# 72 Nature Gifts in A Year

## Time-Honored Artistry Melding with the Wisdom of the Twenty-Four Solar Terms

*A journey where ancient wisdom*
*and nature’s rhythm unite*
*in Zhang Ruoai’s “Ink Marvels of the Pearl Forest”;*

*A meticulous fusion of the Twenty-Four Solar Terms*
*and the Seventy-Two Pentads,*
*bringing each detail and insight vividly to life.*

# Seventy-Two Nature Gifts in A Year

Time-Honored Artistry Melding with the
Wisdom of the Twenty-Four Solar Terms

**Yumeng (ILSA) Li**
Zhang Ruoai (Qing Dynasty)

PEAR & MULBERRRY

# 前言

## Preface

China' s profound historical and cultural depth, ancient and enriched through generations, is exemplified in the "二十四節氣" (Twenty-Four Solar Terms), a unique cultural legacy born from the laboring people' s ingenuity. These terms intricately segment the agricultural cycle, encapsulating a comprehensive grasp of meteorological changes, phenological patterns, and farming practices, thereby significantly influencing societal development and civilizational progress.

Originally, the Twenty-Four Solar Terms primarily guided agricultural activities. Farmers adhered to these periods for planting, fertilizing, and harvesting, where timing was critical for crop yields. Delays could drastically affect harvests. However, the reach of the Solar Terms extends beyond agriculture, permeating aspects of daily life like dietary habits, living arrangements, health practices, and even festive traditions and customs. In summer, for example, people adopt cooling measures, while in winter, they focus on staying warm. Importantly, health regimens are adapted to each term, aligning with the natural cycle and bodily needs.

The Seventy-Two Pentads (七十二候) represent one of China' s earliest calendars for guiding agricultural activities, integrating knowledge of astronomy, meteorology, and phenology. Defined by a structure where "以五日為候" (five days form a Pentad), "三候為氣" (three Pentads create a Solar Term), "六氣為時" (six Solar Terms constitute a season), and "四時為歲" (four seasons complete a year), leading to a total of "一歲二十四節氣共七十二候" (seventy-two Pentads across twenty-four Solar Terms within a year). Originating from the Yellow River Basin, the concept of the Seventy-Two Pentads was first documented by 呂不韋 (Lü Buwei) in the "呂氏春秋(Lüshi Chunqiu)" , and was comprehensively recorded in the 2nd century BCE in the "逸周書·時訓解 (Yi Zhou Shu·Shi Xun Jie)". This book invites you on a journey through the Twenty-Four Solar Terms and the Seventy-Two Pentads, a Huaxia cultural treasure. These terms transcend an agricultural calendar, embodying a philosophy of universal interconnectedness. Detailed exploration will reveal natural transitions, plant growth cycles, and animal behaviors behind each term, showcasing ancient wisdom in harmony with nature. Pentads like "東風解凍" (East Wind Melts the Ice) herald spring' s warmth and life' s awakening; "蟄蟲始振" (Hibernating Insects Stir) marks insects awakening from hibernation; "魚陟負冰" (Fish Swim Upward Under the Ice) depicts fish swimming against currents near thin ice. These vivid descriptions transcend mere natural records, reflecting a deep respect and understanding of nature' s laws.

"墨妙珠林" (Ink Marvels of the Pearl Forest), a masterful work by 張若靄 (Zhang Ruoai), comprises twelve volumes of large album leaves, each identified by earthly branches. The 卯冊 (The Fourth Album) stands out with Zhang' s renderings of the Twenty-Four Solar Terms. Integrating climatic elements, customary activities, and ecological features of fauna and flora into landscape paintings, these works vividly portray each term' s distinct natural and societal scenes.

Zhang Ruoai (1713—1746), styled 晴嵐 (Qinglan), from 桐城 (Tongcheng), 安徽 (Anhui), descended from a notable lineage. His grandfather, 張英 (Zhang Ying), and father, 張廷玉 (Zhang Tingyu), were distinguished scholars under Emperors 康熙 (Kangxi) and 雍正 (Yongzheng)/乾隆 (Qianlong), respectively. Zhang' s exemplary performance in the 1733 imperial examination was humbly relinquished for other scholars, and he advanced from a second-tier Jinshi to an influential academician and eventually the Minister of Rites, earning the title of Earl.

Known for his superb painting skills, particularly in landscapes and bird-and-flower motifs, Zhang was heavily influenced by the Wu School. In "墨妙珠林" 卯冊, he captures the Twenty-Four Solar Terms, from "立春" (Beginning of Spring) to "大寒" (Greater Cold). The ancient adage "春山艷冶而如笑，夏山蒼翠而如滴，秋山明淨而如妝，冬山慘淡而如睡" (Spring mountains vibrant as if smiling, summer lush as if dripping, autumn clear as if adorned, winter bleak as if sleeping) highlights the challenge of depicting these seasonal landscapes, a complexity that made Zhang's works particularly beloved by Emperor Qianlong.

We intricately align and dissect the Twenty-Four Solar Terms and "七十二候" (Seventy-Two Pentads) with "墨妙珠林," inviting readers into a realm where nature, art, and culture intertwine. As you delve into this book, you will be prompted to consider the significance of these natural phenomena and their influence on ancient life and customs. This deep engagement will inspire reverence for nature and an appreciation for how the ancients coexisted harmoniously with their environment, passing down their wisdom. It's an opportunity to rediscover life's "小確幸" (small, certain happiness) amidst modernity's pace.

# 目錄

## Table of Contents


春

Spring

立春 (Lìchūn) - Beginning of Spring
雨水 (Yǔshuǐ) - Rain Water
驚蟄 (Jīngzhé) - Waking of Insects
春分 (Chūnfēn) - The Spring Equinox
清明 (Qīngmíng) - Pure Brightness
穀雨 (Gǔyǔ) - Grain Rain

夏

Summer

立夏 (Lìxià) - Beginning of Summer
小滿 (Xiǎomǎn) - Lesser Fullness of Grain
芒種 (Mángzhòng) - Grain in Beard
夏至 (Xiàzhì) - The Summer Solstice
小暑 (Xiǎoshǔ) - Lesser Heat
大暑 (Dàshǔ) - Greater Heat

秋

Autumn

立秋 (Lìqiū) - Beginning of Autumn
處暑 (Chǔshǔ) - End of Heat
白露 (Báilù) - White Dew
秋分 (Qiūfēn) - The Autumn Equinox
寒露 (Hánlù) - Cold Dew
霜降 (Shuāngjiàng) - Frost’ s Descent

冬

Winter

立冬 (Lìdōng) - Beginning of Winter
小雪 (Xiǎoxuě) - Lesser Snow
大雪 (Dàxuě) - Greater Snow
冬至 (Dōngzhì) - The Winter Solstice
小寒 (Xiǎohán) - Lesser Cold
大寒 (Dàhán) - Greater Cold

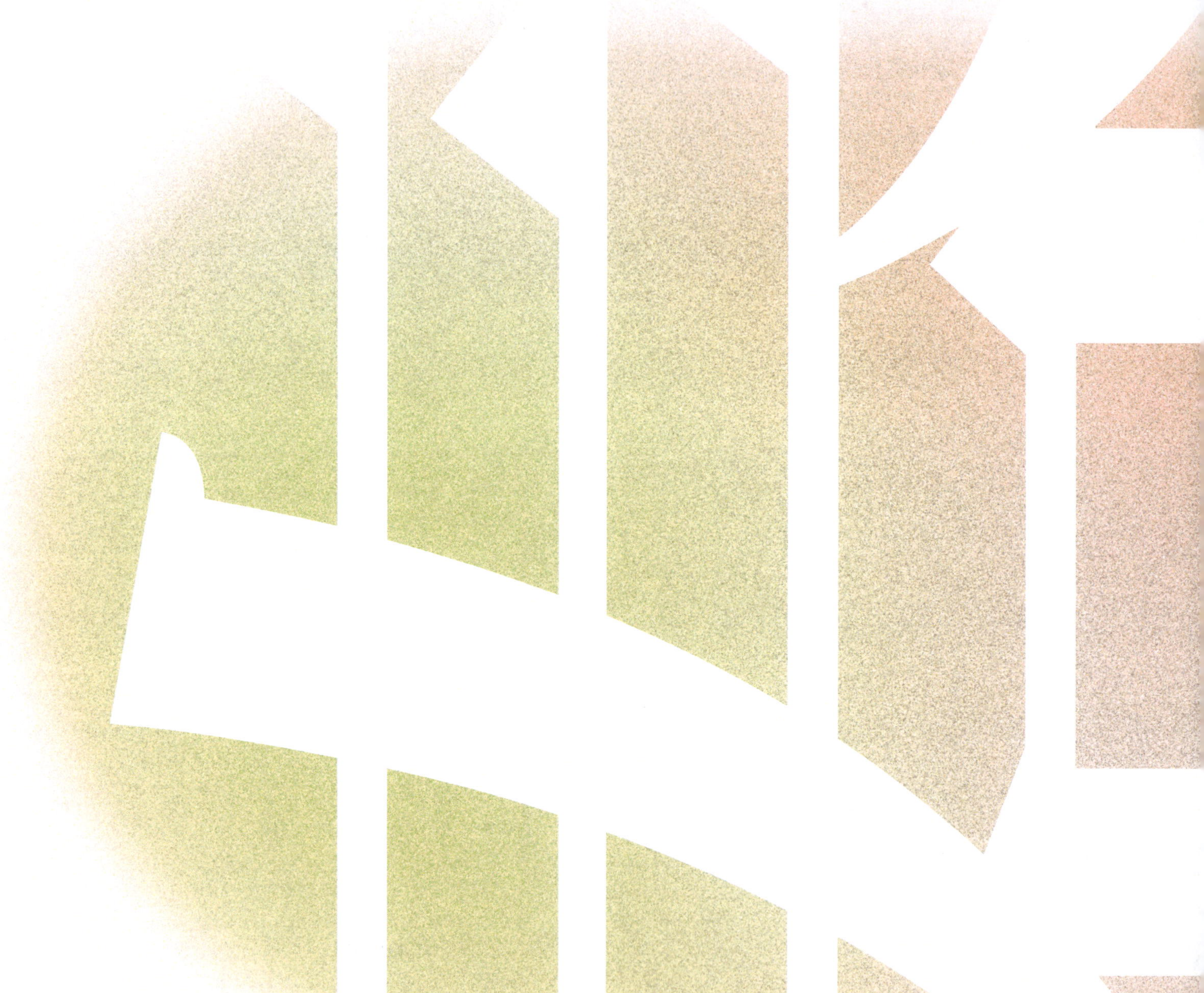

# Spring

# 春

Chūn

Lì

立

# BEGINNING OF SPRING

February 4 (± 1)

# 立春

## Beginning of Spring

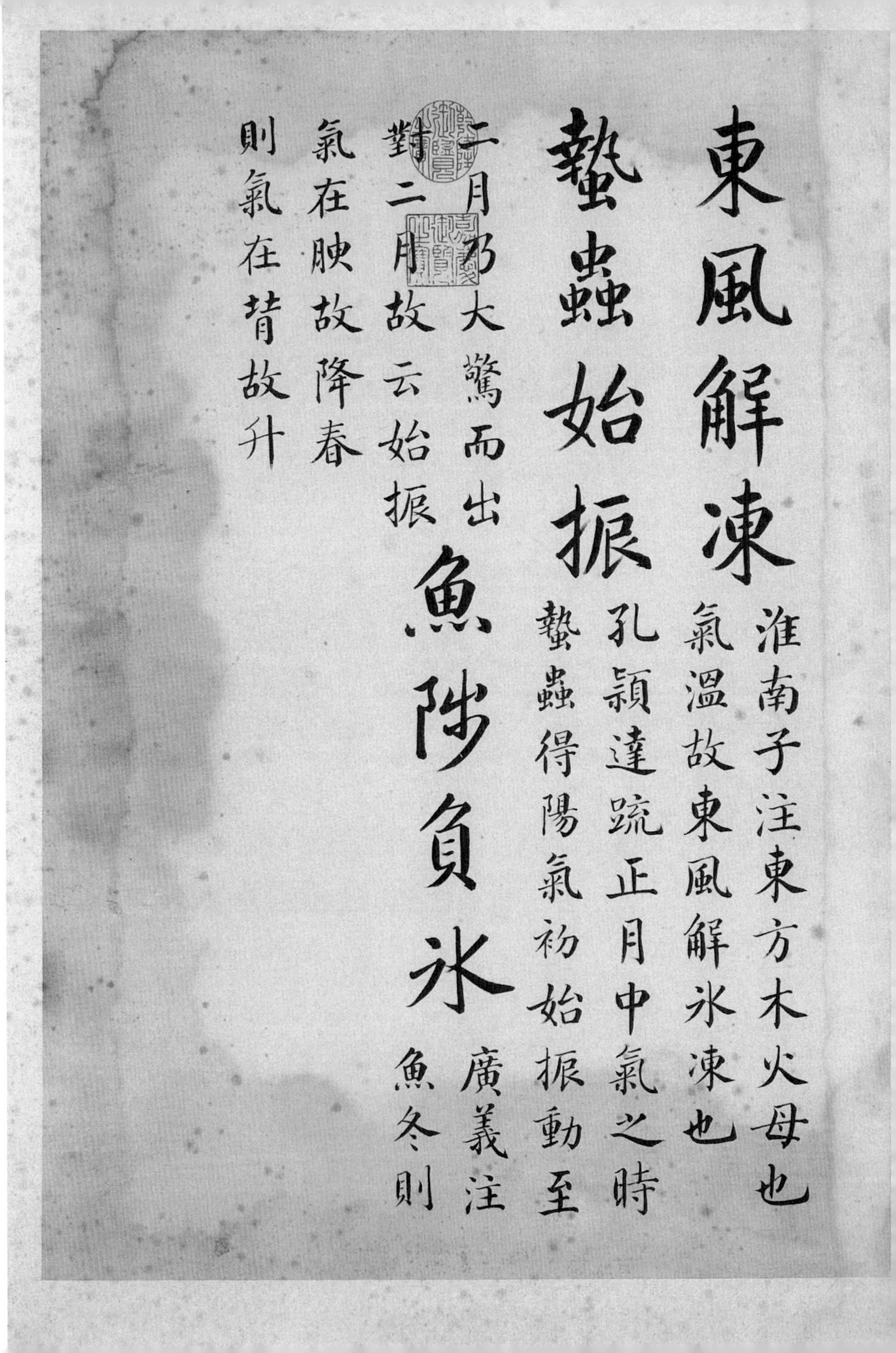


Lìchūn : The beginning of spring, symbolizing the first day of spring. "立 (Lì)" implies emergence, indicating a transition from the end of winter to the start of spring. It's a time when the Big Dipper points east, making east the direction of spring.

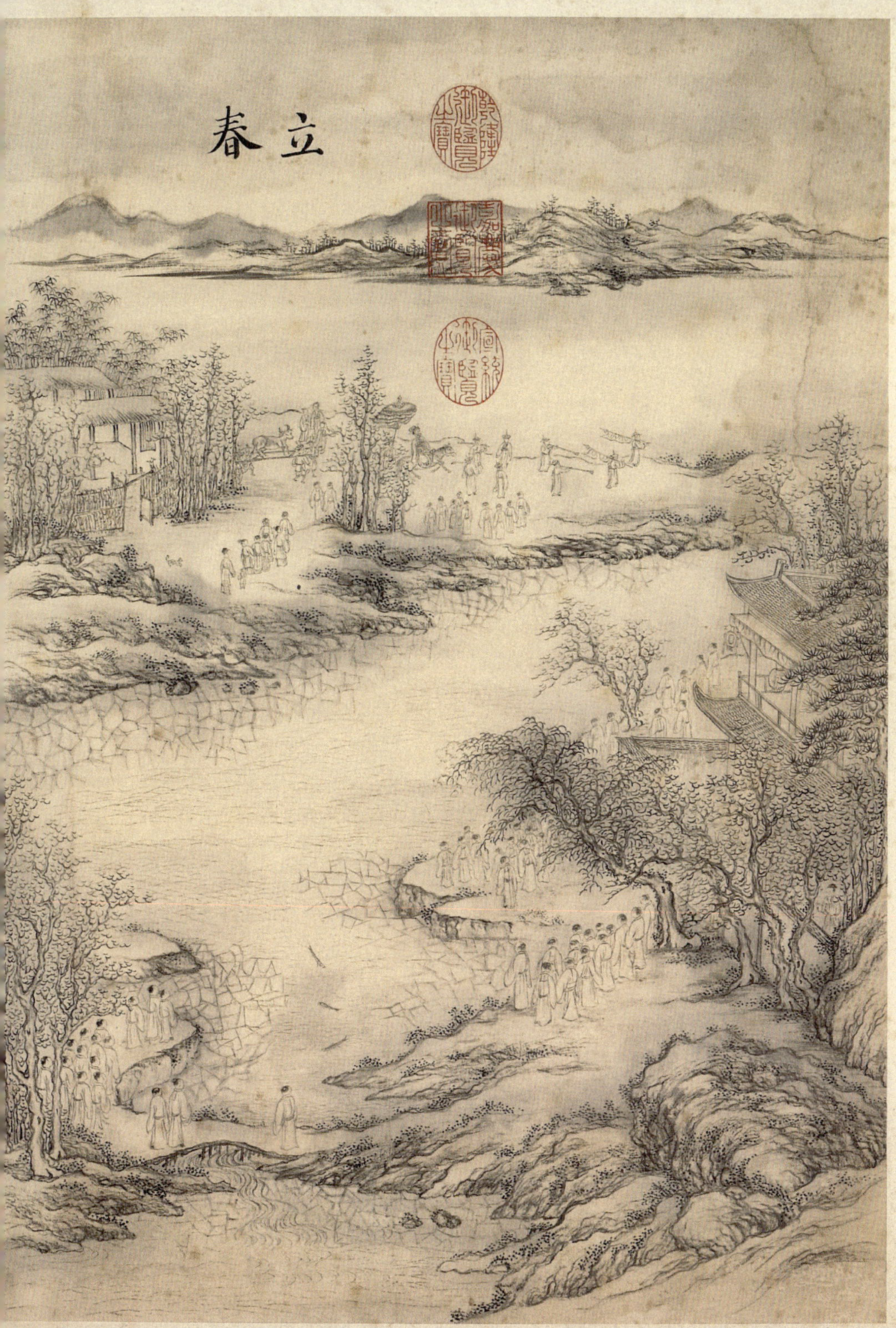


初候

## 東風解凍
## Eastern Winds Thaw

二候

## 蟄蟲始振
## Hibernating Insects Awaken

三候

## 魚陟負冰
## Fish Rise to the Ice

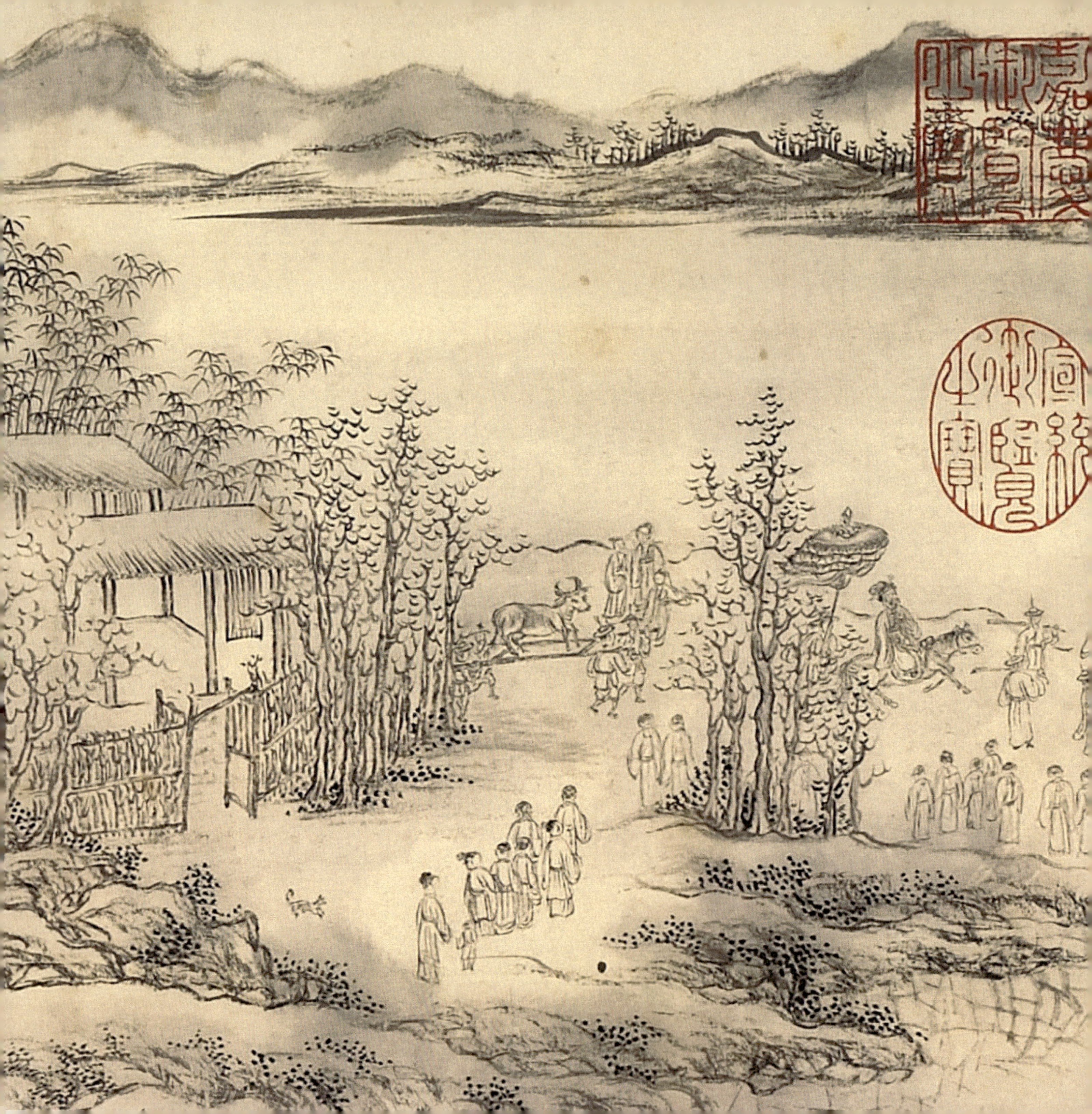
嘉慶御覽之寶
宣統御覽之寶

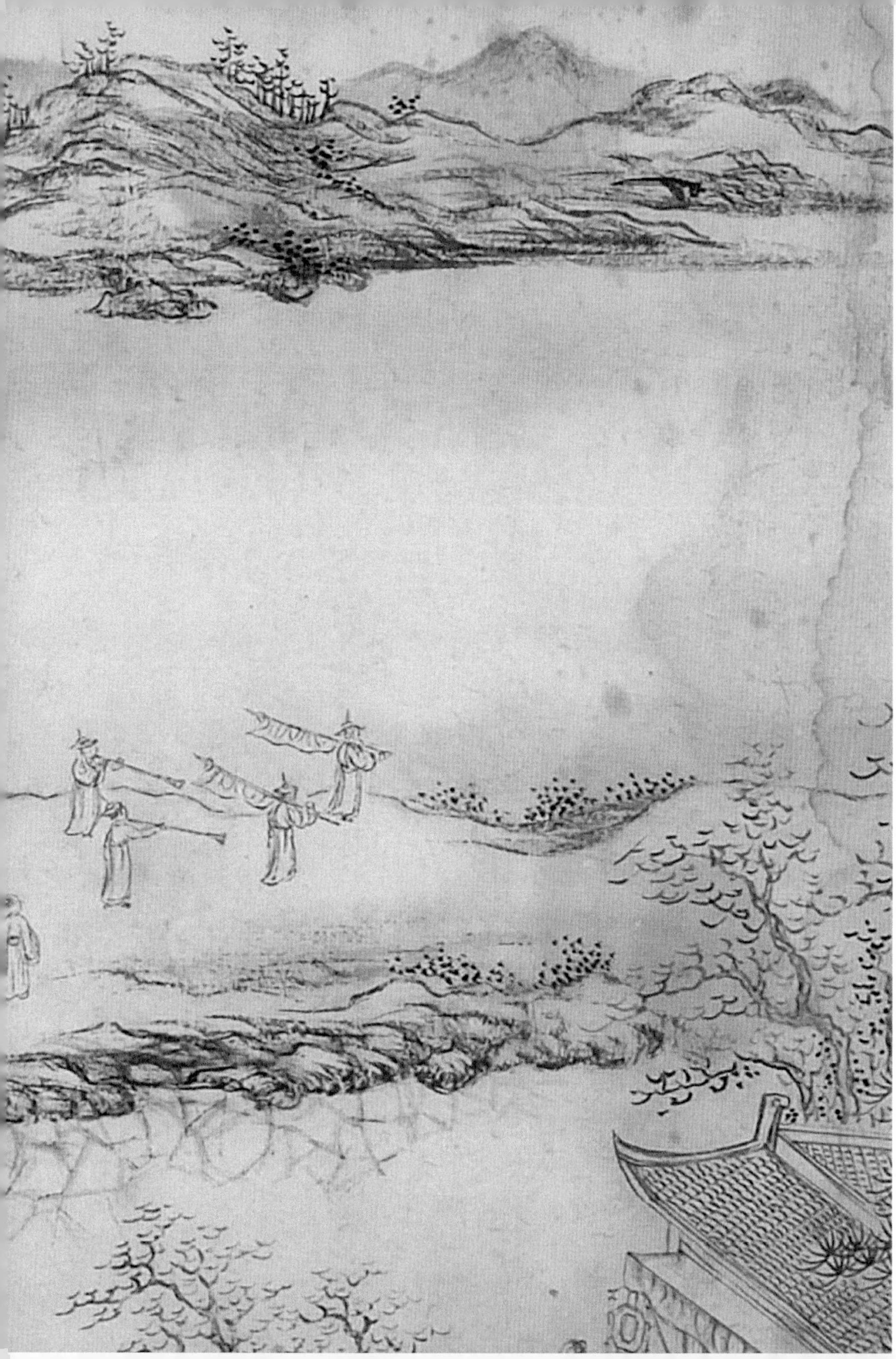

初候

# 東風解凍

## Eastern Winds Thaw

The first herald of Lìchūn is "Eastern Winds Thaw" (**東風解凍** Dōng Fēng Jiě Dòng), where the east wind signifies the melting of ice and the promise of spring.

二候

# 蟄蟲始振

## Hibernating Insects Awaken

The second sign of Lìchūn is "Hibernating Insects Awaken" (蟄蟲始振 Zhé Chóng Shǐ Zhèn):
As the ground slowly warms, hibernating insects start to stir from their winter slumber. This sign, "Hibernating Insects Awaken," reflects the increasing vibrancy and activity beneath the surface. It' s an indication of the earth coming back to life, with the once quiet and still undergrowth beginning to buzz with the movement of awakening creatures, heralding the coming abundance of spring.

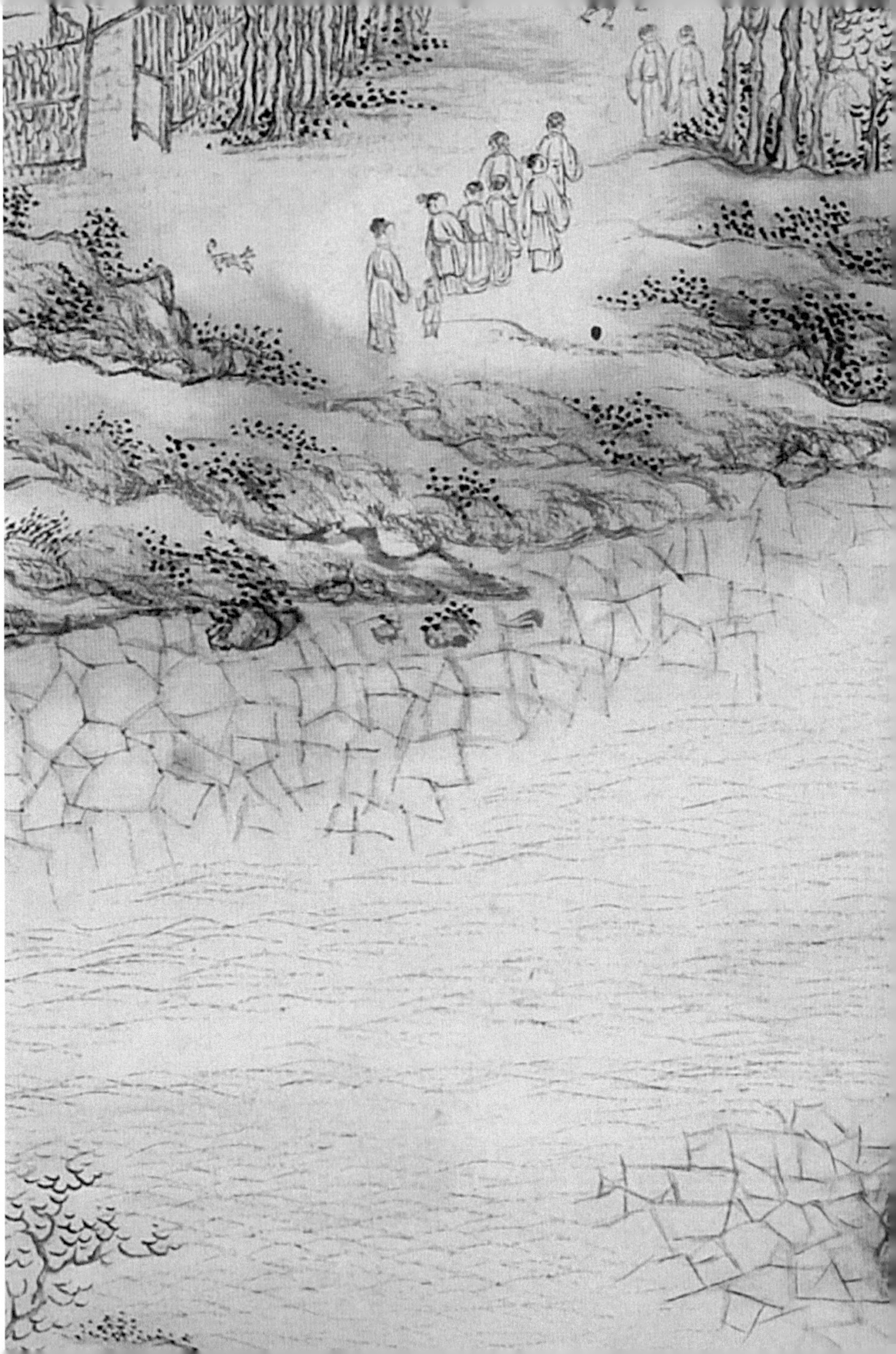

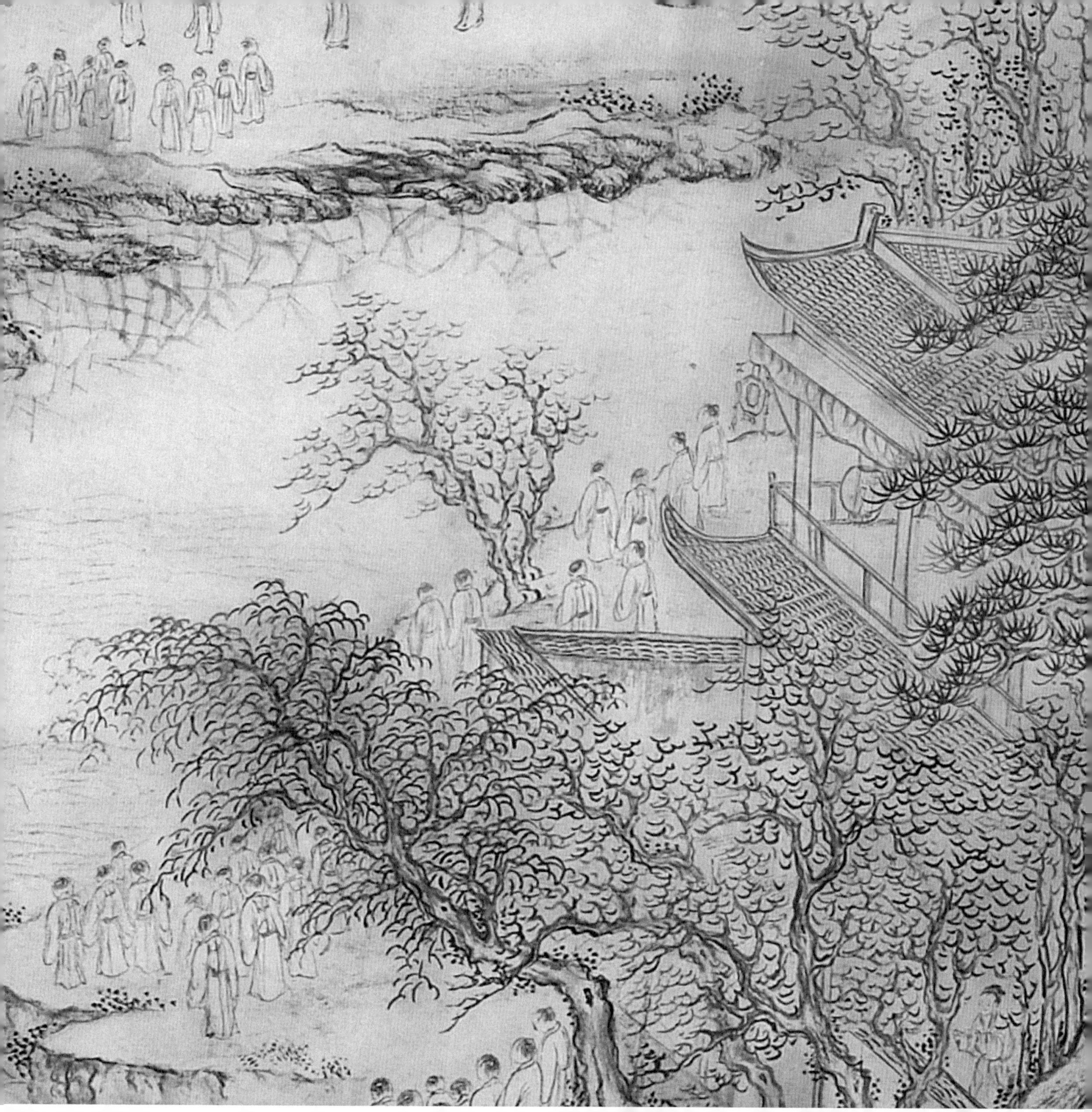

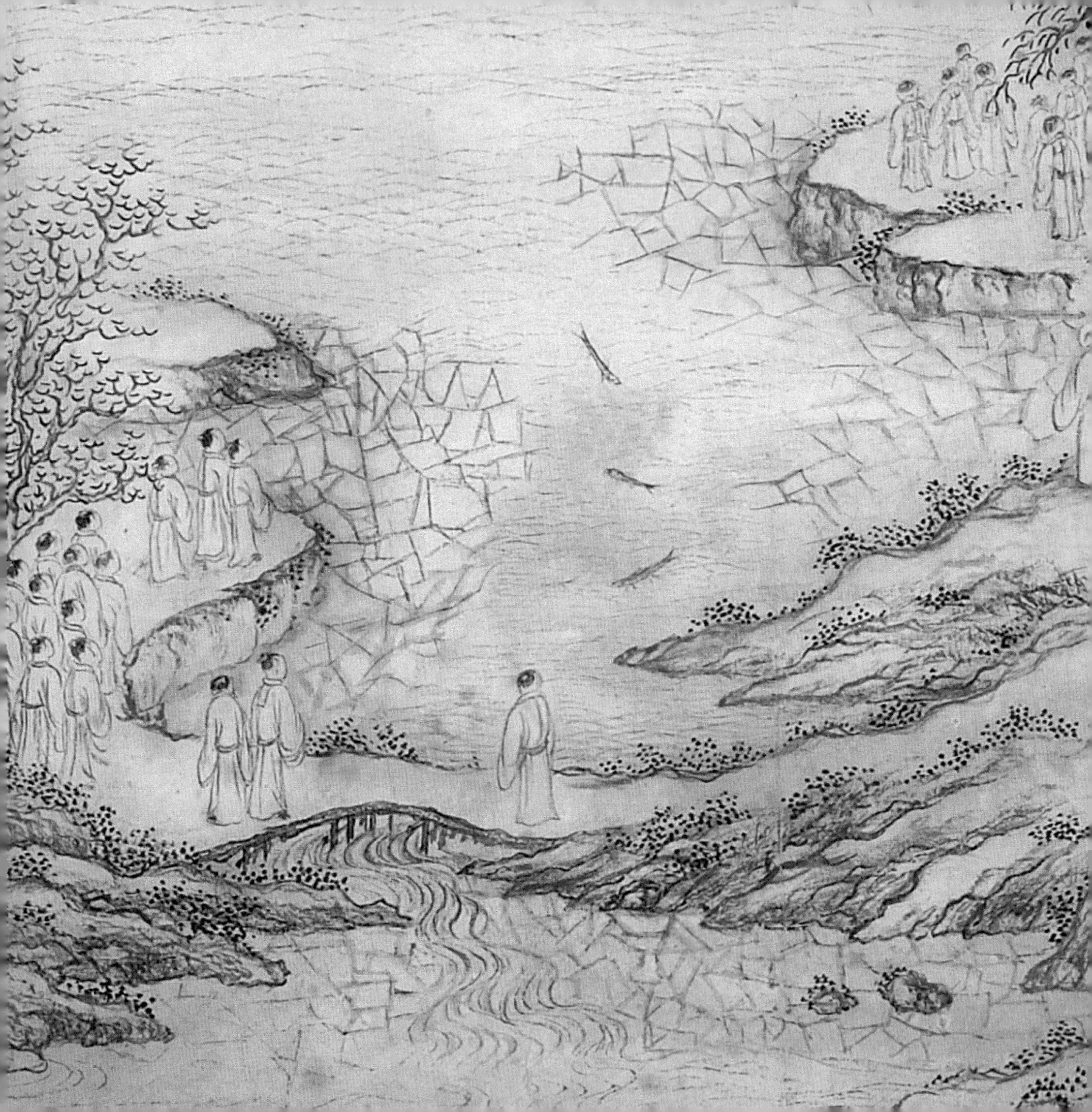

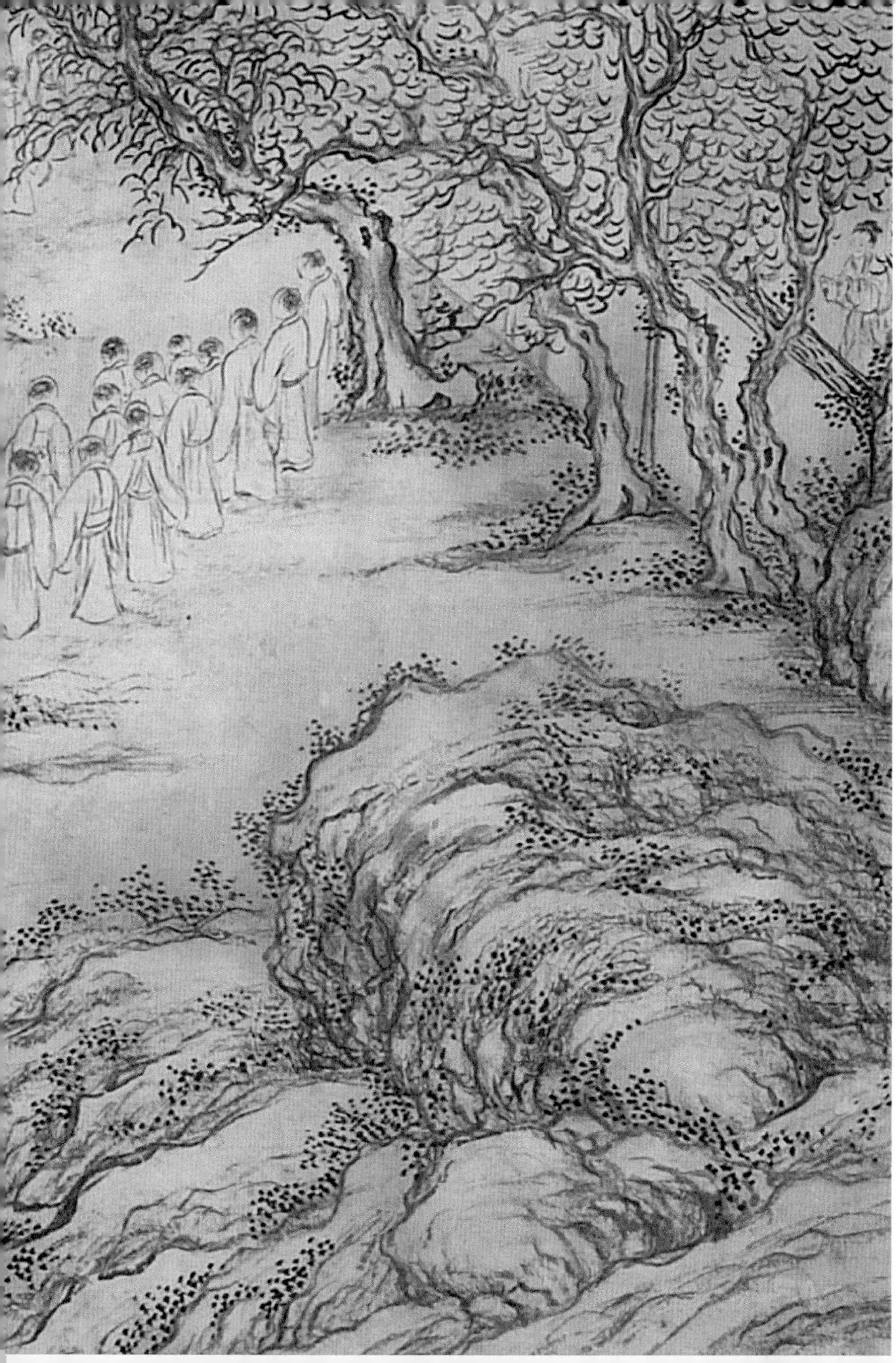

三候

# 魚陟負冰

## Fish Rise to the Ice

The third sign is "Fish Rise to the Ice" (魚陟負冰 Yú Zhì Fù Bīng), this term refers to the time when fish begin to rise to the surface of the water as temperatures gradually warm, swimming amidst the still present fragments of melting ice. From an observer's perspective on the shore, it appears as if the fish are carrying pieces of ice on their backs as they navigate through the water. This sign is a poetic depiction of life resuming in the aquatic world, marking a period of transition from the cold of winter to the promise of spring, with the fish's movements symbolizing the awakening and vitality of nature as it responds to the changing seasons.

# 水

Shuǐ

Yǔ

雨

# RAIN WATER

February 19 (± 1)

# 雨水

## Rain Water

Yǔshuǐ : Literally means "Rain Water," signifying the rainy and cold weather of early spring. It's a time for moisture and preparation for agriculture, as described in ancient texts.

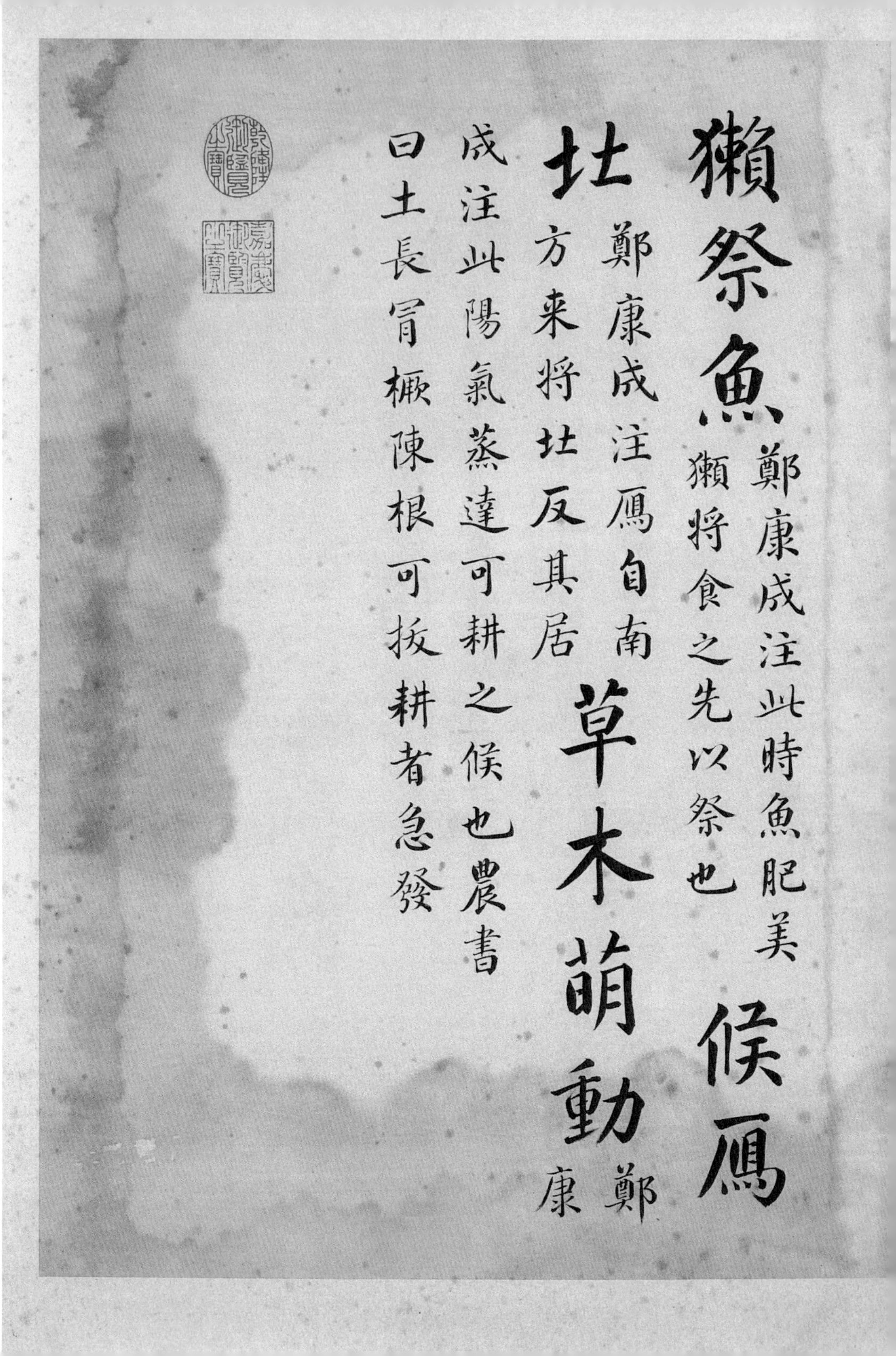

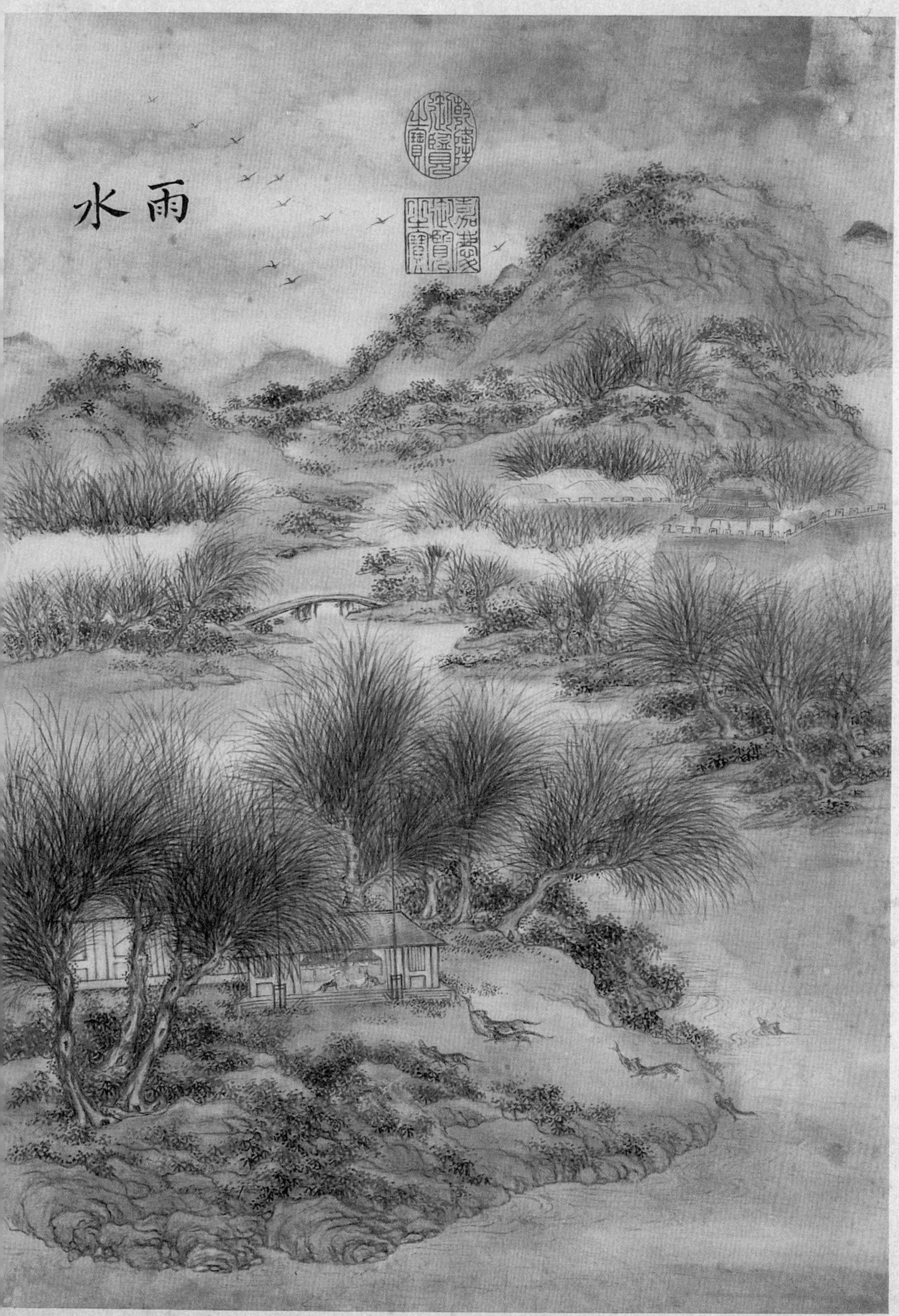


初候

# 獺祭魚

# Otters Perform Fish Ritual

二候

# 候雁北

# Geese Head North

三候

# 草木萌動

# Plants Begin to Sprout

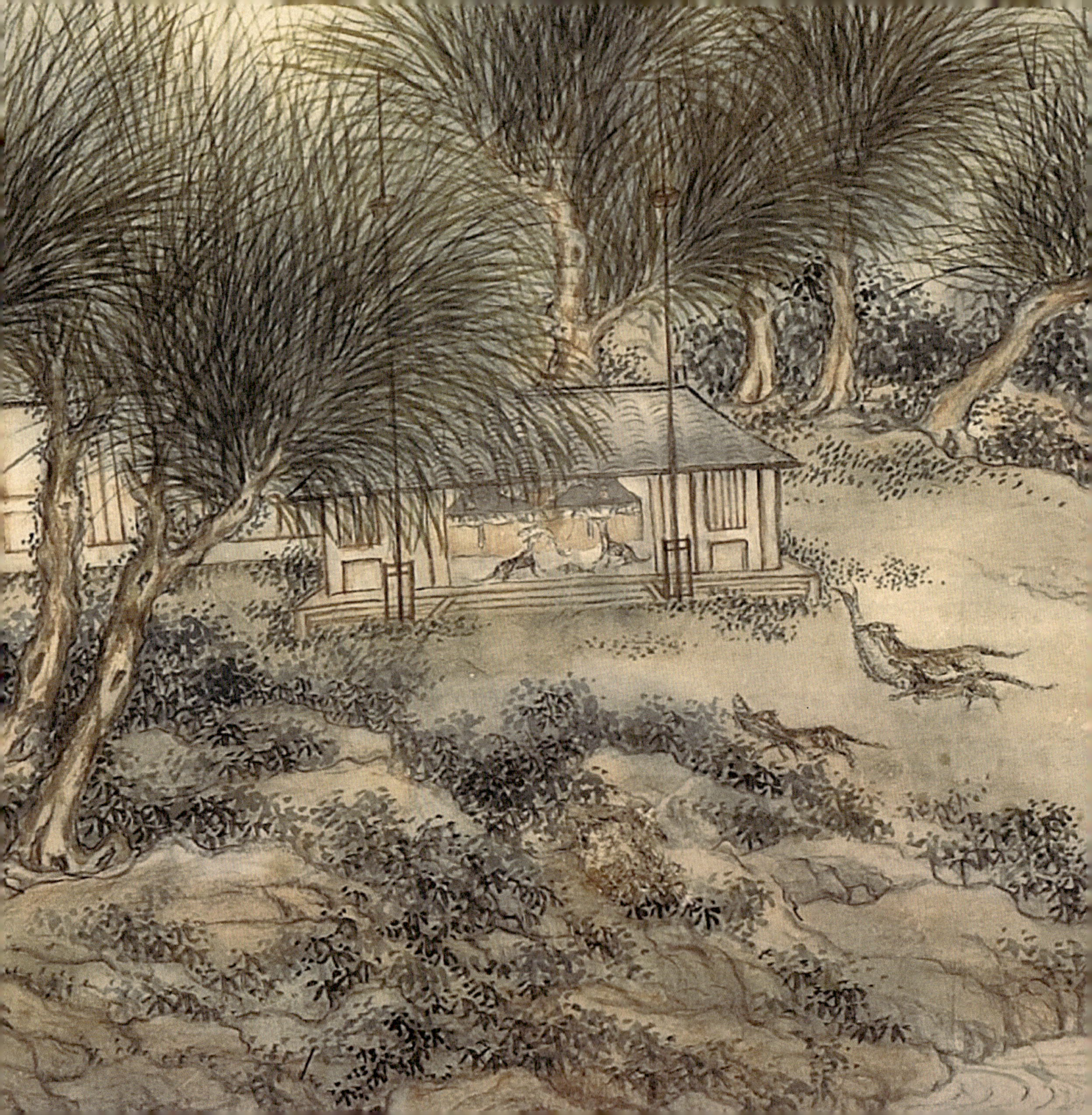

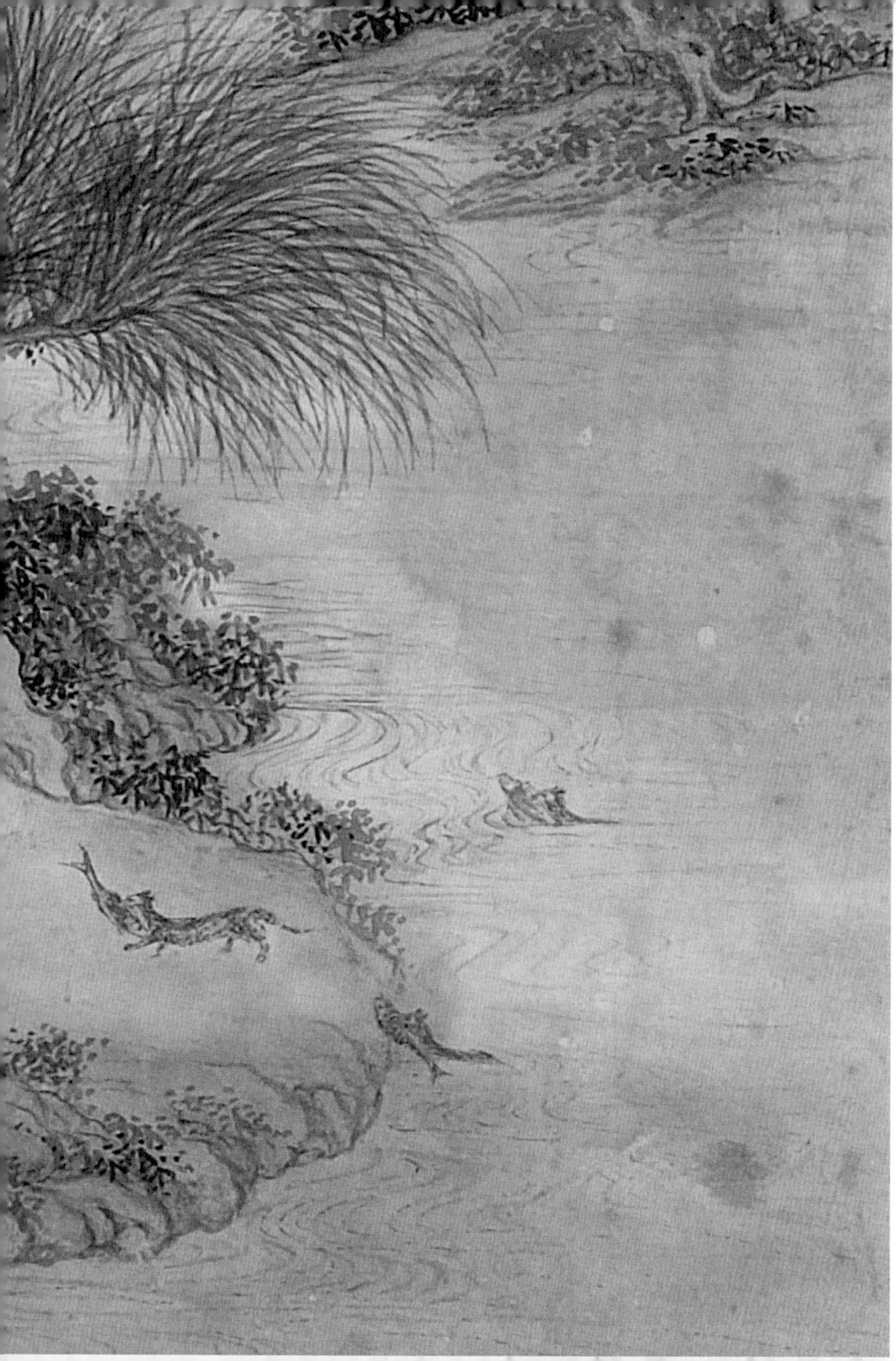

初候

# 獺祭魚

## Otters Perform Fish Ritual

The inaugural indication of Yǔshuǐ is "Otters Perform Fish Ritual" (**獺祭魚 Tǎ Jì Yú**). This sign portrays otters meticulously arranging their caught fish along the riverbank, a behavior reminiscent of a ritualistic offering. The otters' actions seem to hon- or their prey before consumption, symbolizing a deep-rooted natural instinct intertwined with the season's shift. It's a captivating display of how life adapts and respects the cyclical nature of existence.

二候

# 候雁北
## Geese Head North

The second cue is “Geese Head North” (候雁北 Hòu Yàn Běi). As the climate warms in the south, migratory geese commence their northward voyage. These birds, moving from the warmer southern regions of China to the cooler northeast or even to Siberia, are responding to the shifting Yin and Yang energies of nature. Their migration marks a broader environmental transition, signaling a change in temperature and the progression of the season.

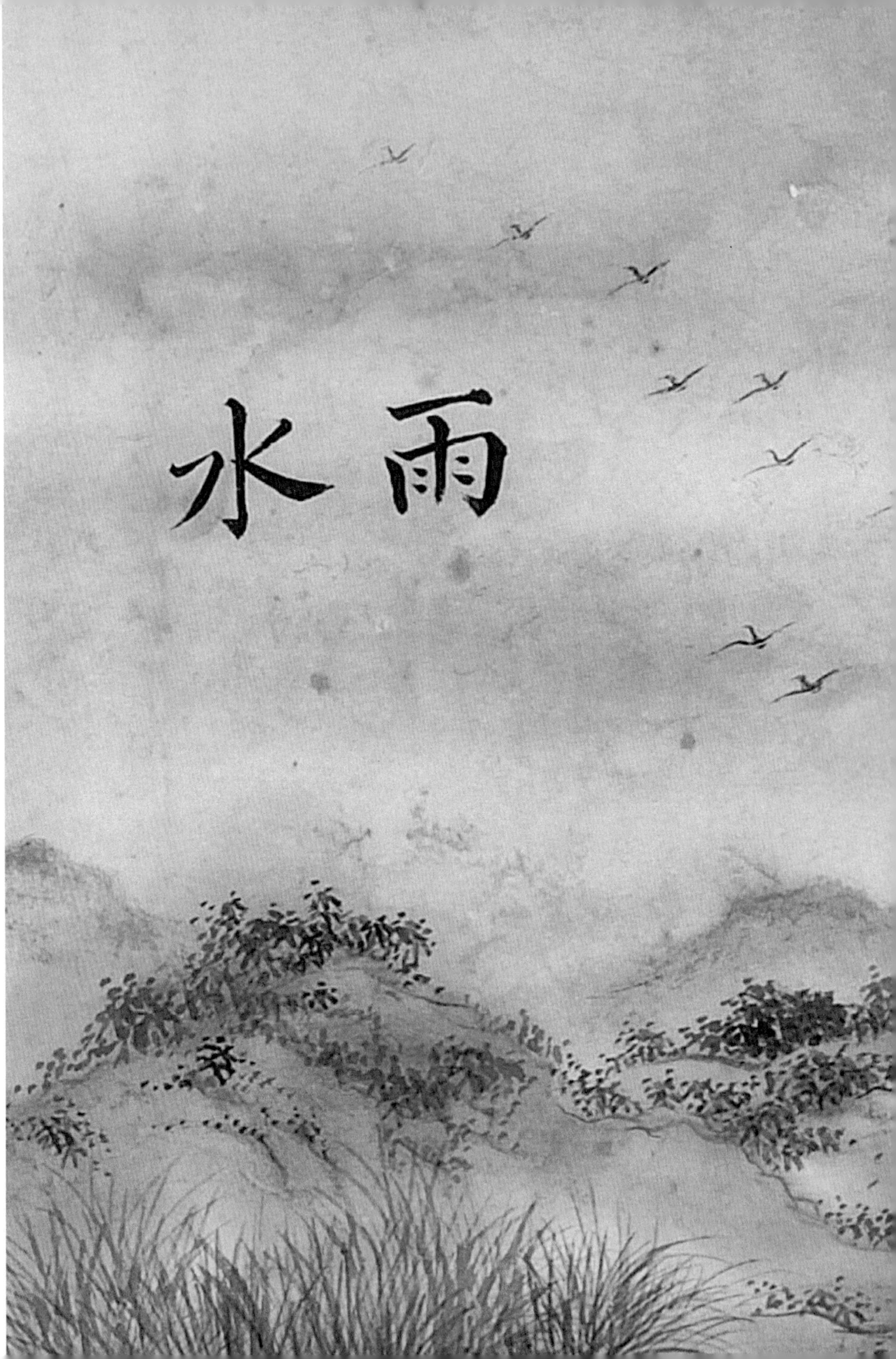

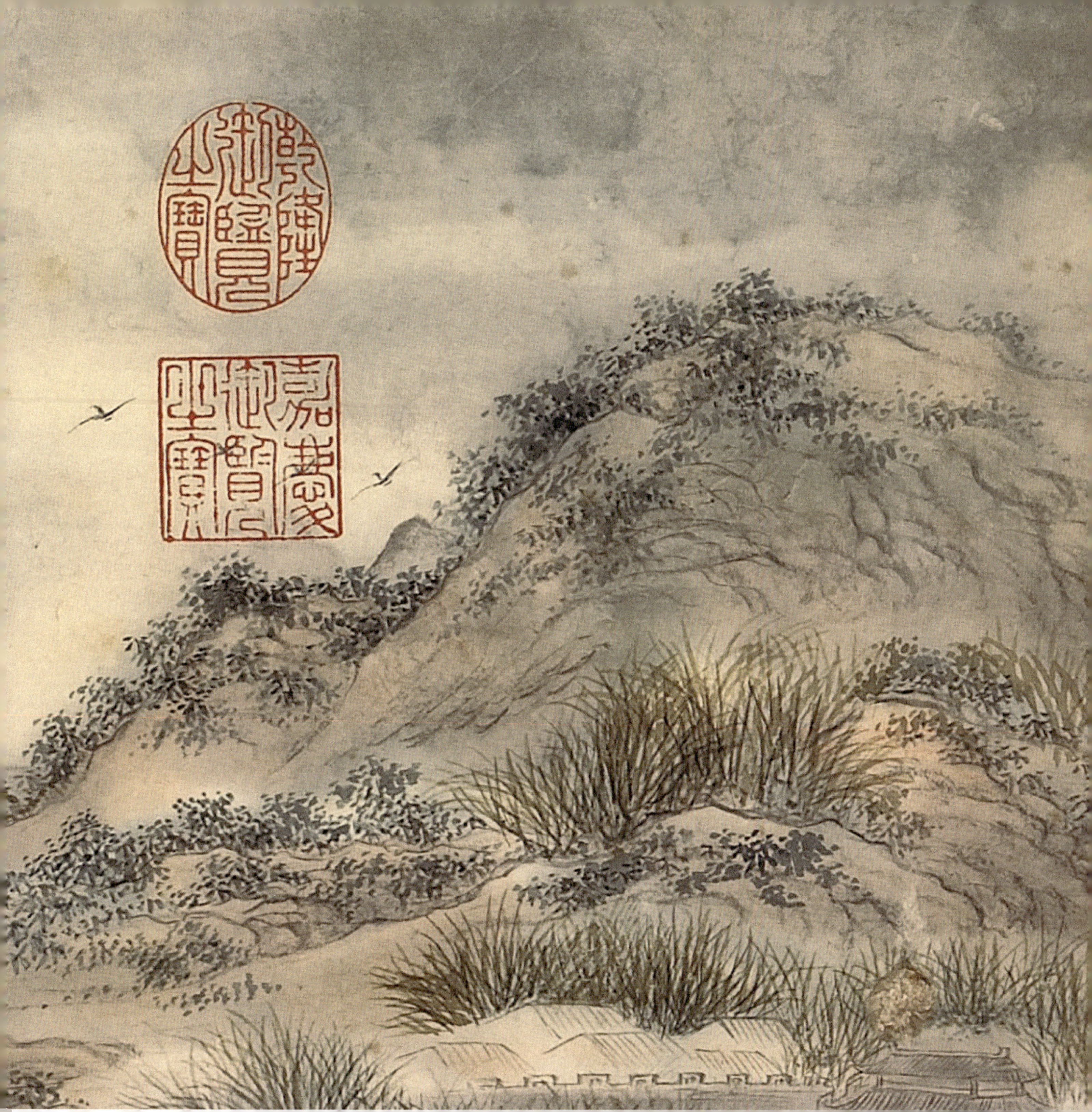

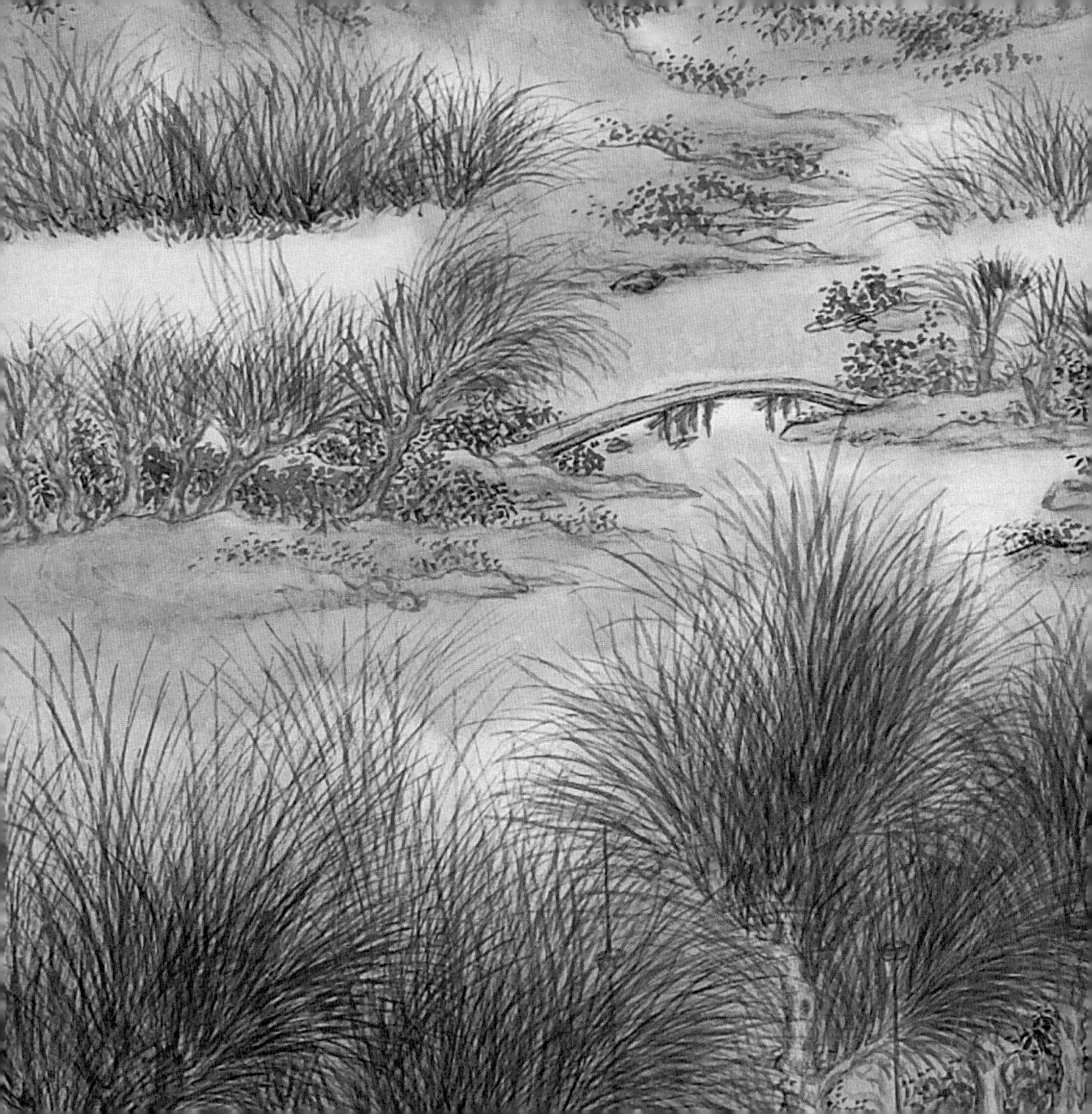

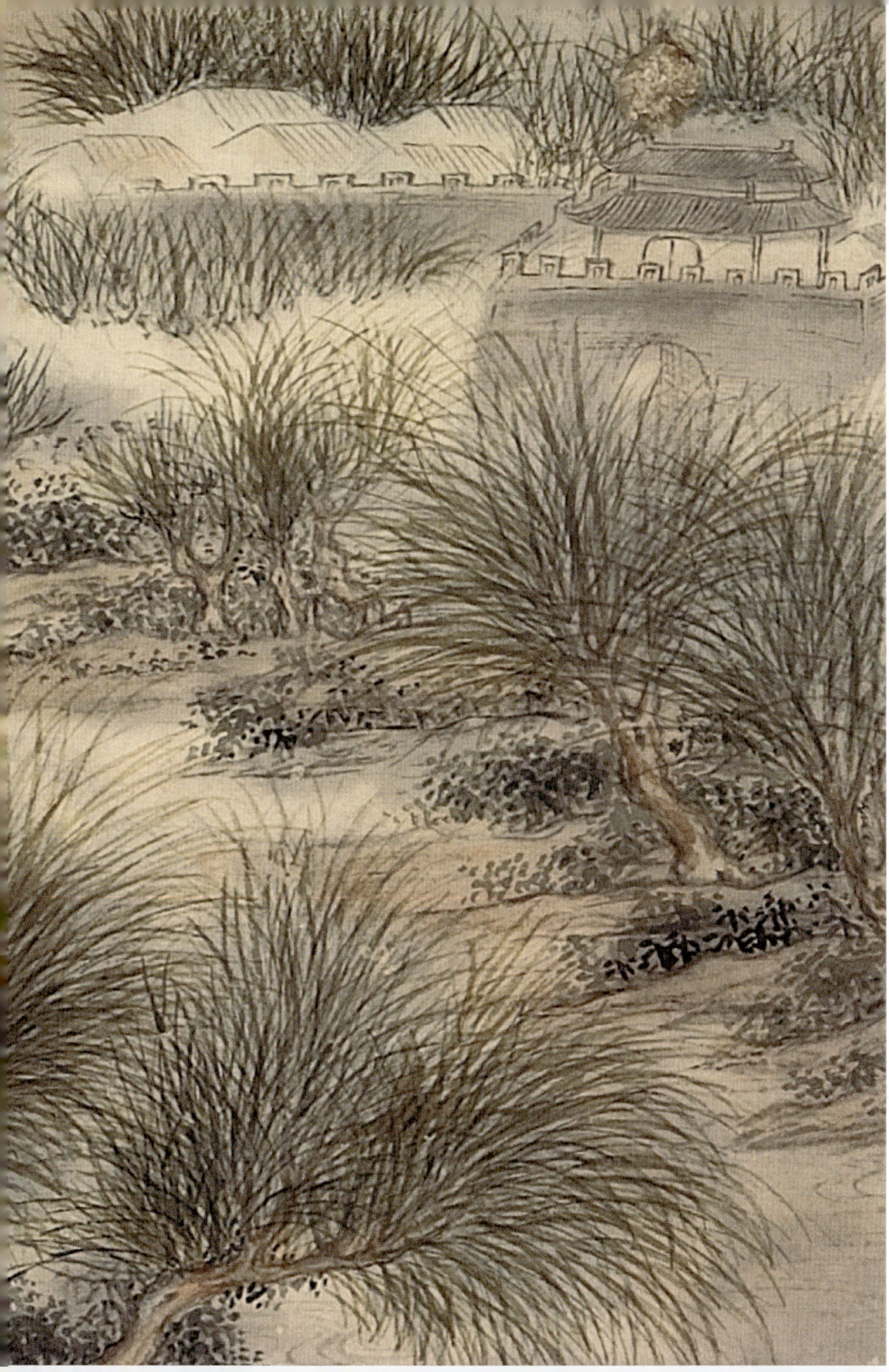

三候

# 草木萌動

## Plants Begin to Sprout

The third marker is “Plants Begin to Sprout” (草木萌動 Cǎo Mù Méng Dòng), which captures the burgeoning of plant life, with grasses and trees sprouting anew. It’s a vivid sign of nature’s revival and growth, as the balanced interplay of Yin and Yang energies nurtures the reawakening of the natural world. The sprouting of plants heralds a period of renewal, symbolizing the earth’s reinvigoration and the lushness soon to envelop the landscape.

# 蟄

Zhé

Jīng

驚

# WAKING OF INSECTS

March 6 (± 1)

# 驚蟄

## Waking of Insects

Jīngzhé : Also known as "Awakening of Insects," is deeply rooted in the legend of Pangu creating the world, where his breath became the wind and his voice became thunder. This term marks the time when the thunder, hidden within the earth during autumn and winter, bursts forth in spring, breaking the soil with its powerful roar. This thunderous awakening signifies the end of hibernation for all beings, whether beneficial or harmful insects, stirring them into activity. It is also a critical time for farmers, signaling the need to fertilize and commence planting various crops, aligning agricultural activities with the awakening energies of earth and sky.

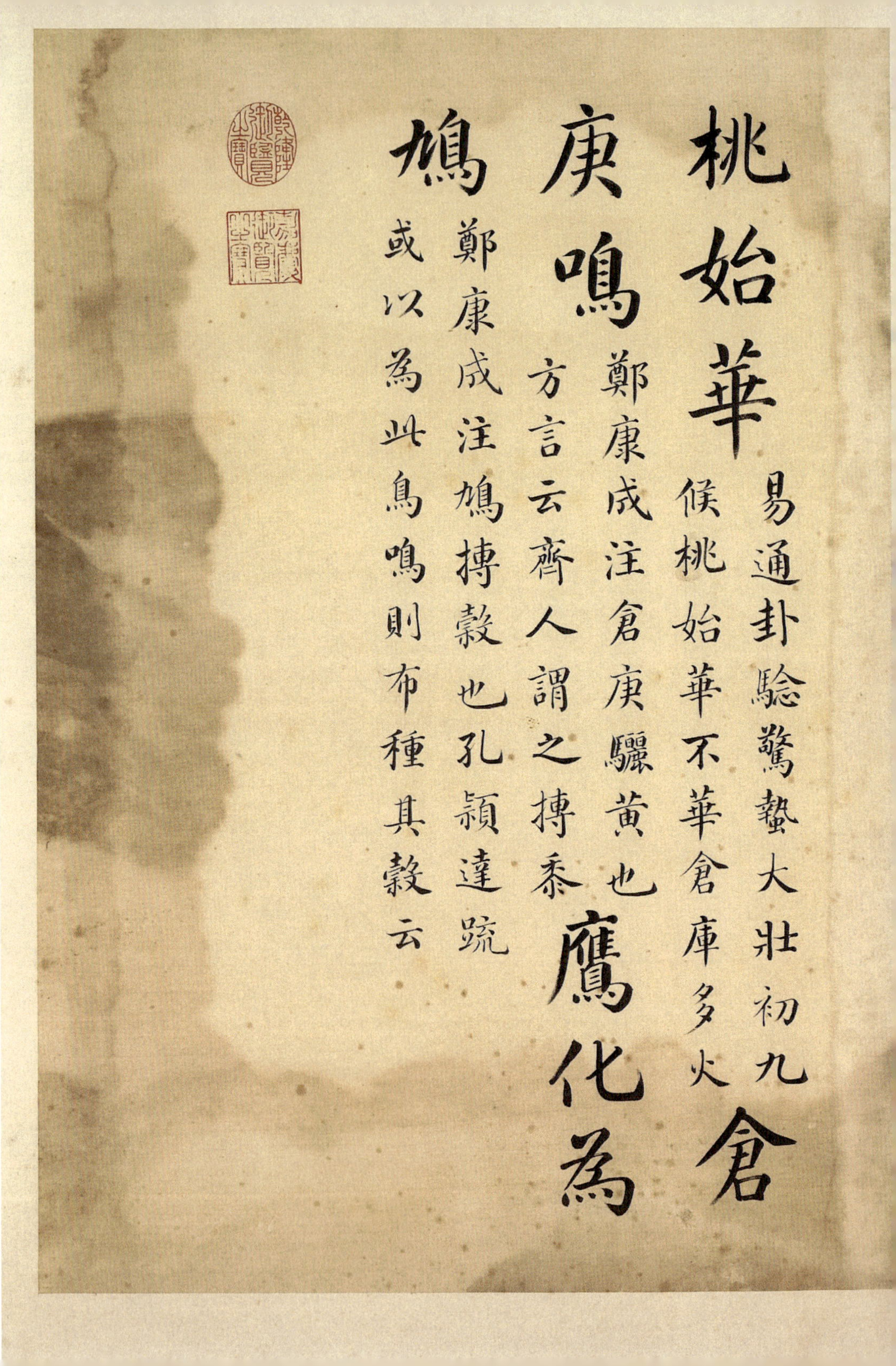

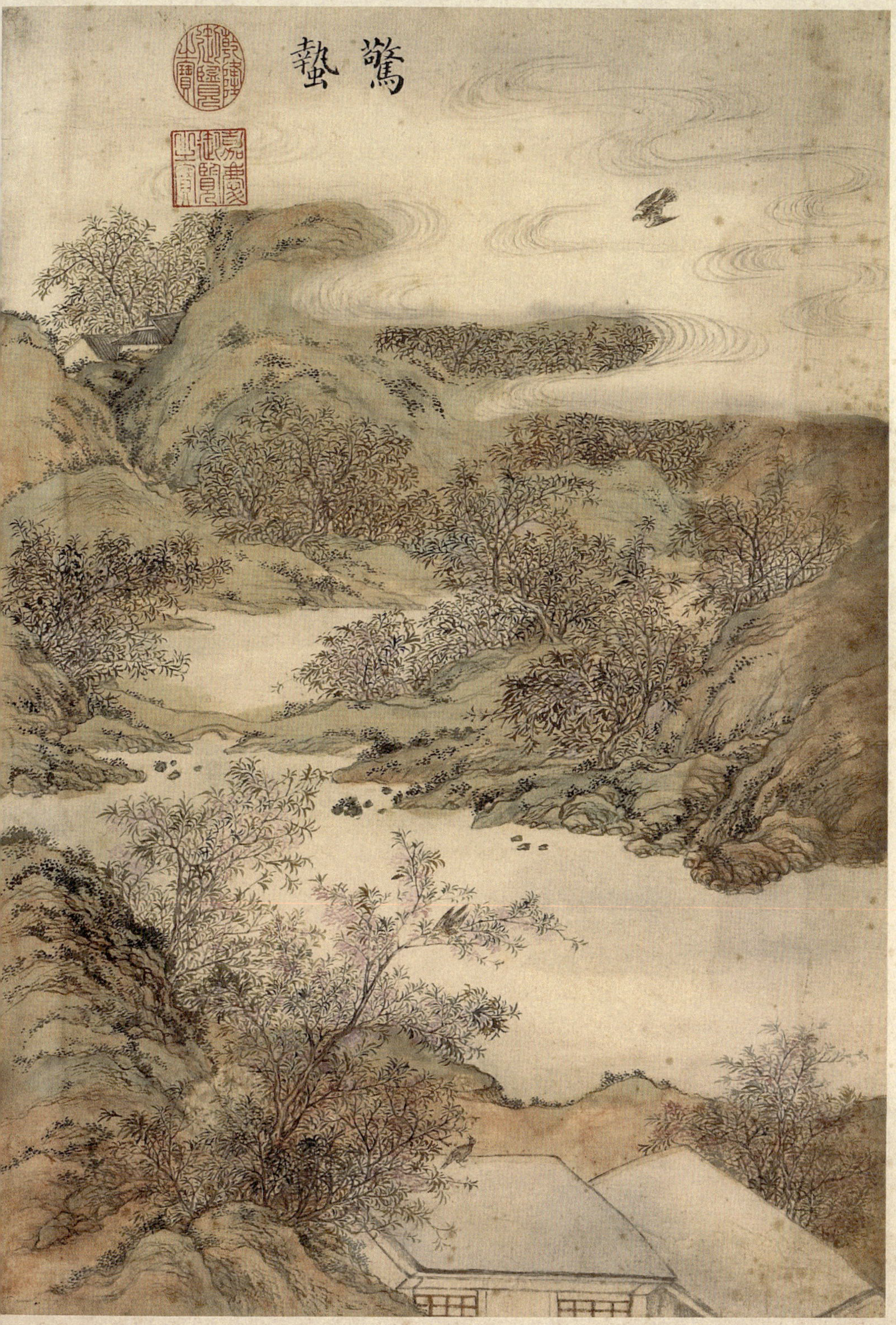


初候

# 桃始華
# Peaches Begin to Bloom

二候

# 倉庚鳴
# Orioles Start Singing

三候

# 鷹化為鳩
# Hawks Transform into Doves

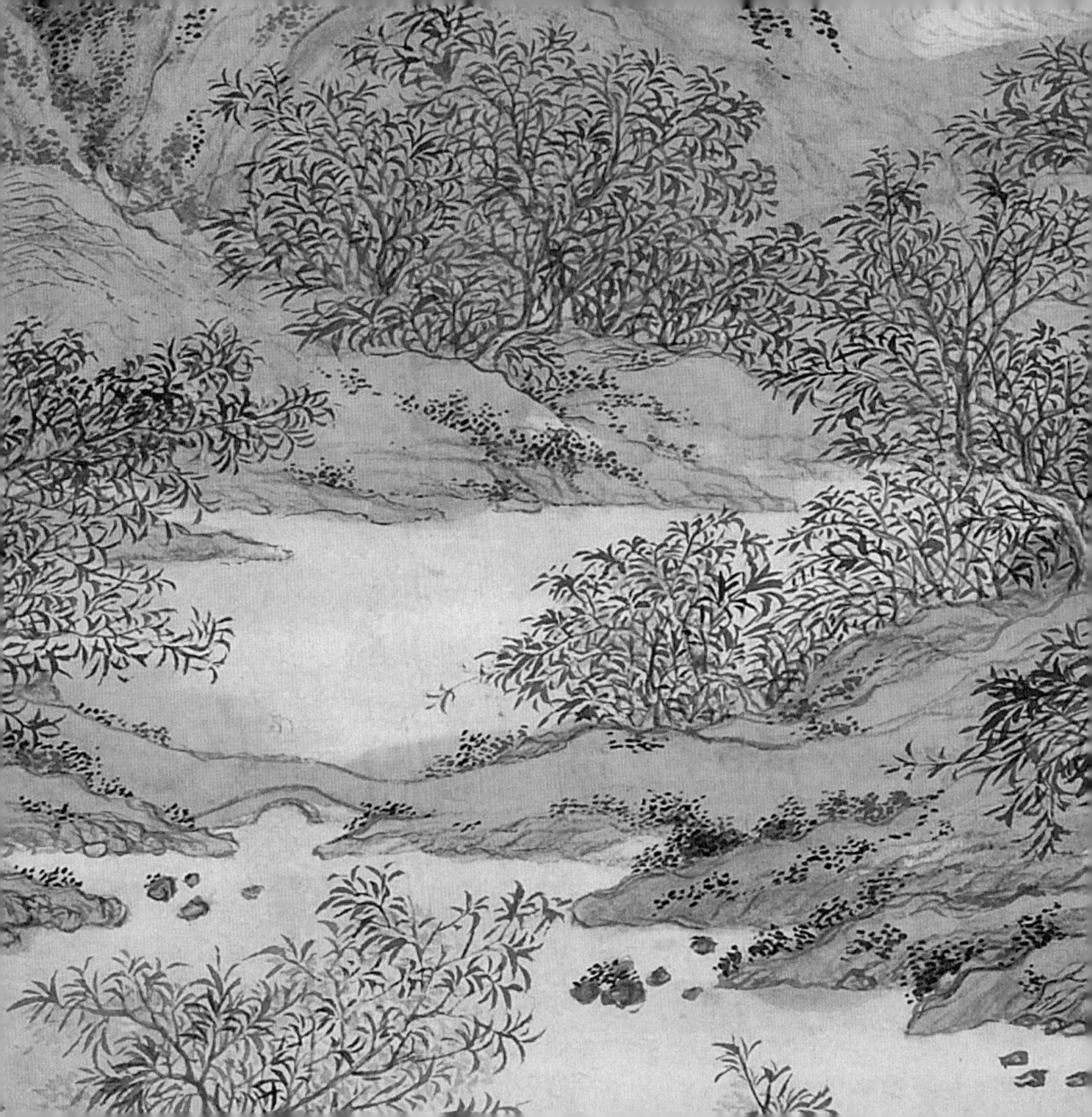

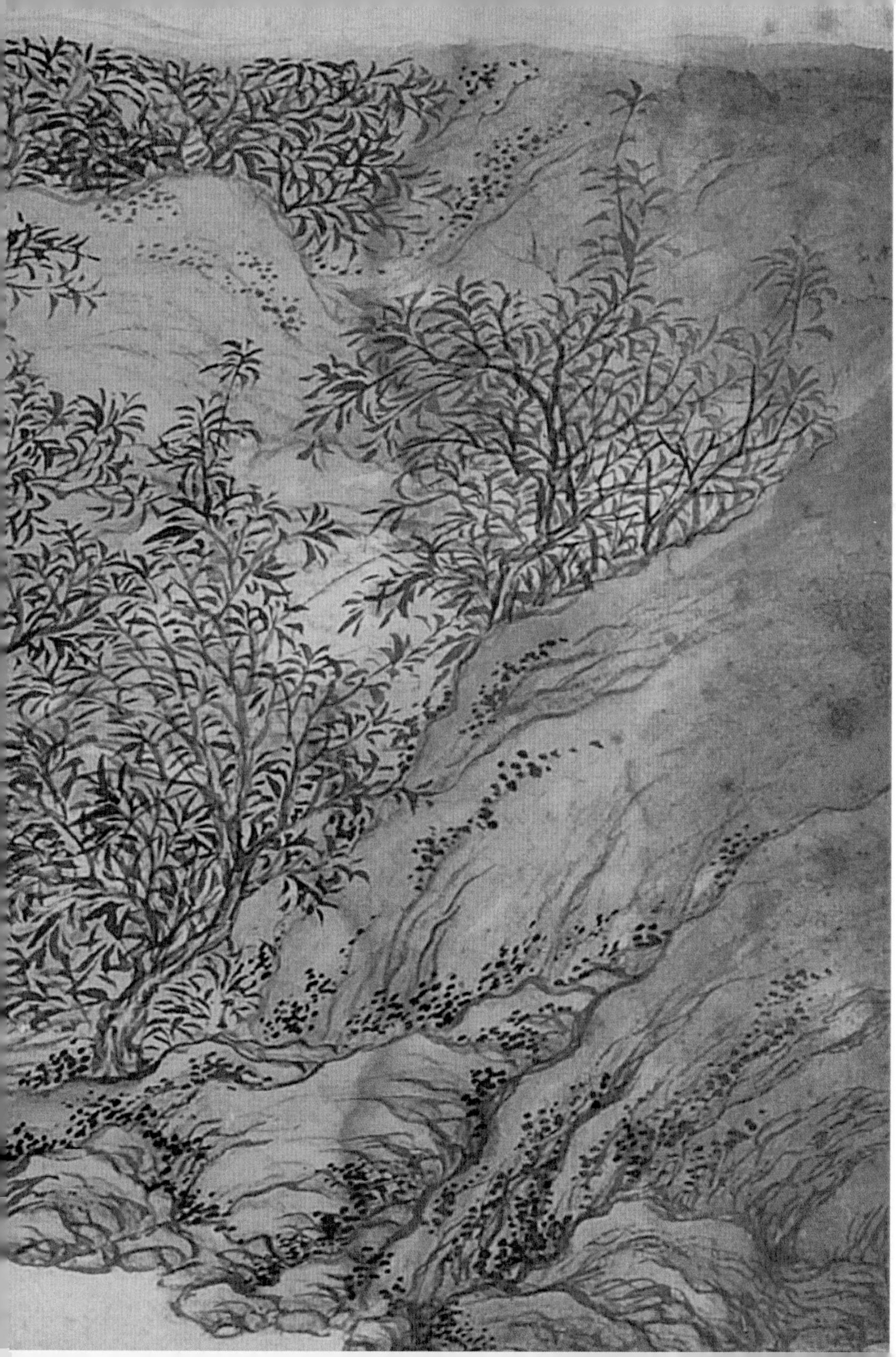

初候

# 桃始華

## Peaches Begin to Bloom

The first glimpse is "Peaches Begin to Bloom" (桃始華 Táo Shǐ Huá), which is marked by the blossoming of peach flowers, which have lain dormant through the harsh winter. As the climate warms, the buds break open in a vibrant display of renewal and beauty, signaling the definitive arrival of spring and the reawakening of life.

二候

# 倉庚鳴

## Orioles Start Singing

The second whisper is "Orioles Start Singing" (倉庚鳴 Cāng Gēng Míng), which refers to the oriole's song heralding the arrival of spring. "倉庚," representing the oriole, is a sign of clearness and newness as these birds begin their melodious calls. Their vibrant singing fills the air, marking a time of warmth and renewal. The oriole's song is a symbol of the world coming back to life, as nature dons its colorful mantle and creatures large and small rejoice in the resurgence of growth and vitality.

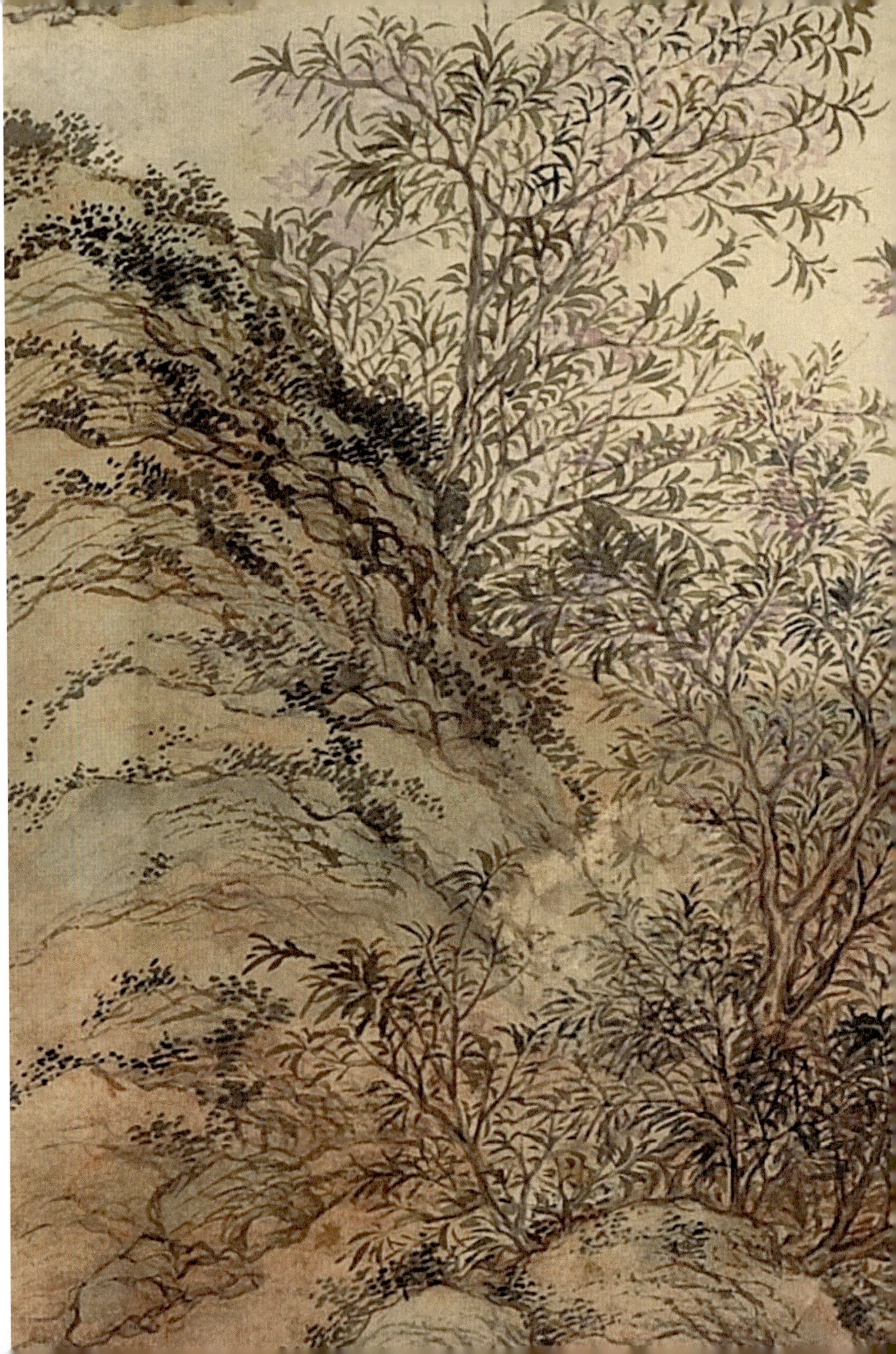

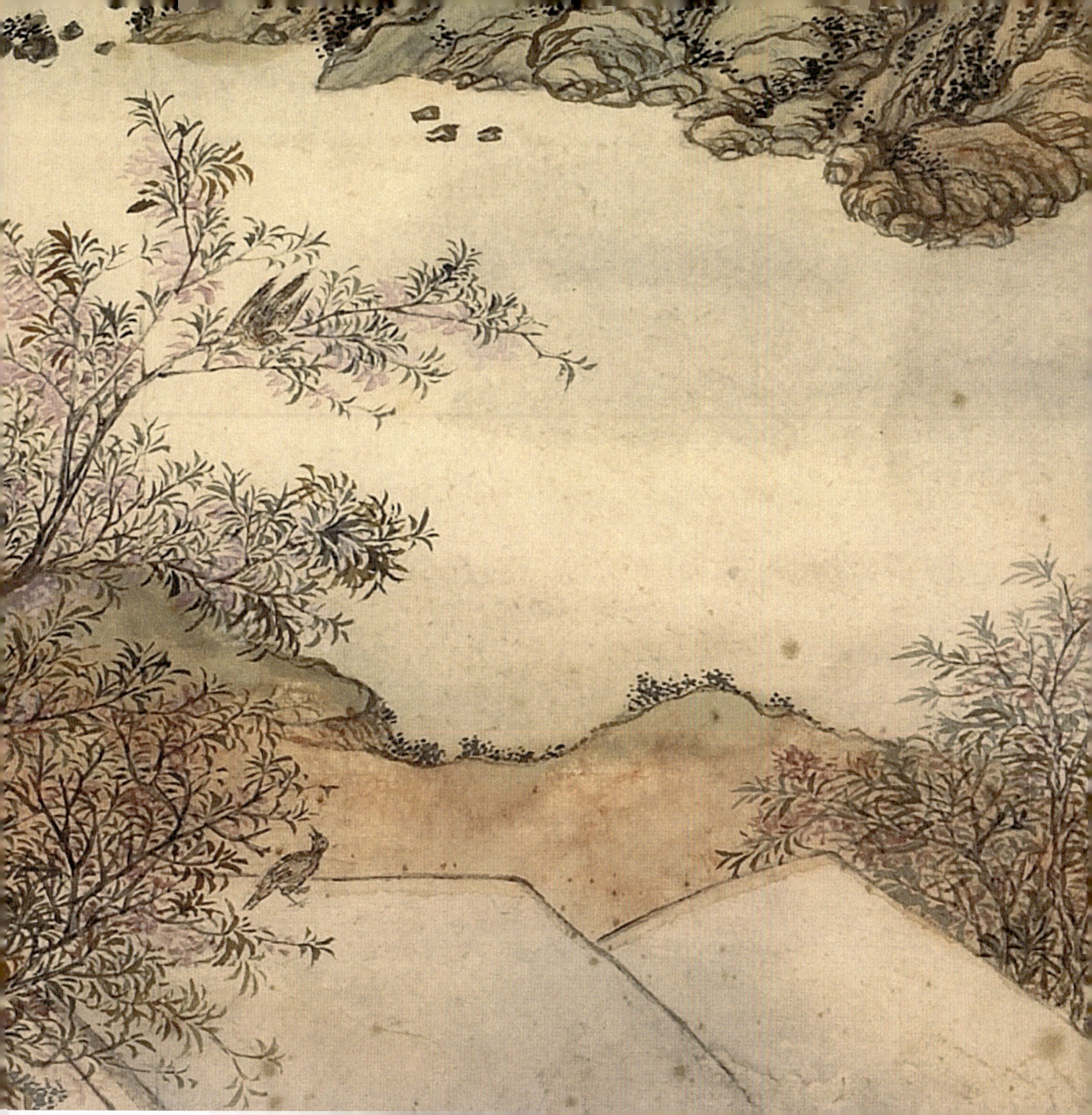

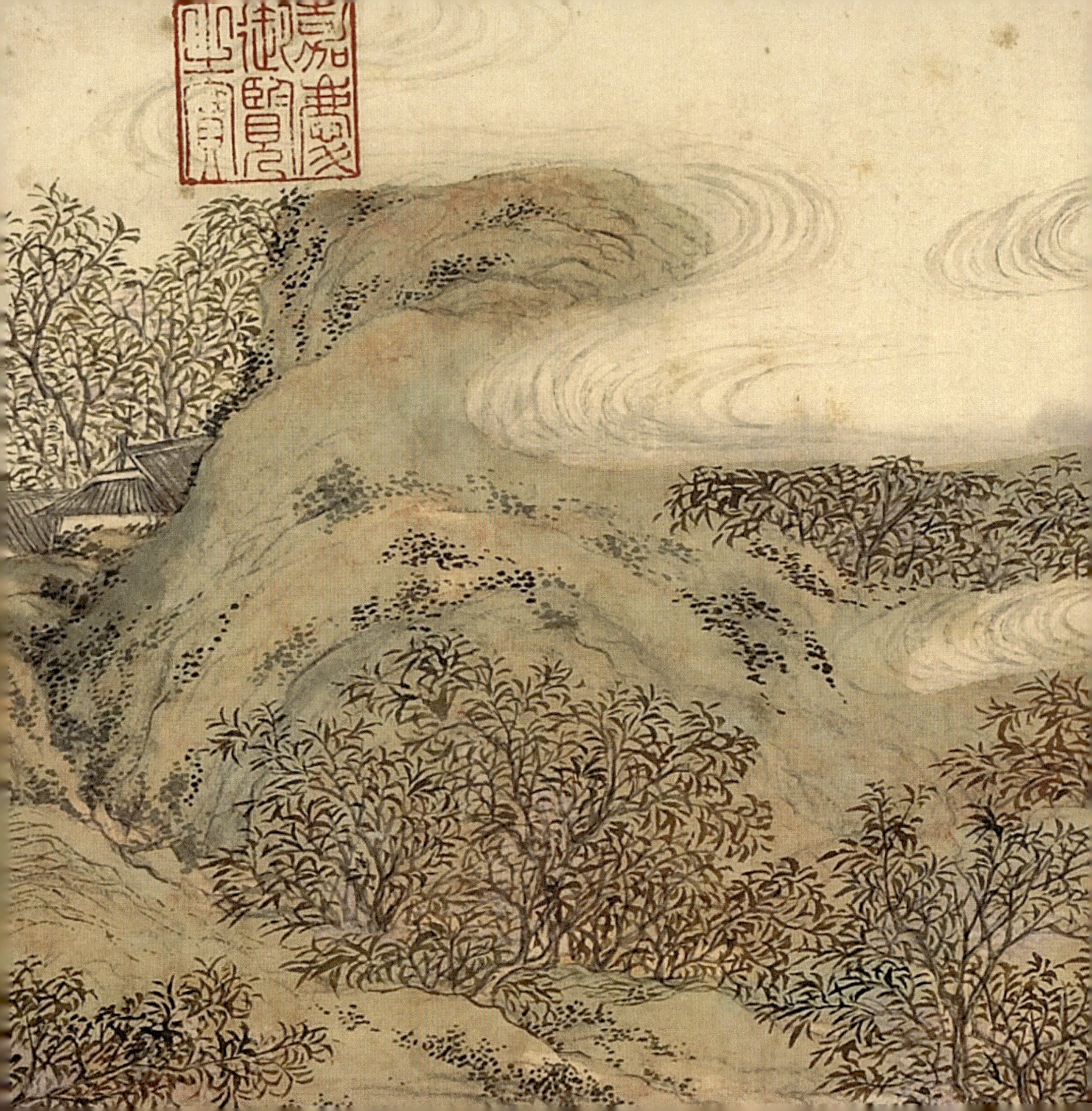
嘉慶御覽之寶

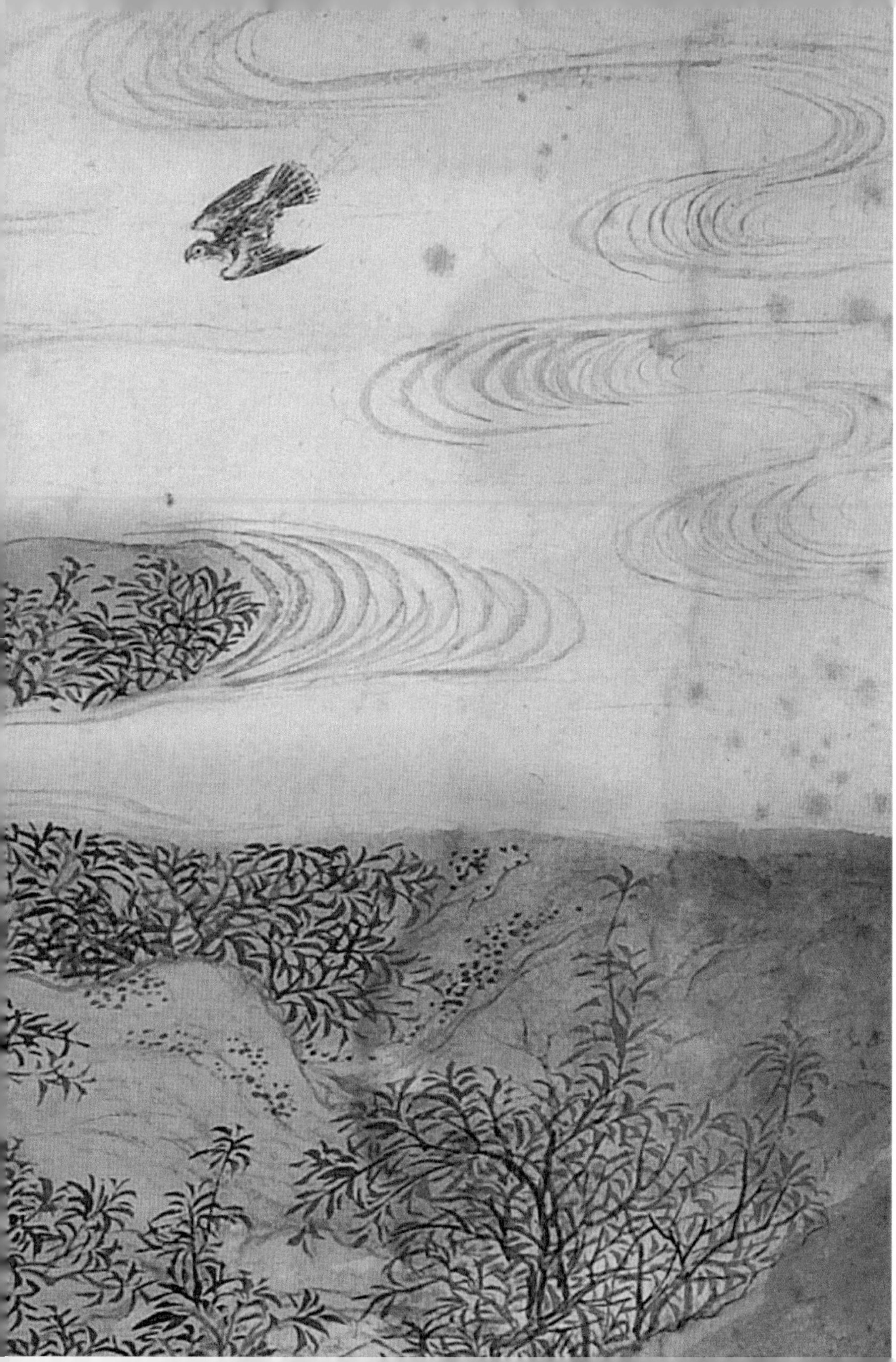

三候

# 鷹化為鳩

## Hawks Transform into Doves

The third hint is “Hawks Transform into Doves” (鷹化為鳩 Yīng Huà Wéi Jiū), which is a metaphorical representation of transformation and new beginnings. As hawks migrate away, doves and other gentle birds become more prominent, symbolizing a shift from the harsher aspects of nature to a more peaceful and nurturing environment. It signifies a time of harmony and rebirth, as the harshness of winter gives way to the gentleness of spring.

# 分

## Fēn

Chūn 春

# THE SPRING EQUINOX

March 21 (± 1)

# 春分

## The Spring Equinox

Chūnfēn : The Spring Equinox, known as "Day and Night Divide," indicating equal length of day and night. It's a time for balance and harmony in nature.

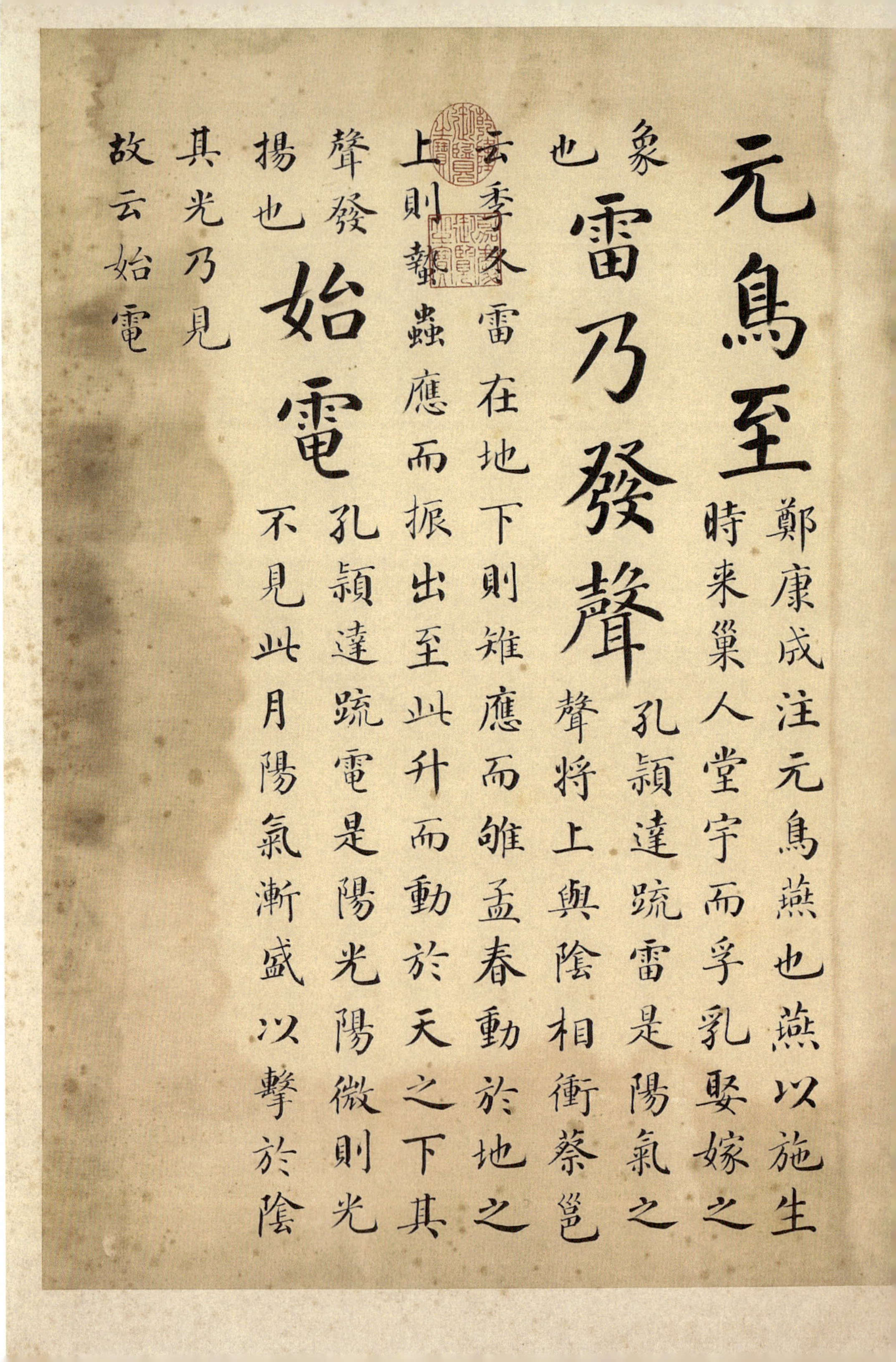
元鳥至
鄭康成注元鳥燕也燕以施生時来巢人堂宇而孚乳娶嫁之象也
雷乃發聲
孔穎達疏雷是陽氣之聲将上與陰相衡蔡邕云季冬雷在地下則雉應而雊孟春動於地之上則蟄蟲應而振出至此升而動於天之下其聲發揚也
始電
孔穎達疏電是陽光陽微則光不見此月陽氣漸盛以擊於陰其光乃見故云始電

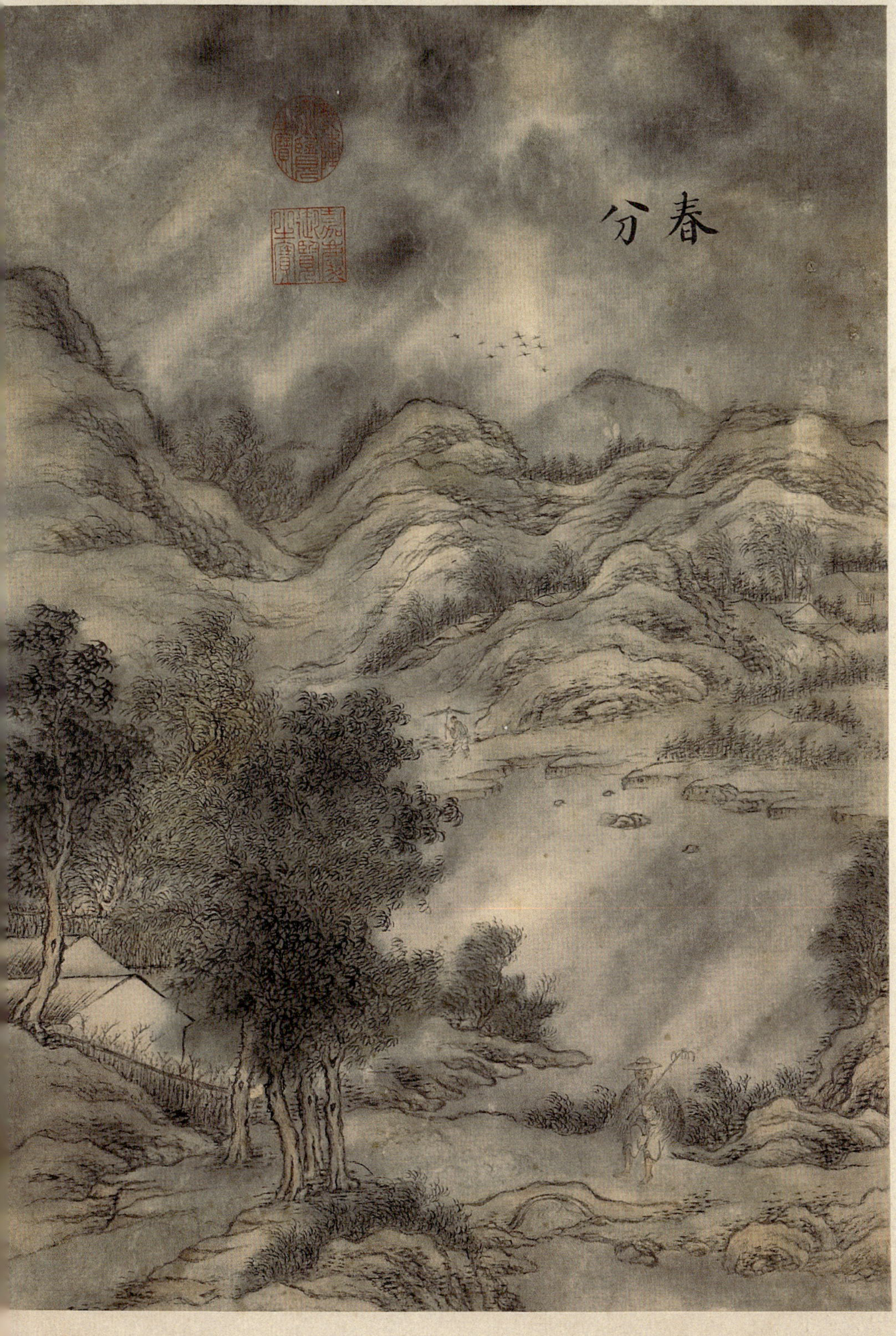


初候

## 元鳥至
## Swallows Arrive

二候

## 雷乃發聲
## Thunder Begins to Sound

三候

## 始電
## Lightning Commences

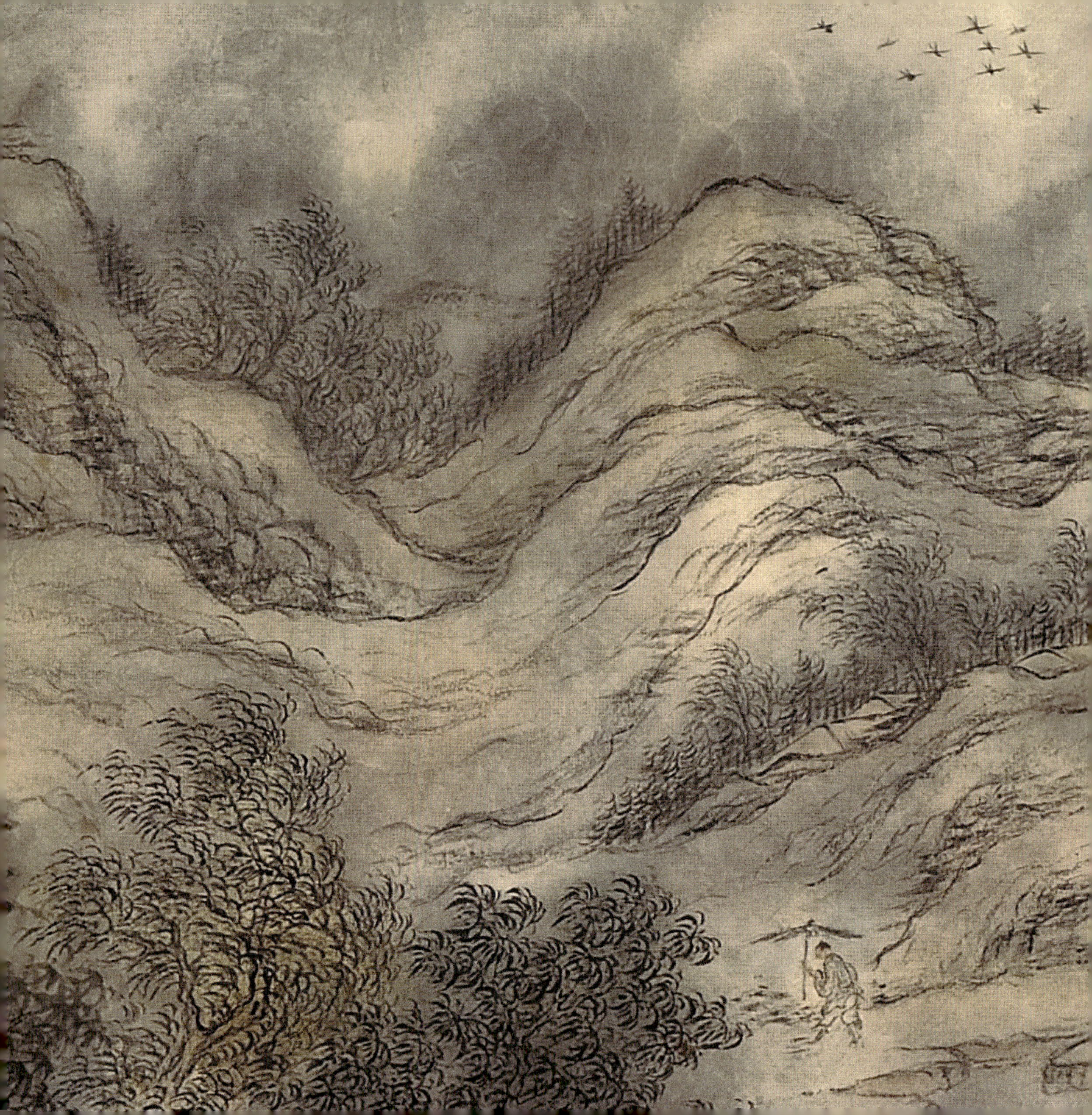

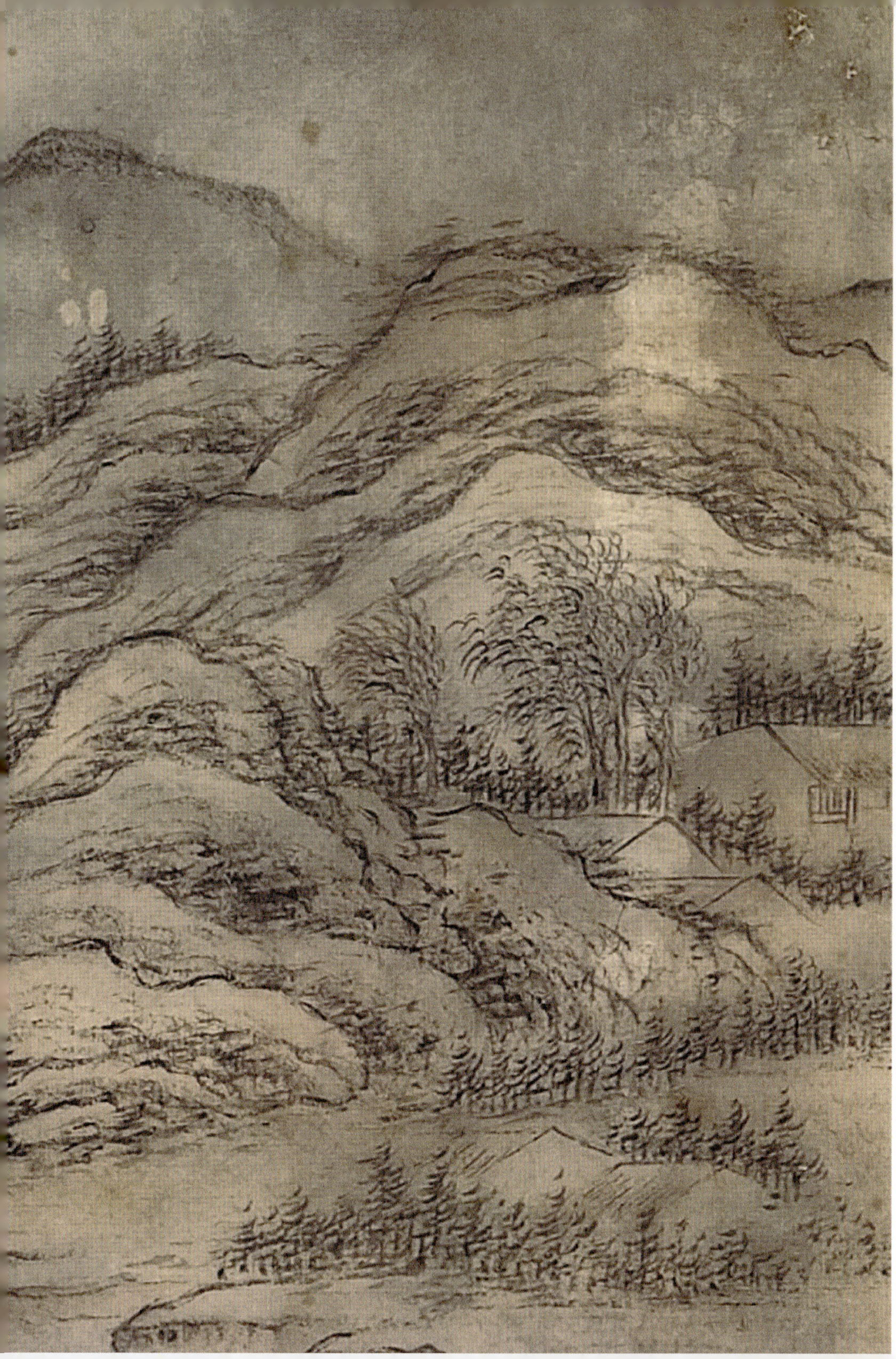

初候

# 元鳥至

## Swallows Arrive

The initial marker is "Swallows Arrive" (元鳥至 Yuán Niǎo Zhì), which is when swallows, also known as "元鳥Yuán Niǎo" or "玄鳥Xuán Niǎo," return from the South to signal the arrival of spring. These migratory birds are a symbol of auspiciousness and harmony, known for their arrival at the Spring Equinox and departure at the Autumn Equinox. As they build nests under the eaves of homes, they begin preparing to nurture their next generation, embodying a sign of good fortune and the renewal of life.

二候

# 雷乃發聲

## Thunder Begins to Sound

The second sign is "Thunder Begins to Sound" (雷乃發聲 Léi Nǎi Fā Shēng), which is marked by the first sounds of thunder in the season, symbolizing the awakening of yang energy and the dynamic force of nature. Thunder, representing sound and yang energy, is seen as the harbinger of the fertile and active season to come. It reflects the ancient understanding of the interconnectedness of natural phenomena, where thunder and lightning are part of the grand orchestration of the cosmos.

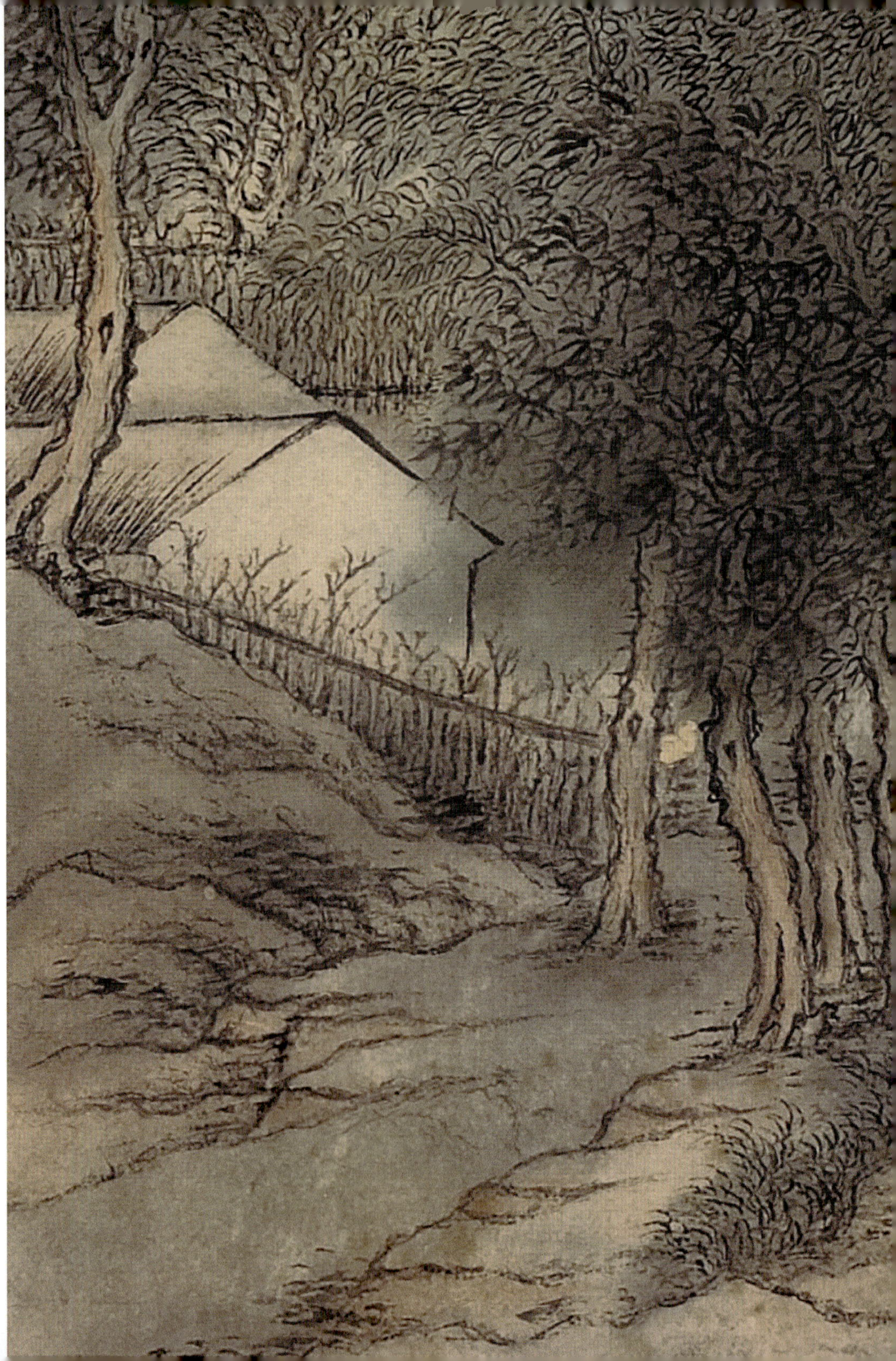

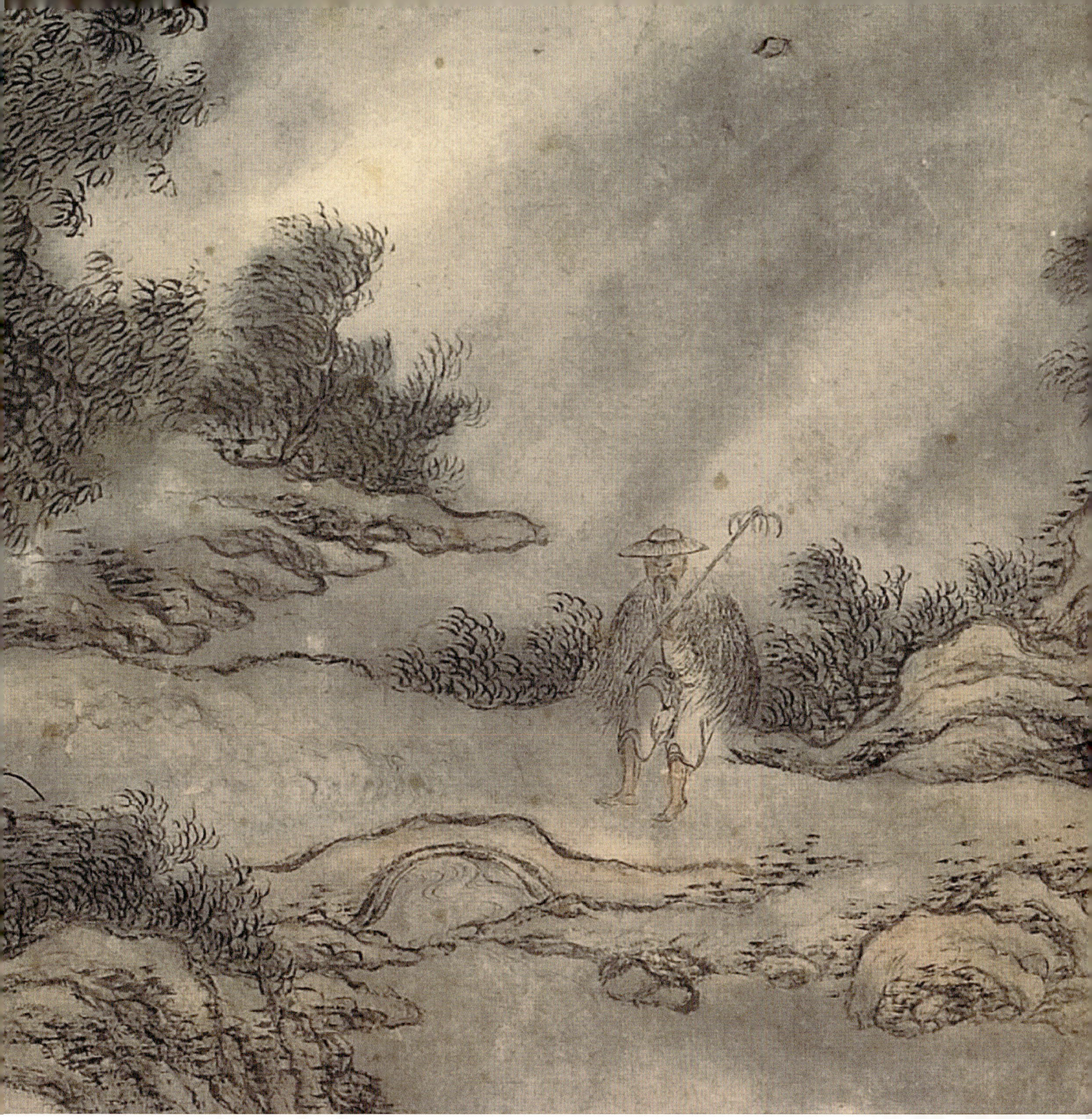

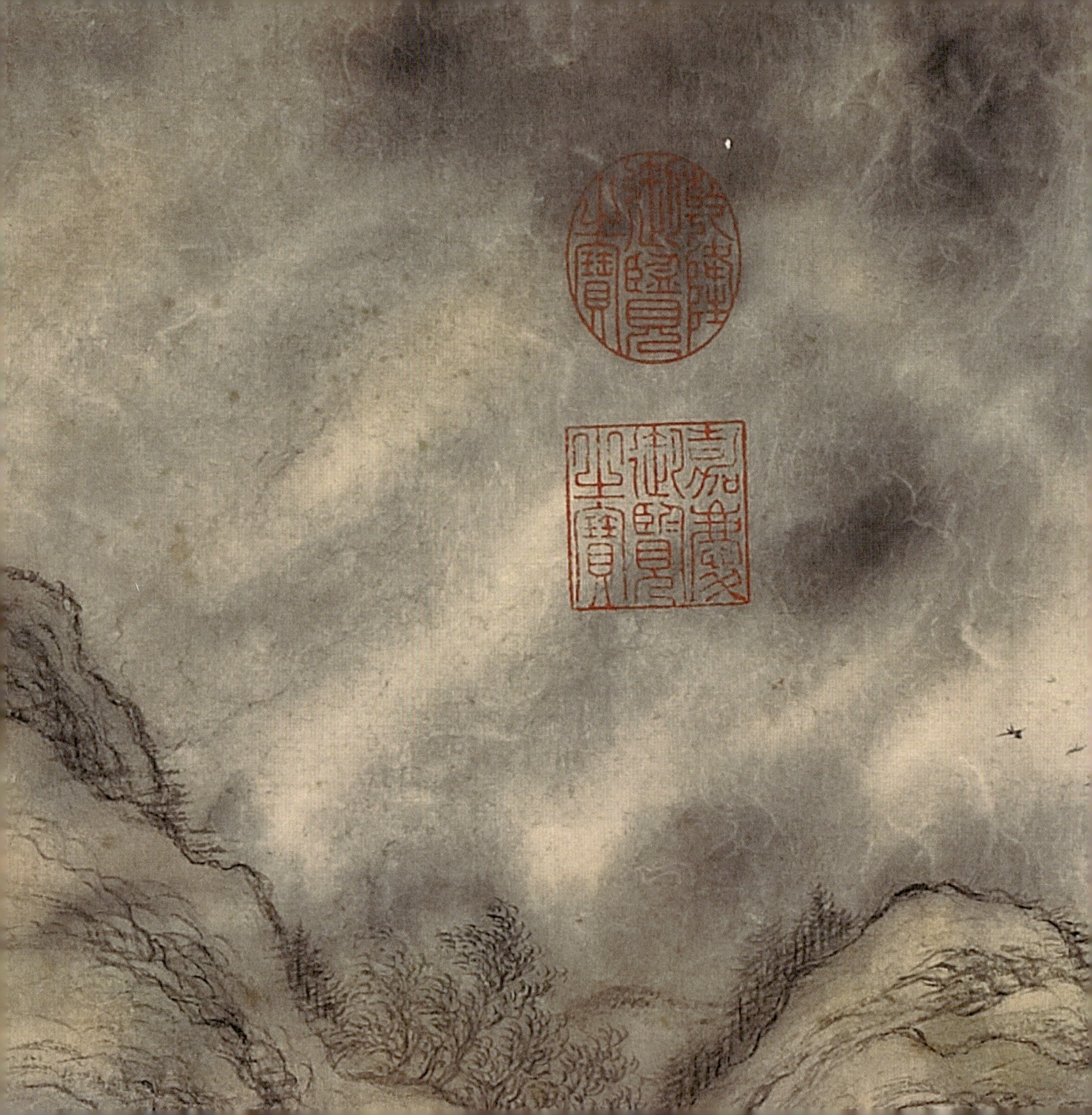
乾隆御覽之寶
嘉慶御覽之寶

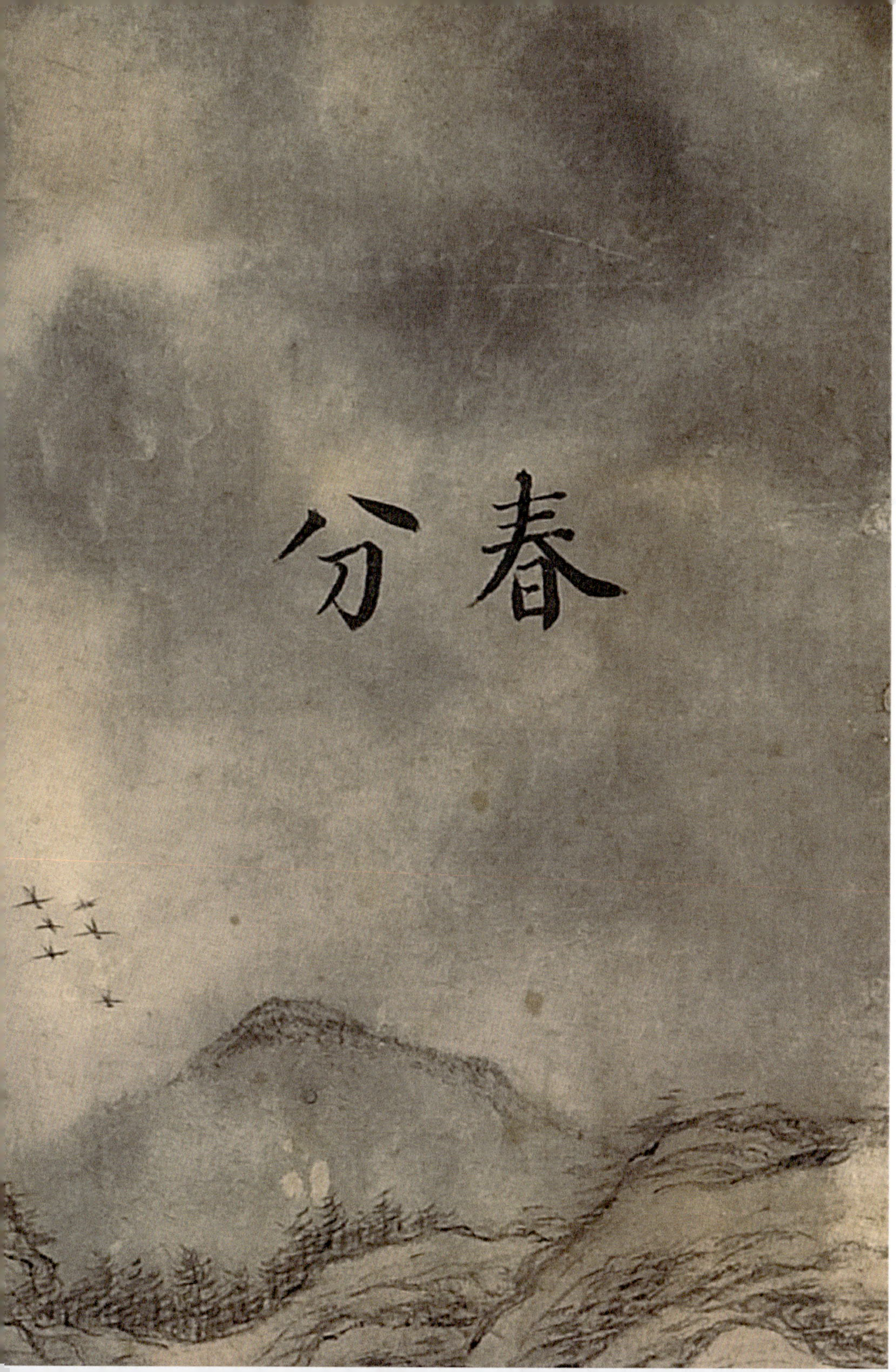


三候

# 始電

## Lightning Commences

The third marker, "Lightning Commences" (始電 Shǐ Diàn), represents the commencement of lightning as a significant phenomenon in the natural world. Ancient understanding acknowledged that while lightning visually precedes thunder, it is conceptually sequenced after thunder due to the belief in the precedence of yang (thunder) over yin (lightning). Thunder, embodying the more vigorous yang energy, is thus the focal point of the preceding marker, while lightning, associated with the more subtle yin energy, takes prominence in this marker. This division reflects a profound respect for the interplay of yin and yang forces in nature, with lightning and thunder together symbolizing the dynamic and powerful changes of the season.

明

Míng

Qīng

清

# PURE BRIGHTNESS

April 5 (± 1)

# 清明

## Pure Brightness

Qīngmíng : Meaning “Clear and Bright,” derives its name from the “Eight Winds” described in the Huainanzi, a classic text of philosophy. These winds, starting from the Winter Solstice, include the Tiao Wind, Ming Shu Wind, Qingming Wind, and others, each symbolizing different elements of change and atmosphere. Qingming is specifically associated with the Qingming wind, a southeast wind that brings warm and abundant rainfall, nurturing the earth. The designation of Qingming as a solar term dates back to the Warring States period, signifying a time of ancestral reverence and the sweeping of tombs, as well as the appreciation of the thriving life as nature bursts into bloom.

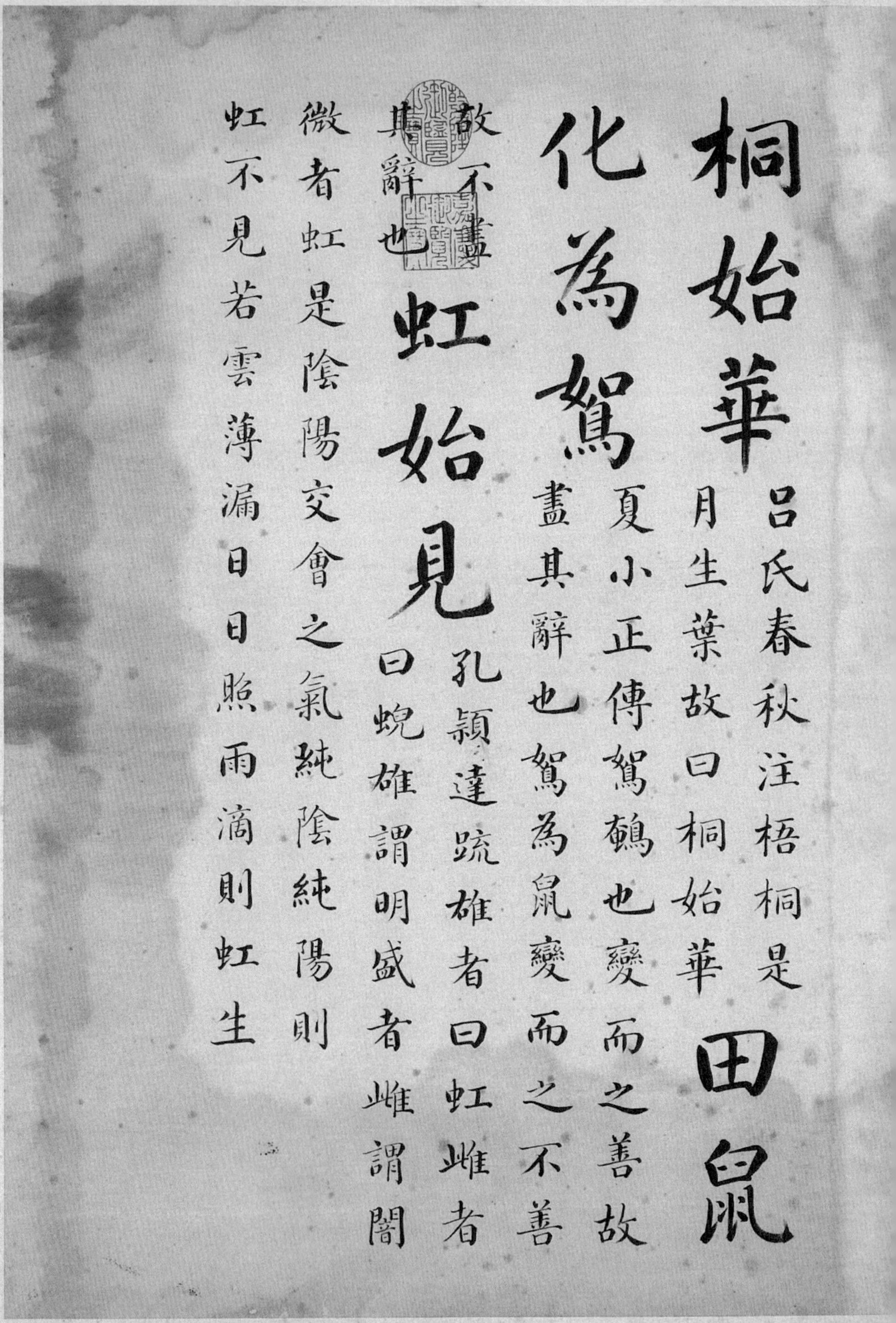

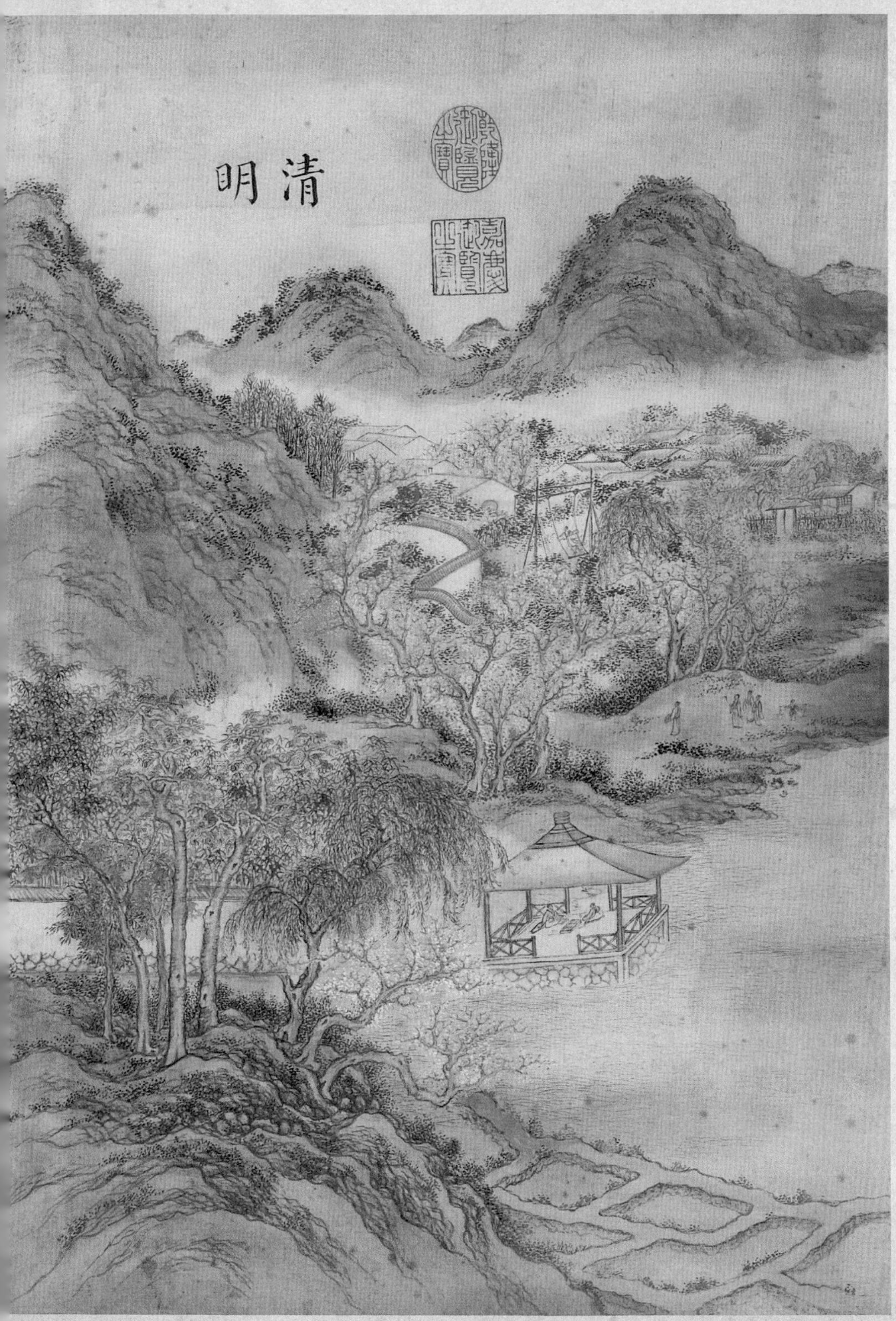


初候

## 桐始華

## Paulownia Blossoms Emerge

二候

## 田鼠化為鴽

## Field Mice Turn to Quails

三候

## 虹始見

## Rainbows Begin to Appear

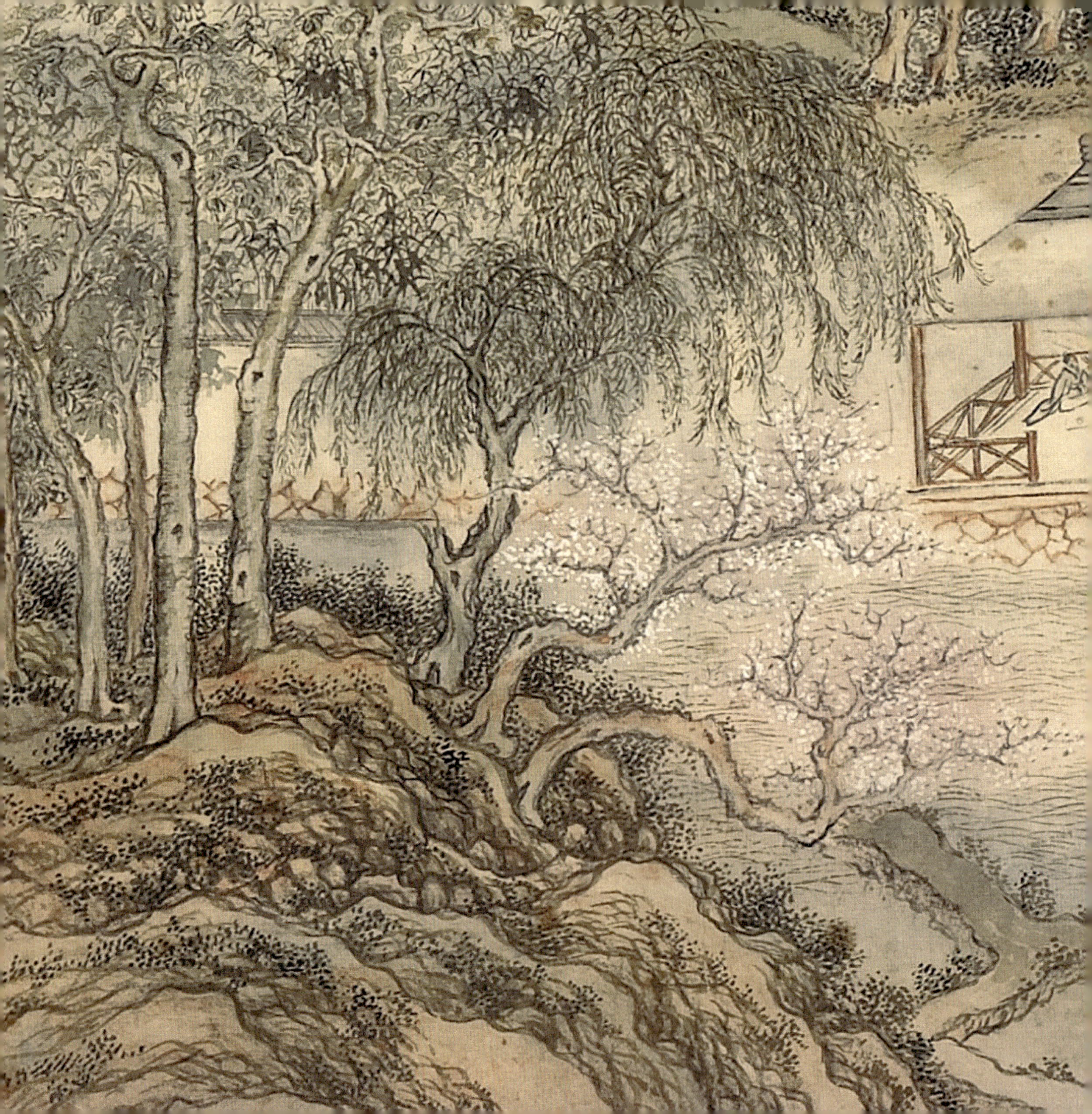

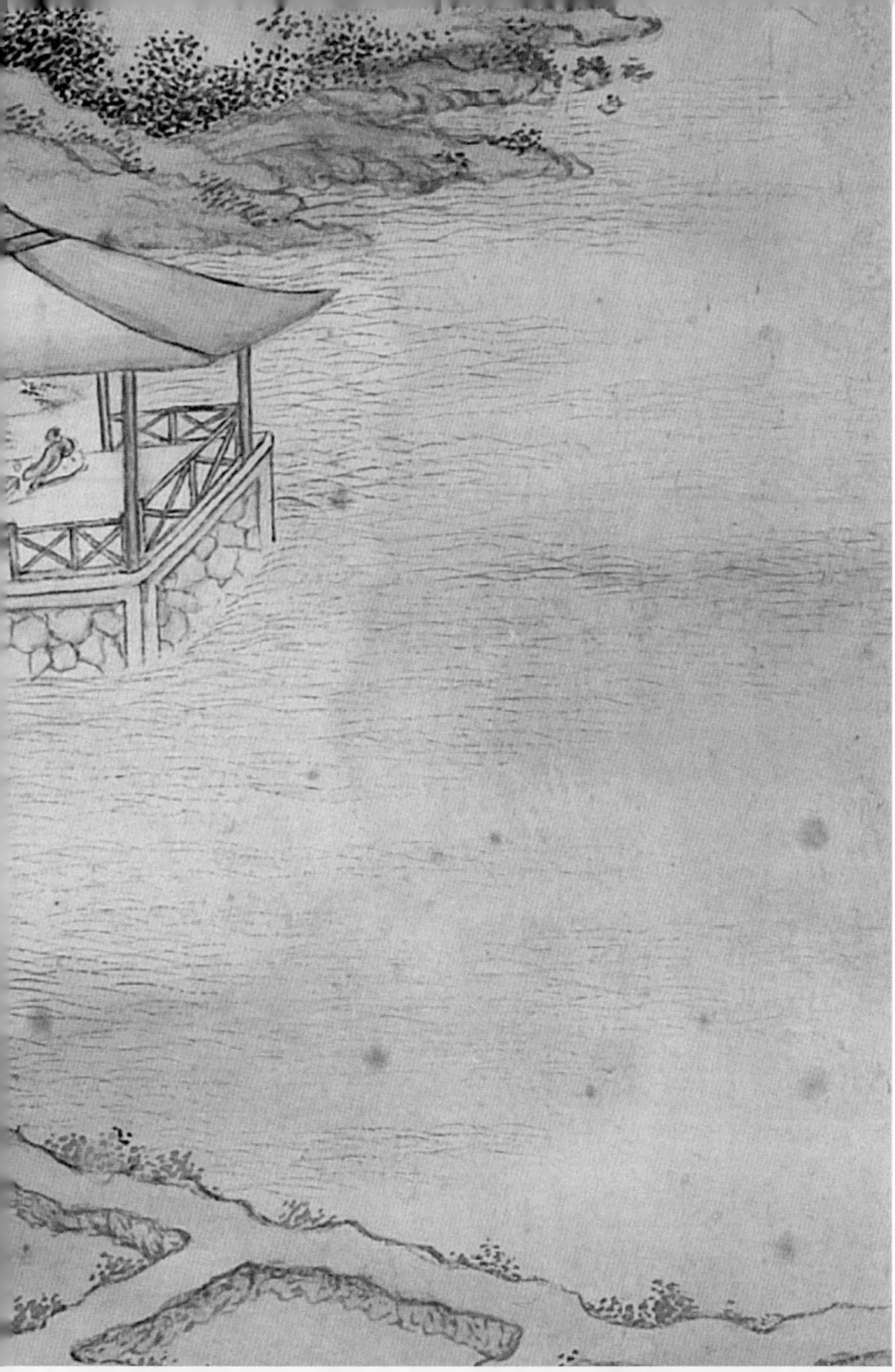

初候

# 桐始華

## Paulownia Blossoms Emerge

"The first sign is ""Paulownia Blossoms Emerge"" (桐始華 Tóng Shǐ Huá). At the beginning of Qīngmíng, Paulownia trees start to bloom, filling the mountains with their delicate white and pale purple flowers. This blooming signifies the further warming of the weather and the full swing of spring, as the landscape is adorned with the gentle hues of the Paulownia blossoms."

二候

# 田鼠化為鴽

## Field Mice Turn to Quails

The second sign is "Field Mice Turn to Quails" (田鼠化為鴽 Tián Shǔ Huà Wéi Rú).
As the sun grows stronger, field mice retreat into their burrows to escape the heat, and quails, which thrive in the Yang energy, become more active. This period symbolizes the shift from Yin to Yang energy, with the diminishing presence of the mice and the emerging activity of the quails reflecting the natural order' s dynamic balance.

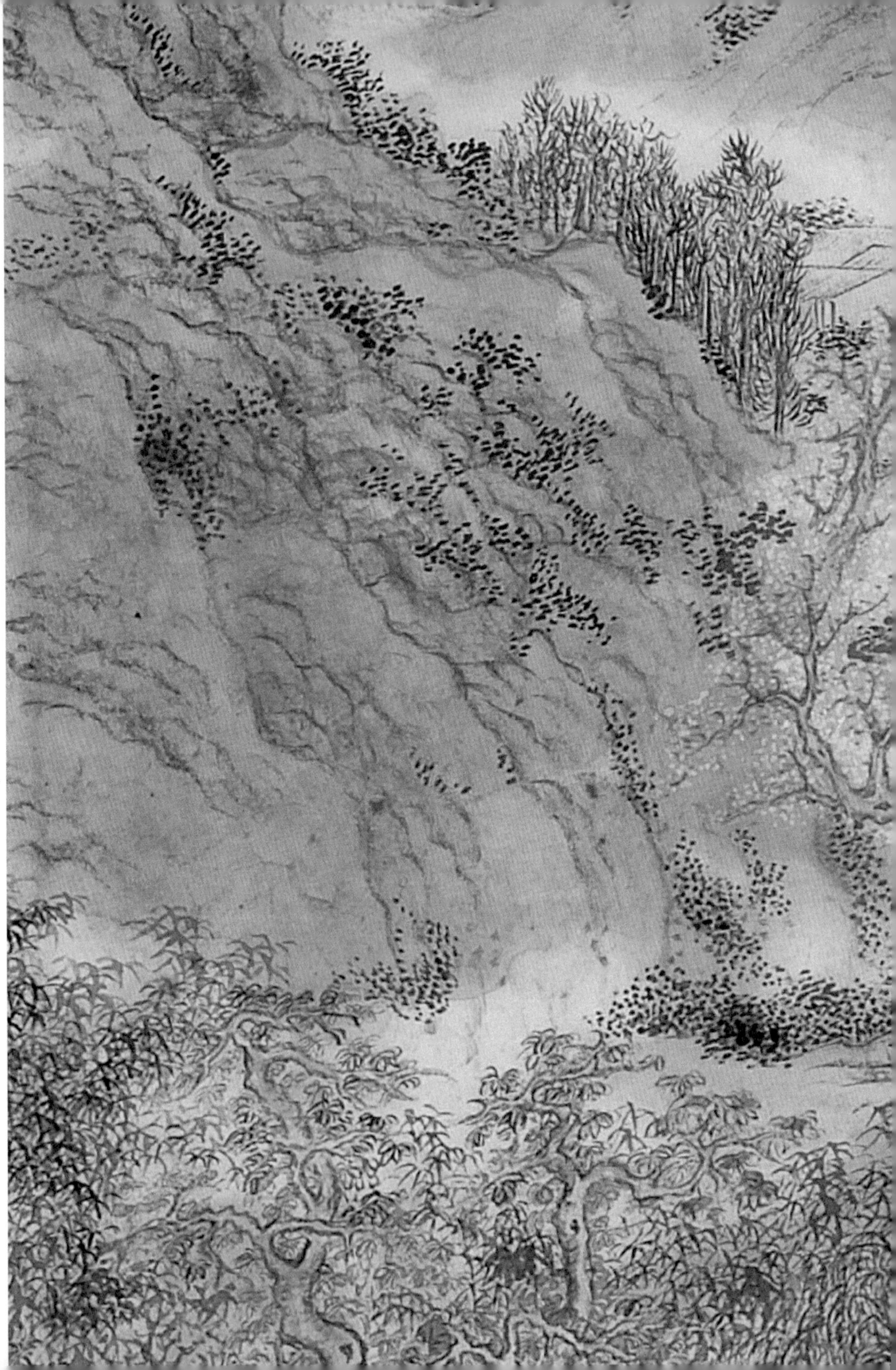

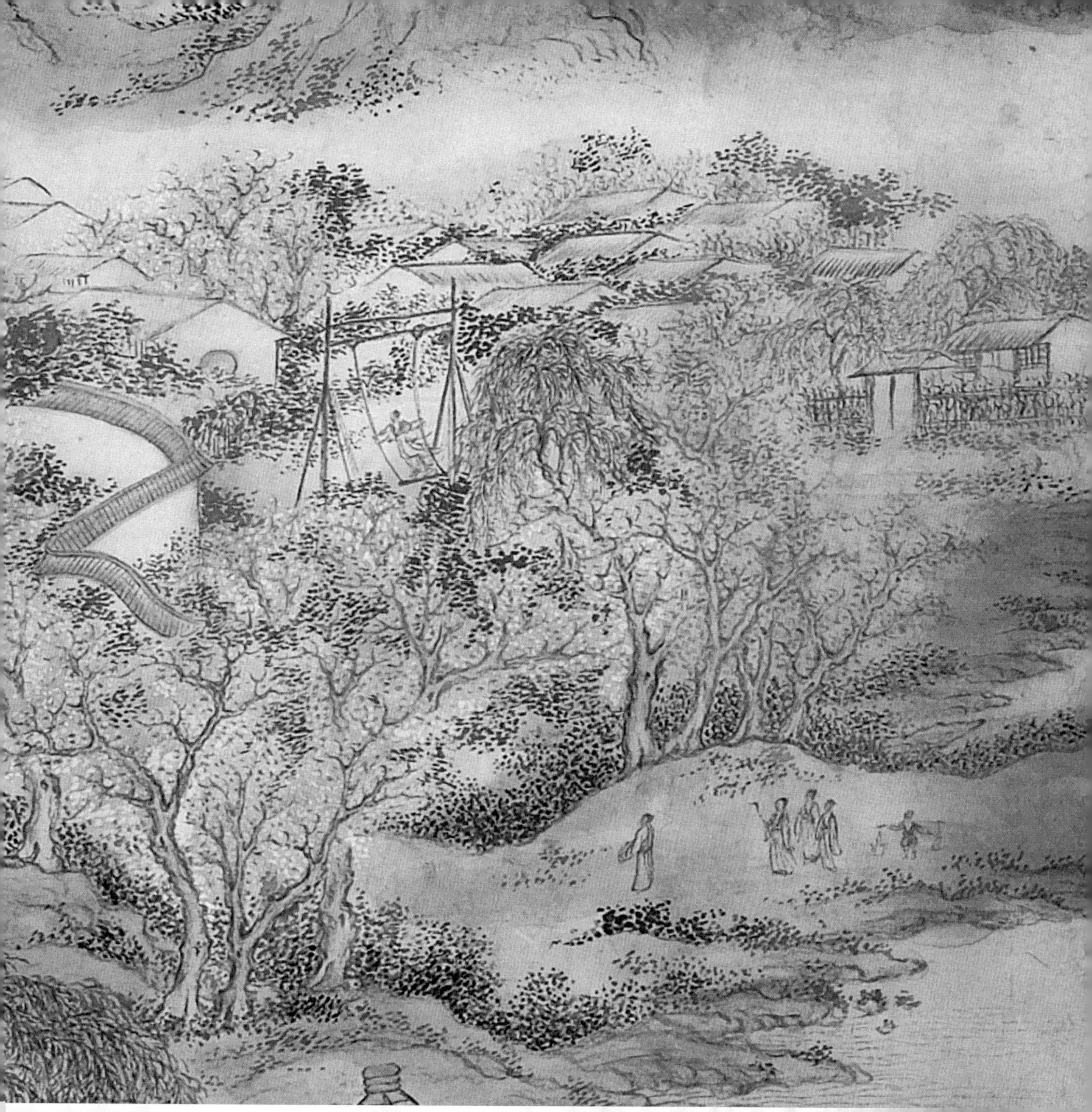

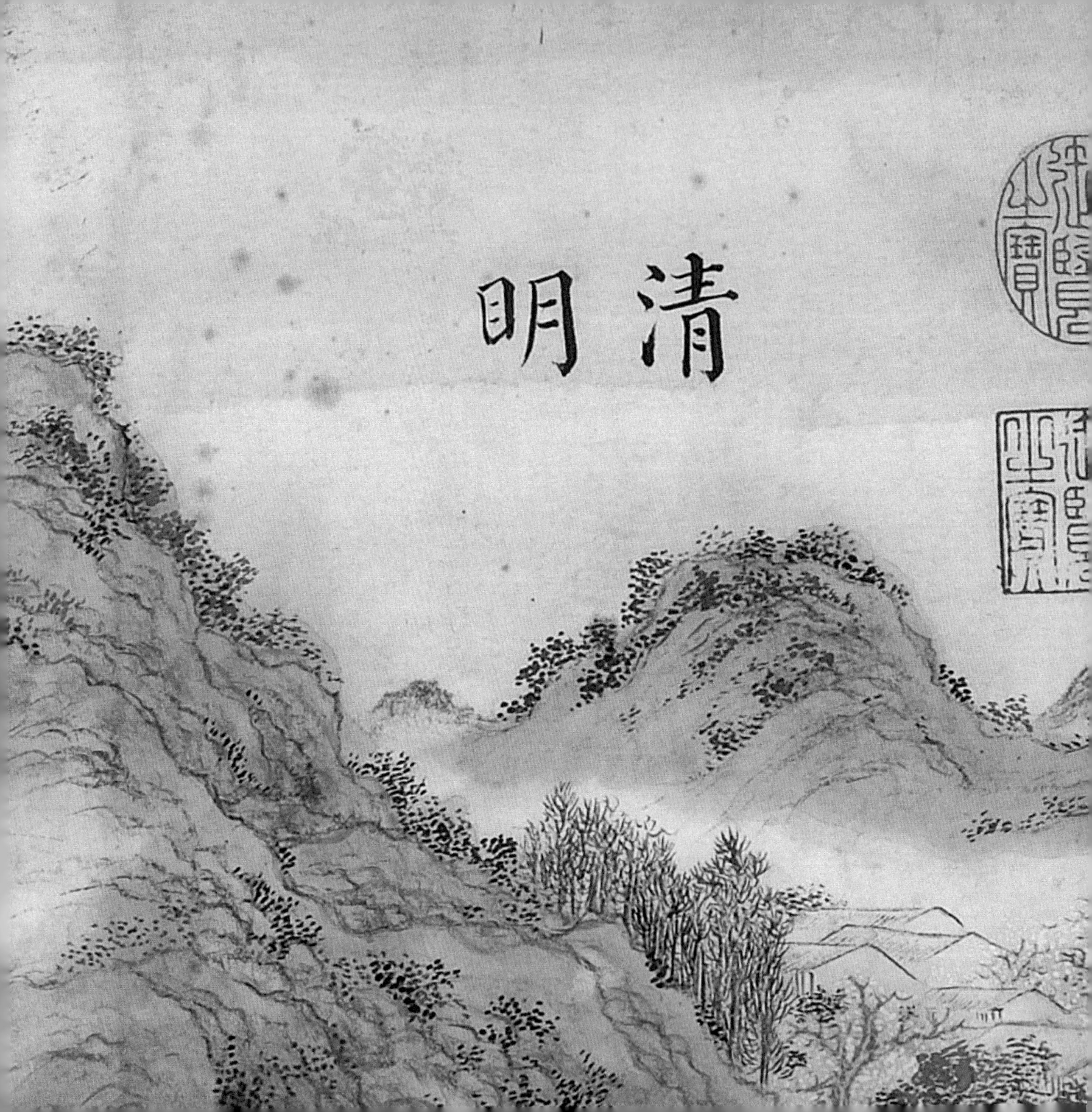
清明

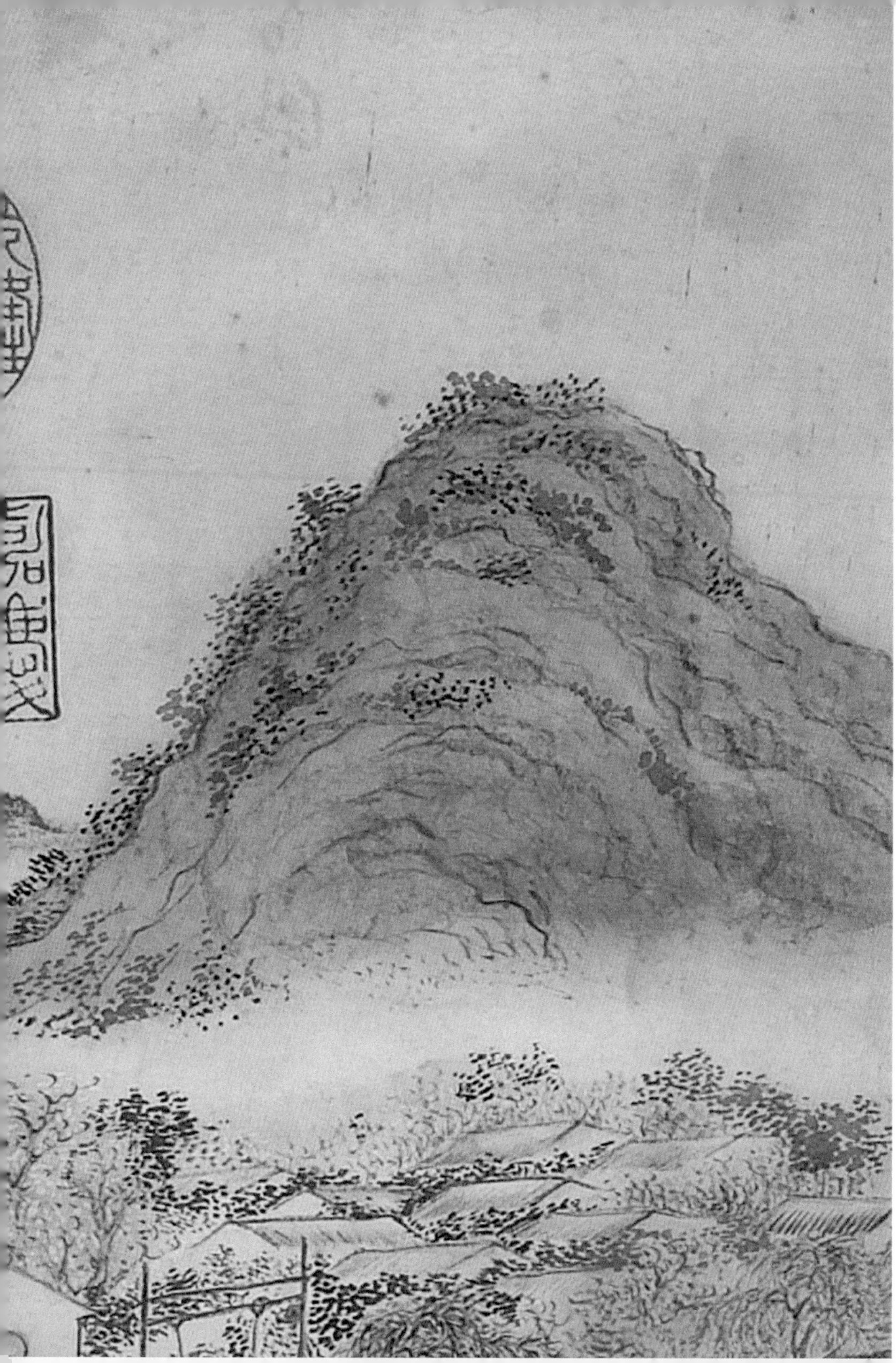

三候

# 虹始見

## Rainbows Begin to Appear

“The third sign is “”Rainbows Begin to Appear”” (虹始見 Hóng Shǐ Jiàn). Rainbows start to make their first appearances during Qīngmíng, often after the spring rains. The presence of rainbows in the sky is a visual testament to the season’ s frequent rains and the subsequent clearing, symbolizing hope, renewal, and the fruitful results of perseverance, much like the clearing of clouds to reveal a bright moon.”

雨

Yǔ

Gǔ

穀

# GRAIN RAIN

April 20 (± 1)

# 穀雨

## Grain Rain

Gǔyǔ : It is emblematic of the invaluable spring showers that foster the growth of crops. Its name is a reminder to farmers of the importance of these rains for the cultivation of grains. The term emphasizes the significance of timely rainfall, crucial for agricultural success and the flourishing of crops. As the last term of spring, it symbolizes the fullness of the season, with nature in its most vibrant and productive stage, urging those who work the land to take heed of the changing climate and prepare accordingly for the growing season ahead.

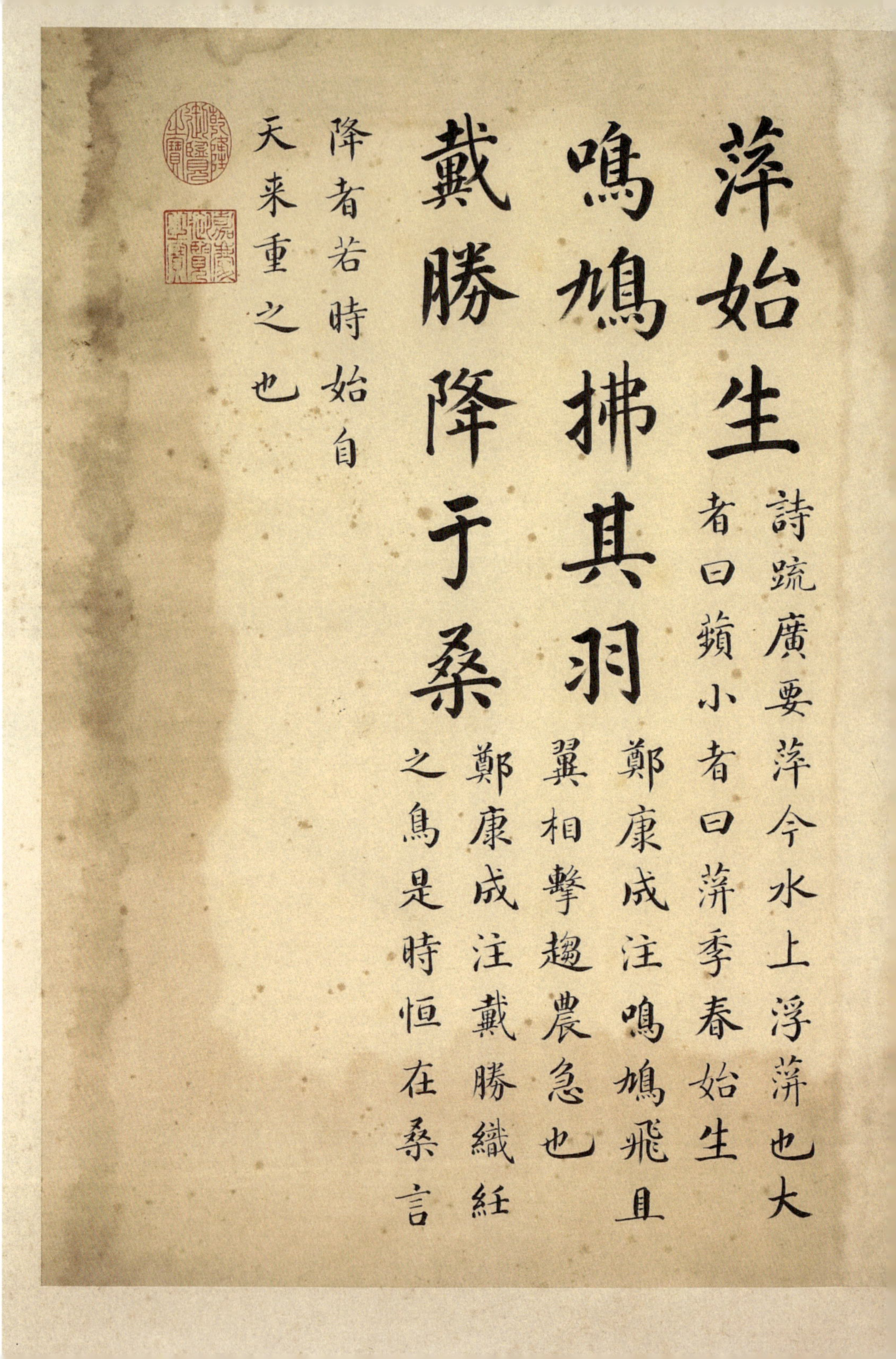
萍始生
詩疏廣要萍今水上浮蓱也大
者曰蘋小者曰蓱季春始生
鳴鳩拂其羽
鄭康成注鳴鳩飛且
翼相擊趨農急也
戴勝降于桑
鄭康成注戴勝織紝
之鳥是時恒在桑言
降者若時始自
天来重之也

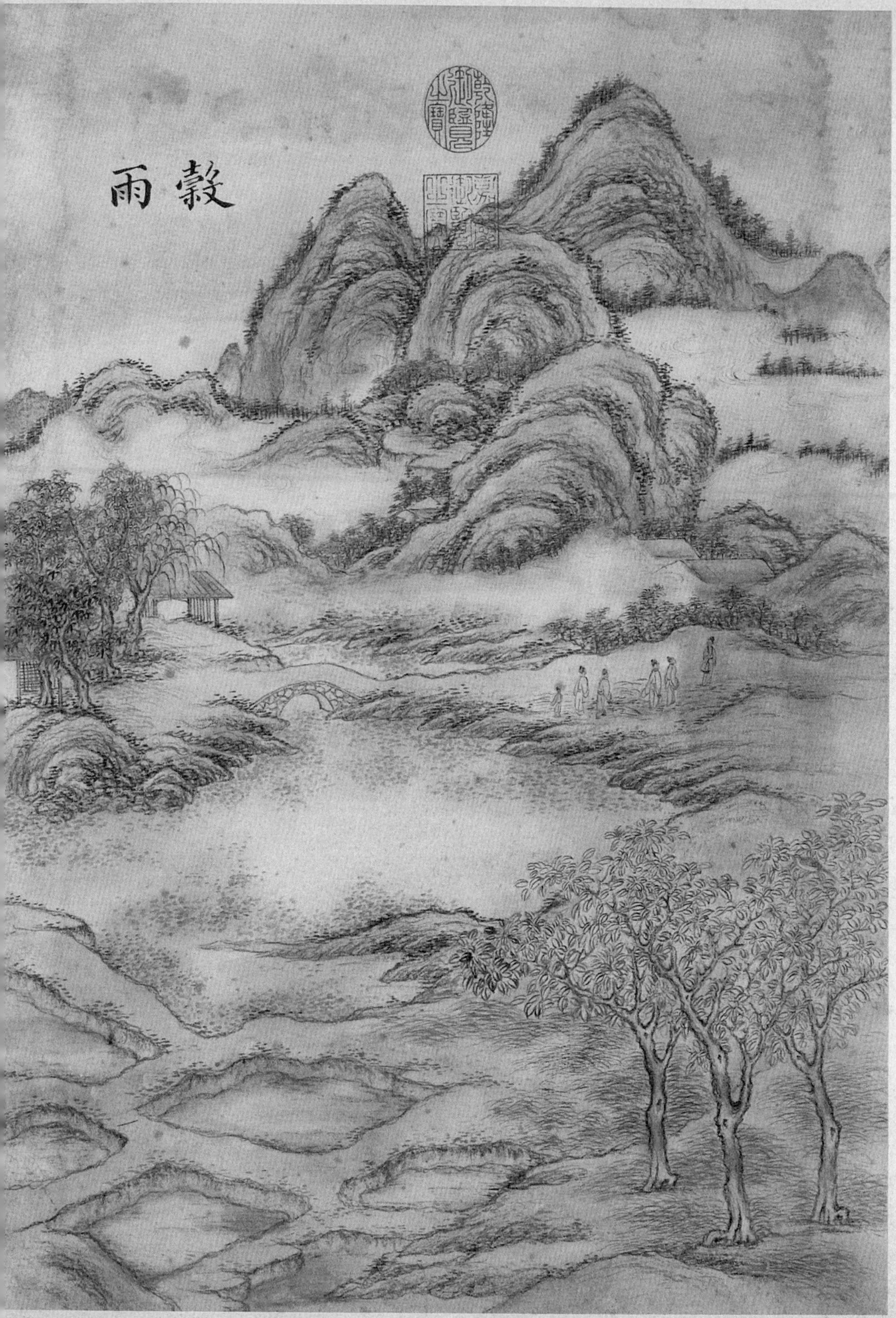


初候

## 萍始生
## Duckweed Begins to Thrive

二候

## 鳴鳩拂其羽
## Doves Coif Their Feathers

三候

## 戴勝降於桑
## Crested Birds Descend on Mulberries

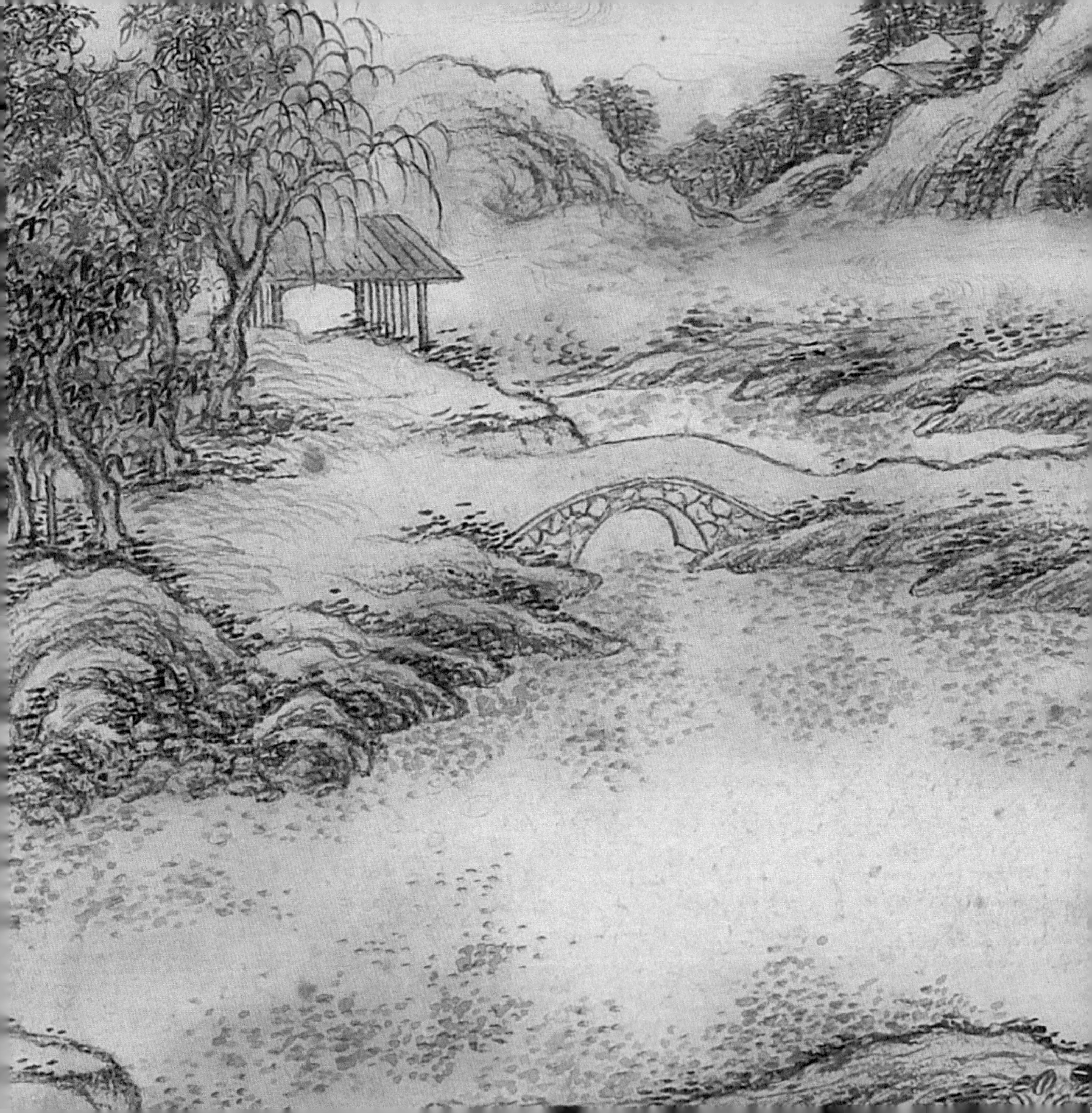

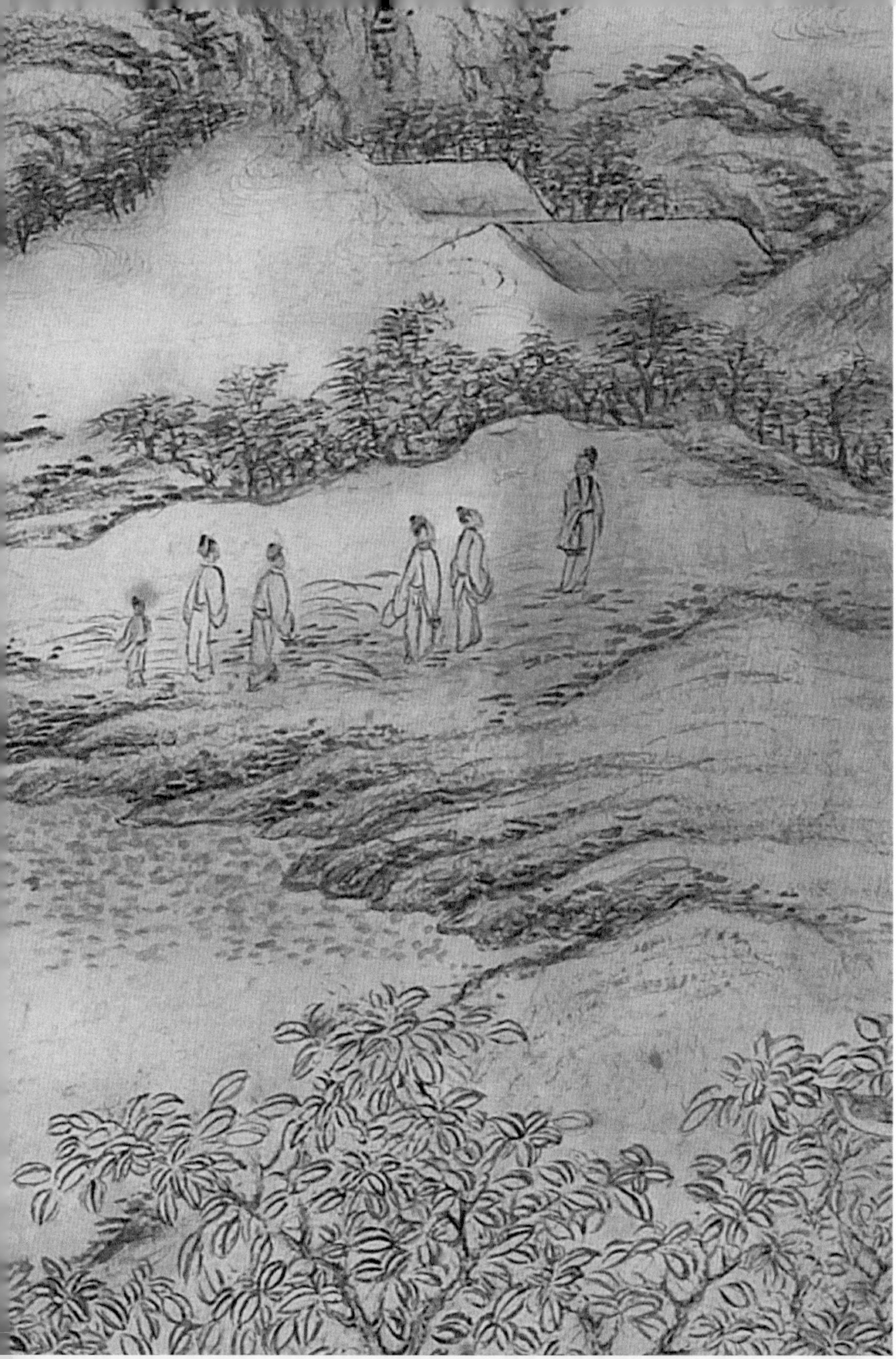

初候

# 萍始生

## Duckweed Begins to Thrive

The first marker is “Duckweed Begins to Thrive” (萍始生 Píng Shǐ Shēng). With the arrival of Gǔyǔ, duckweed starts to proliferate across the water surfaces, signaling the season’s richness and the beginning of a fertile period. Duckweed, a green aquatic plant, thrives with the ample rains of the season, symbolizing the abundance and vitality that Gǔyǔ brings to the natural world.

二候

# 鳴鳩拂其羽

## Doves Coif Their Feathers

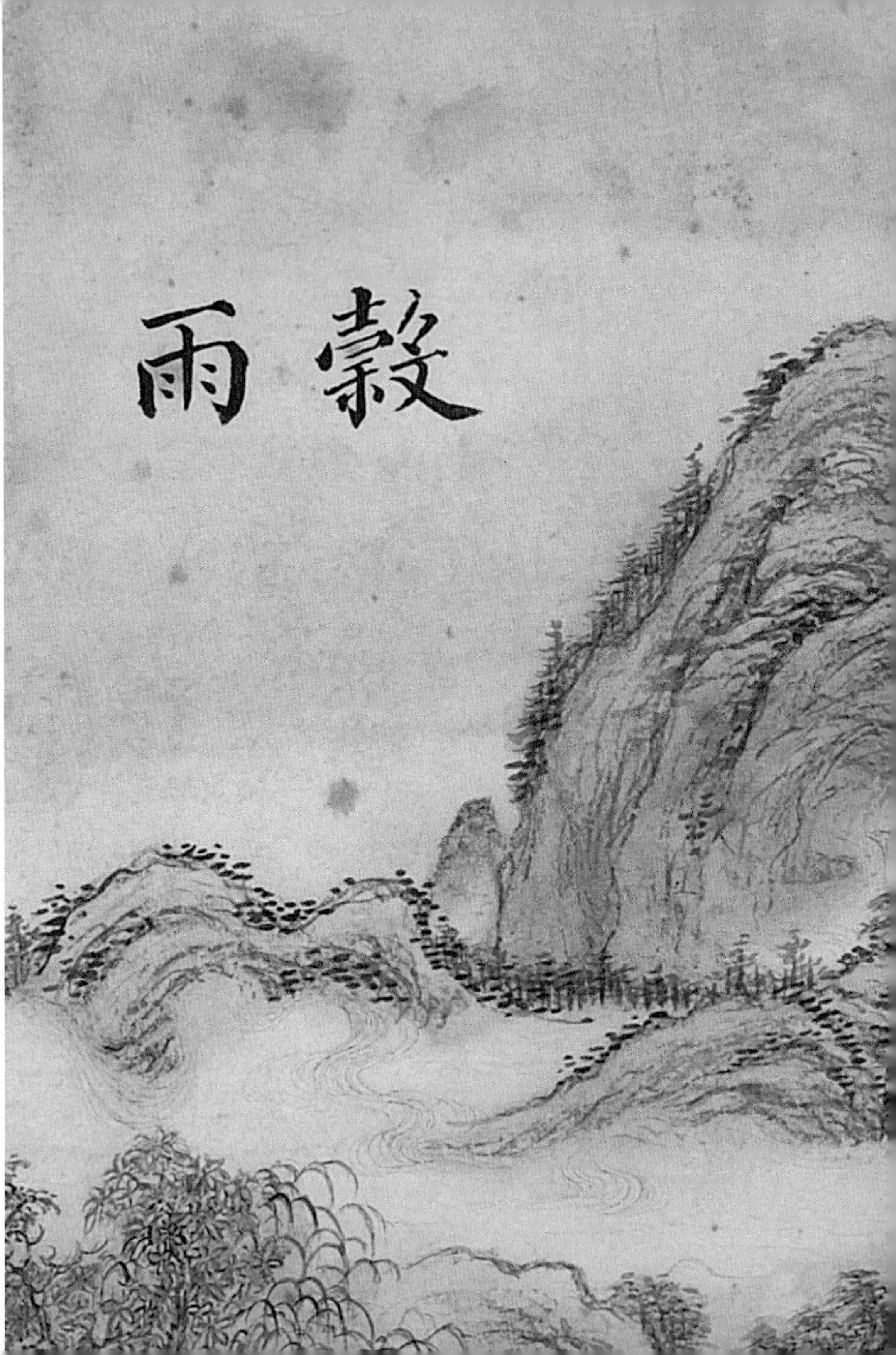


The second indication is “Doves Coif Their Feathers” (鳴鳩拂其羽 Míng Jiū Fú Qí Yǔ).

During this time, doves are seen actively preening and fluffing their feathers, filling the air with their gentle cooing. This behavior is not only part of their natural life cycle but also a reminder to farmers of the ongoing agricultural tasks, as the doves’ activities are intertwined with the rhythms of farming life.

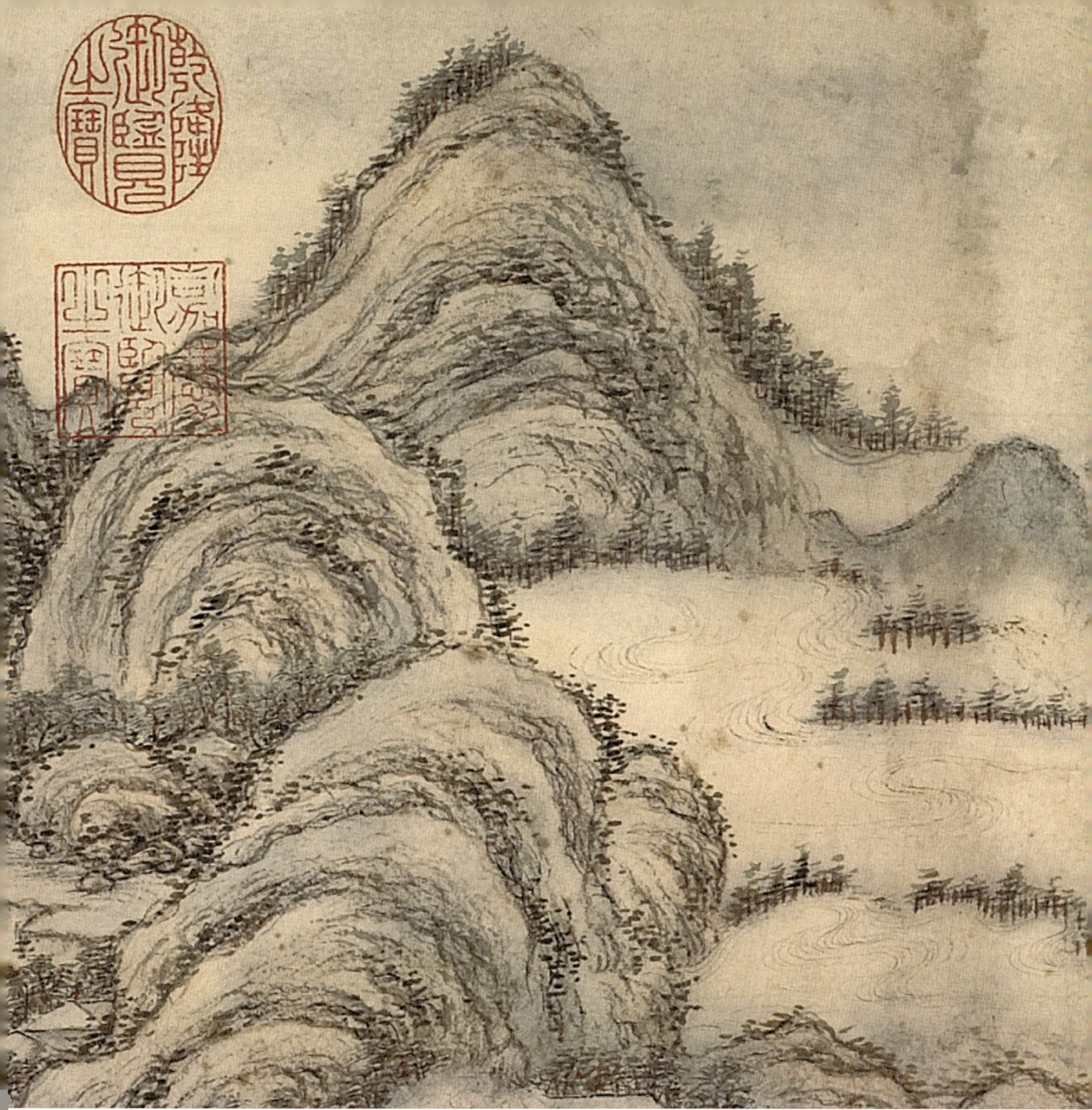

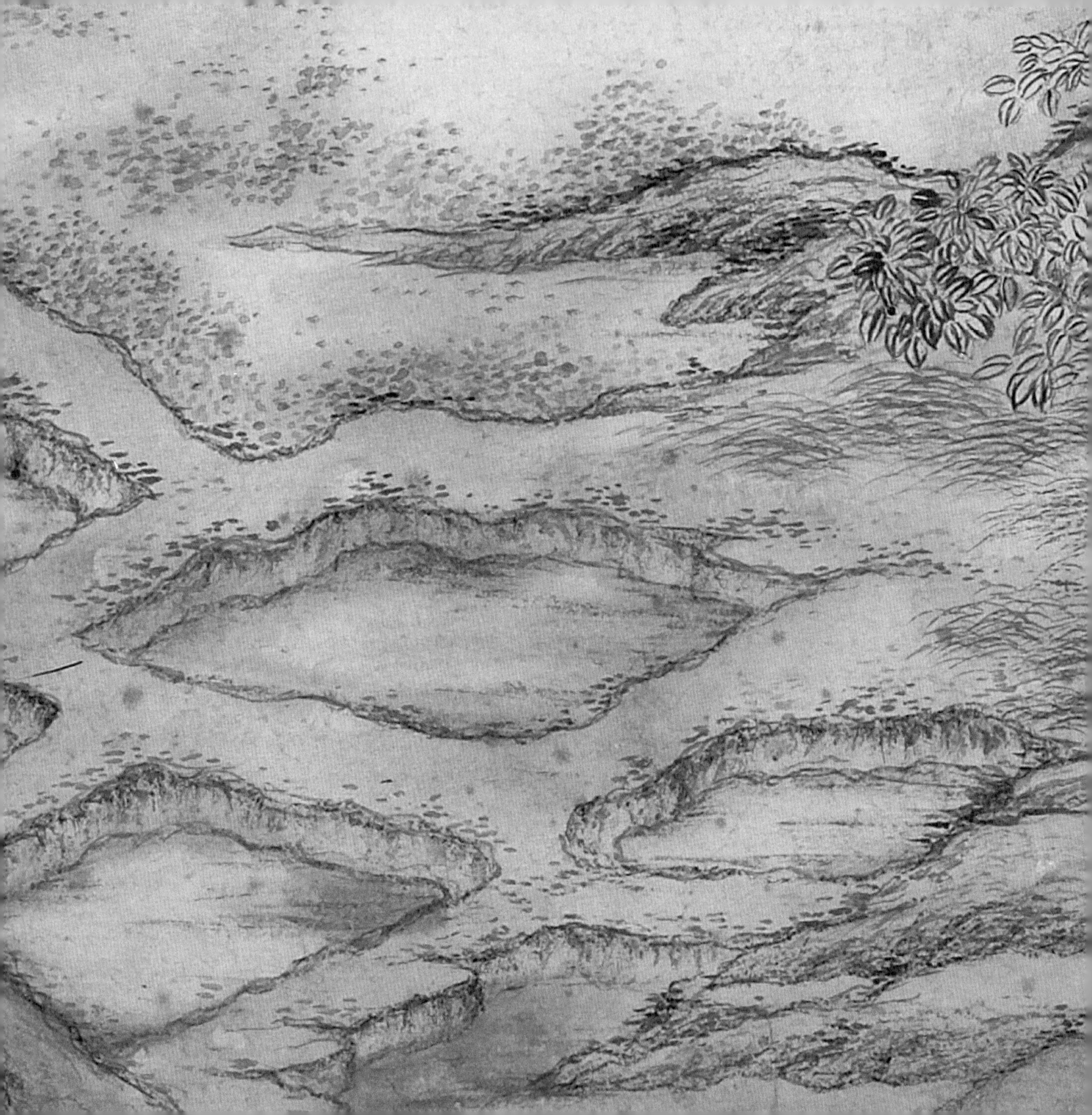

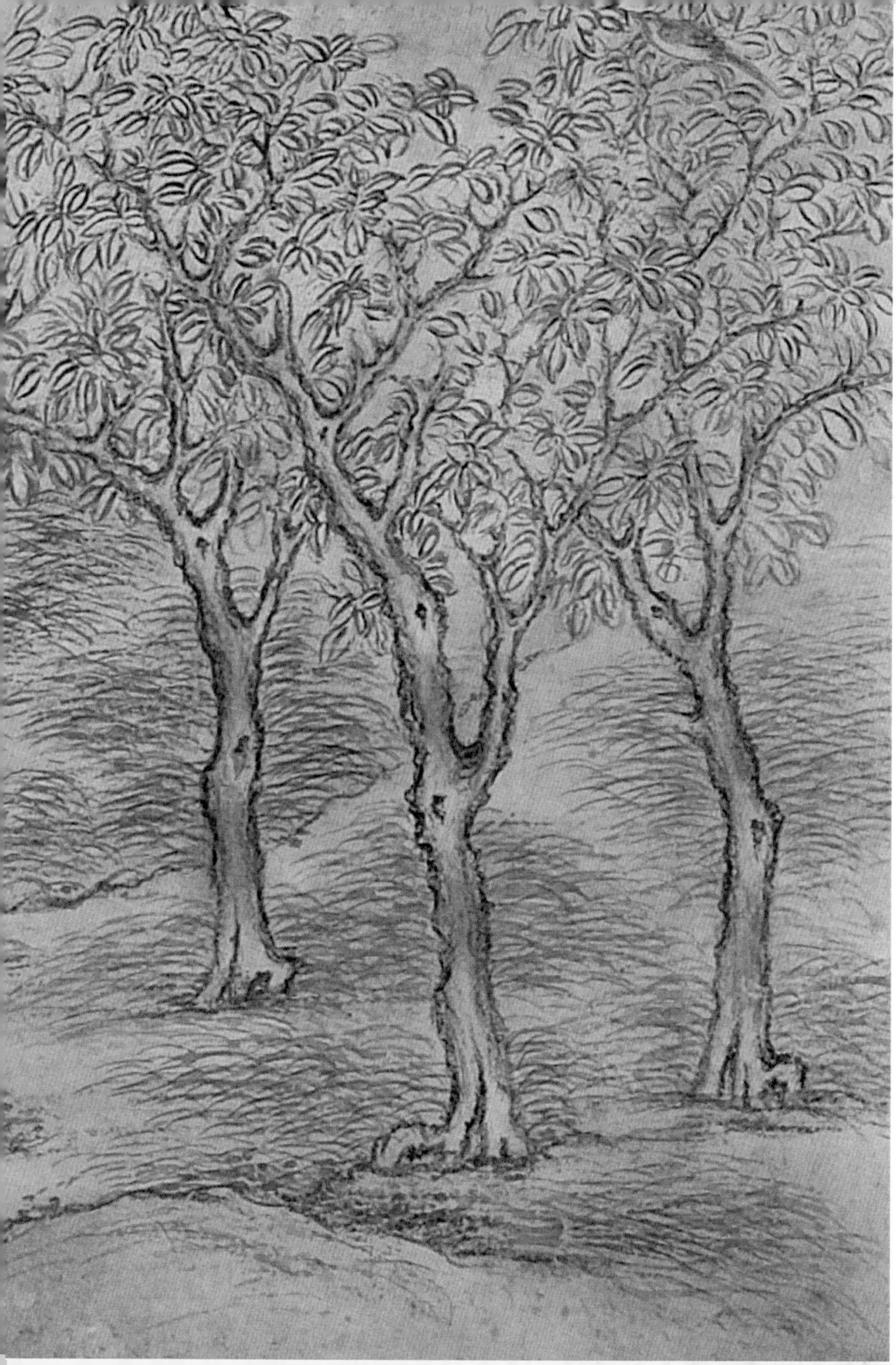

三候

# 戴勝降於桑

## Crested Birds Descend on Mulberries

The third signal is "Crested Birds Descend on Mulberries" (戴勝降於桑 Dài Shèng Jiàng Yú Sāng).
Crested birds, known for their distinctive tufts, migrate to areas along the Yellow and Yangtze rivers, often alighting on mulberry and hemp trees cultivated by farmers. These birds' arrival is a sign of the ongoing cycle of life and the interconnectedness of all beings, as they integrate into the landscape and the agricultural cycle.

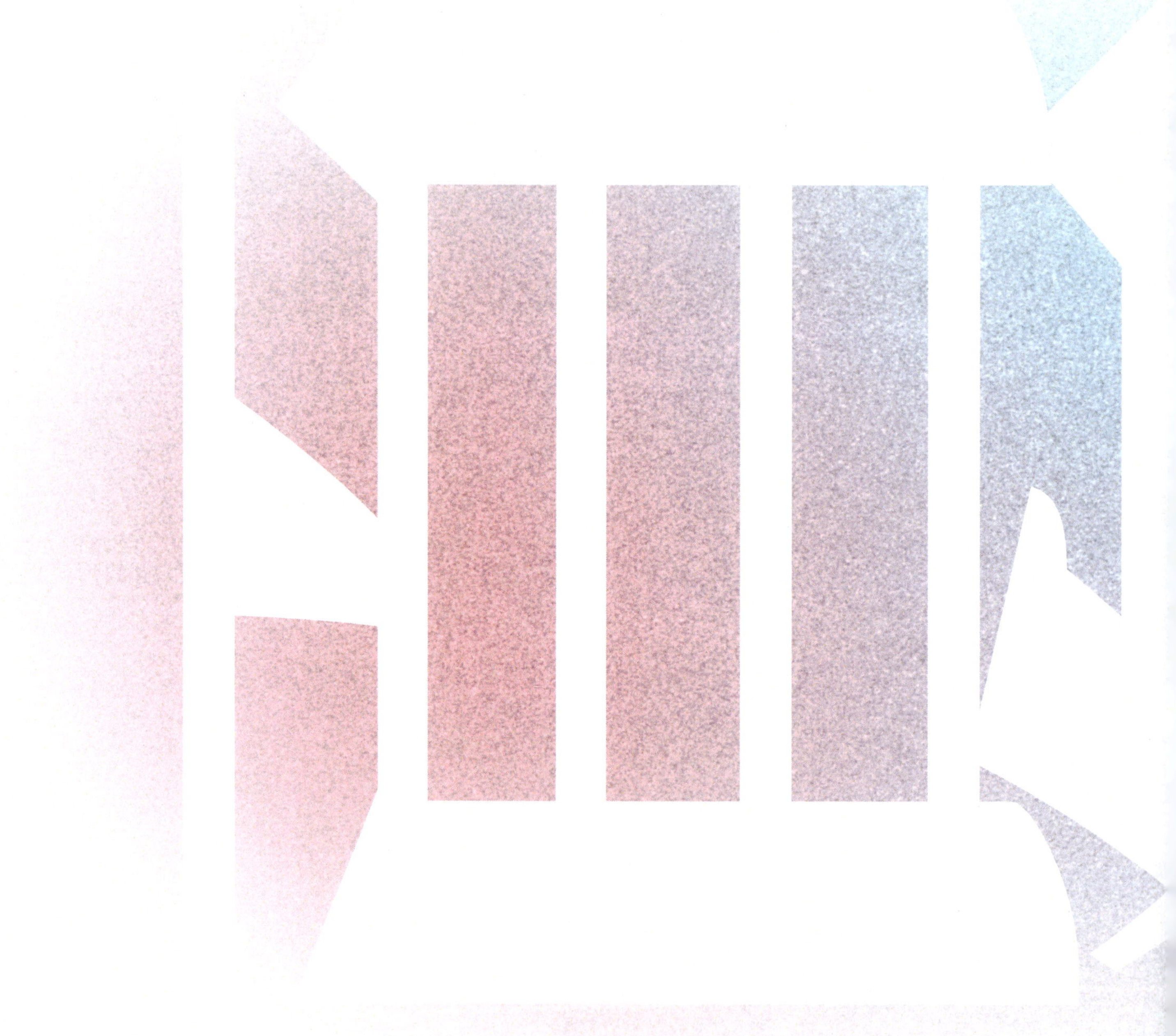

# Summer

夏

Xià

Lì

立

# BEGINNING OF SUMMER

May 6 (± 1)

# 立夏

## Beginning of Summer

Lìxià : The start of summer, interpreted as a period of vigorous growth for all living things. reflecting a period of extensive nurturing and flourishing. In the philosophy of Yin and Yang, Lìxià represents the pinnacle of Yang energy, symbolizing vitality and growth. As the season progresses, the supreme Yang begins to seed the growth of Yin, heralding a transition and balance between the energies.

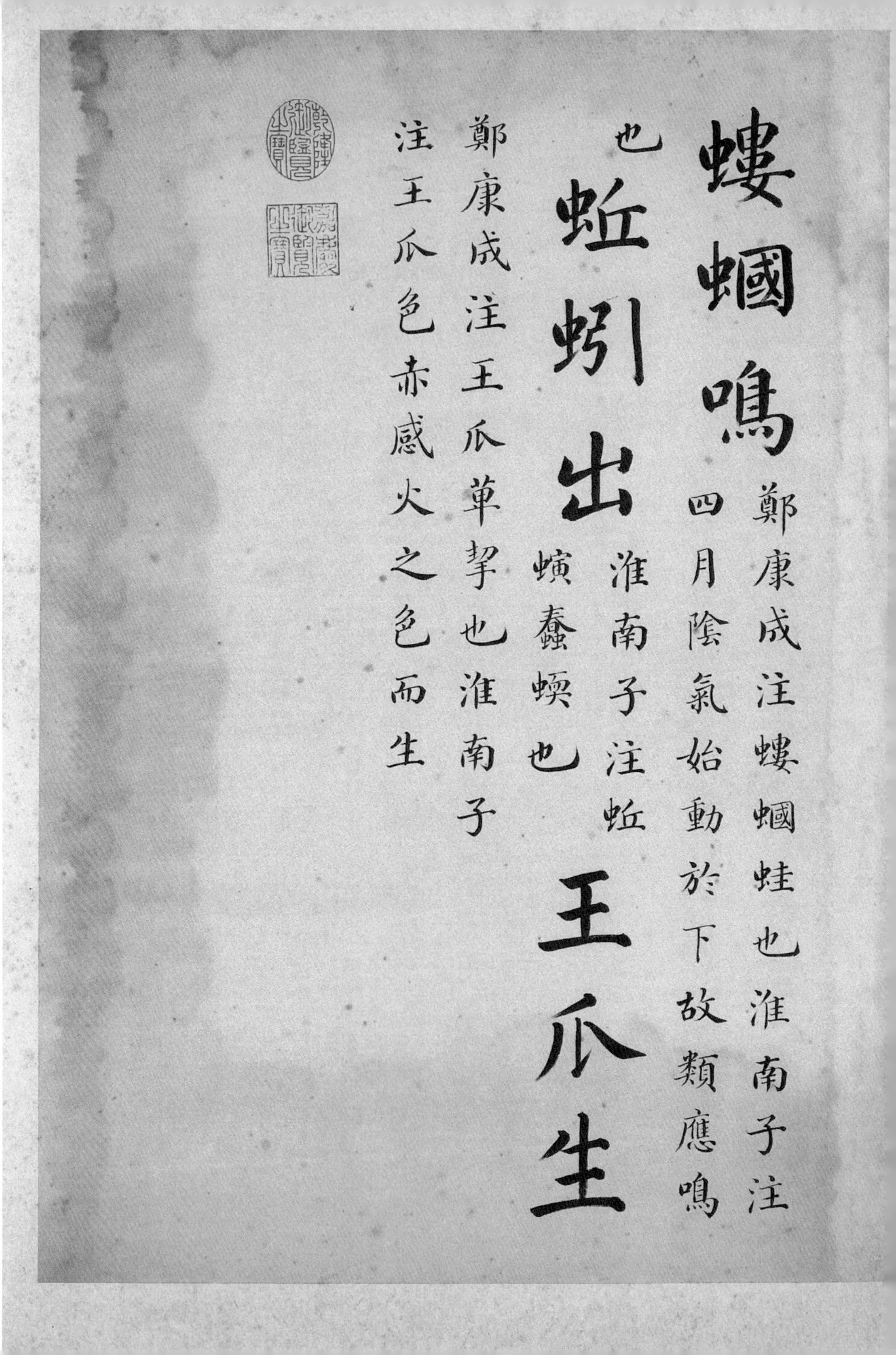

螻蟈鳴
鄭康成注螻蟈蛙也淮南子注
四月陰氣始動於下故類應鳴
也
蚯蚓出
淮南子注蚯
螾蠢蝡也
王瓜生
鄭康成注王瓜萆挈也淮南子
注王瓜色赤感火之色而生

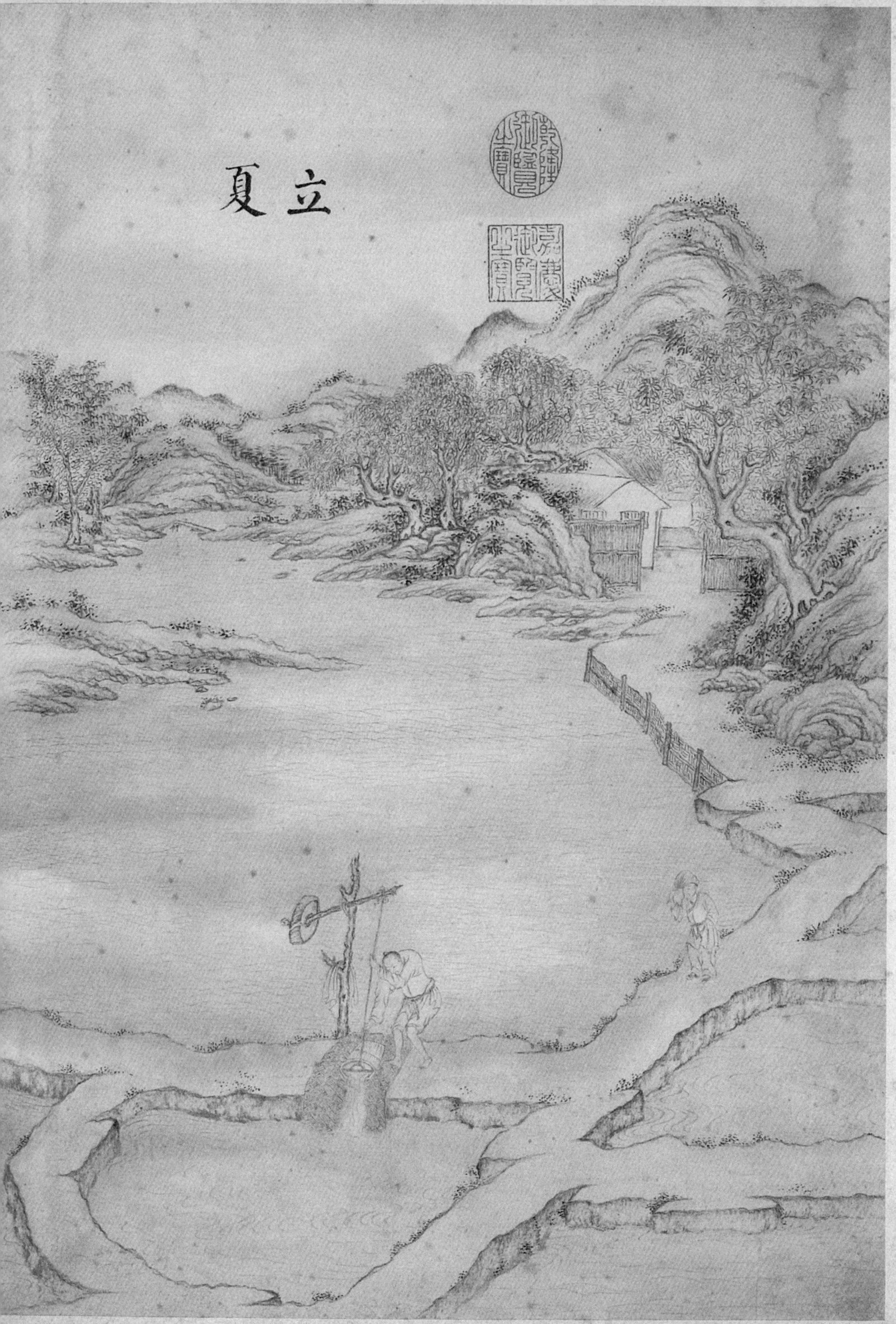


初候

## 螻蟈鳴

## Crickets Chirp

二候

## 蚯蚓出

## Earthworms Emerge

三候

## 王瓜生

## King Melon Grows

# 立夏

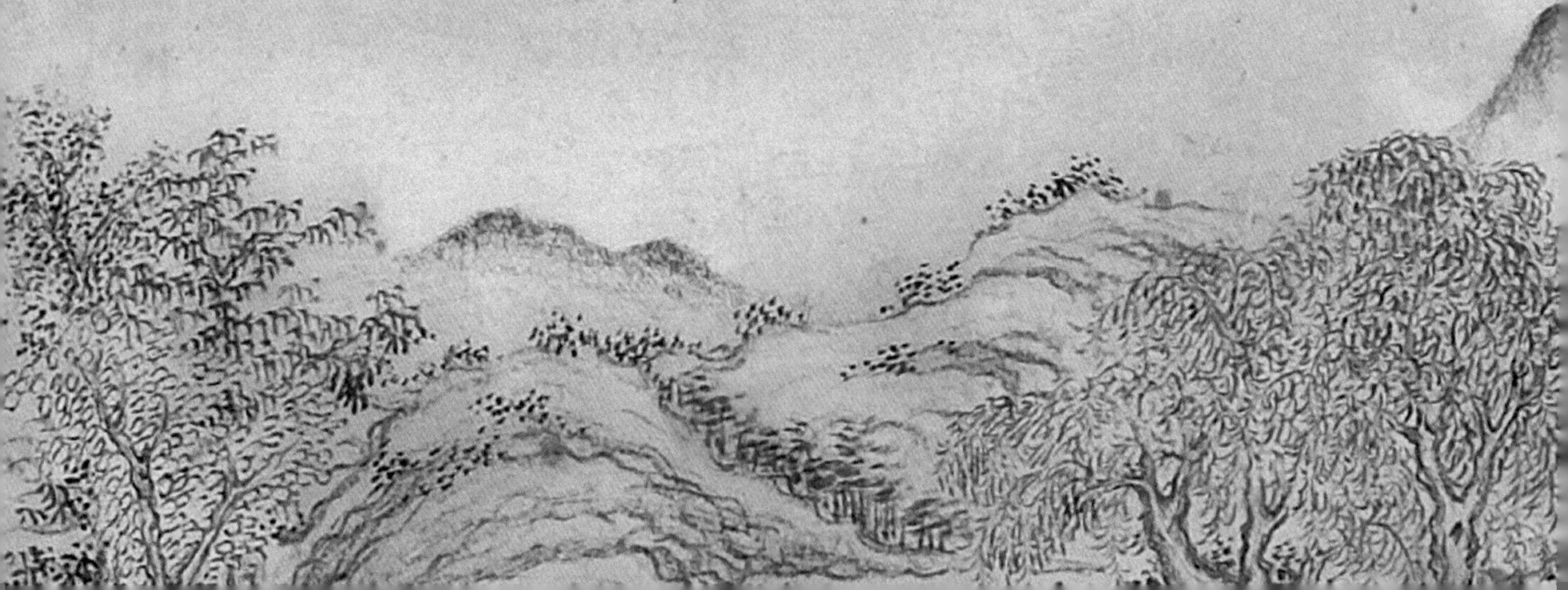

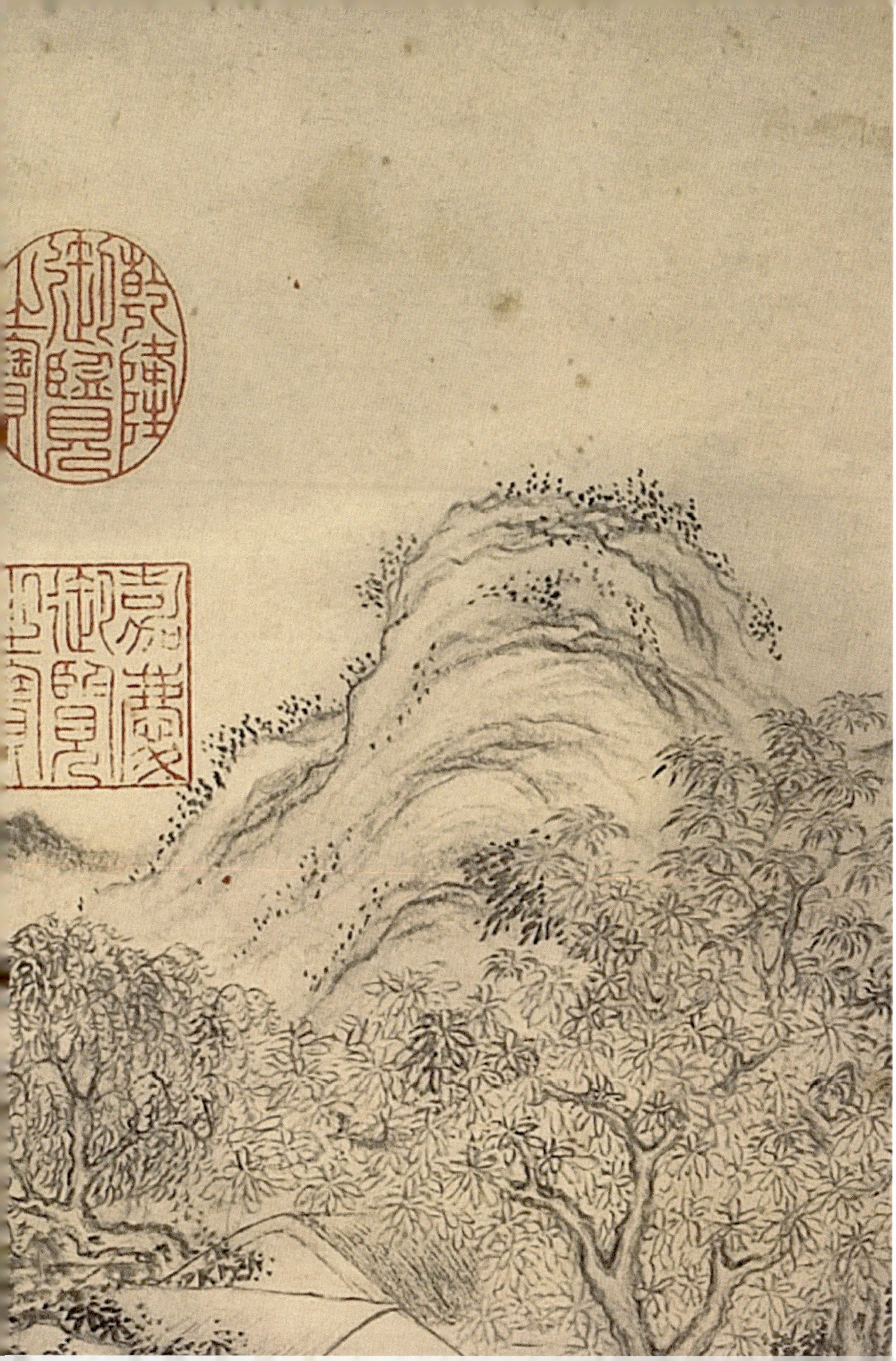

初候

# 螻蟈鳴

## Crickets Chirp

The first harbinger of Lìxià is "Crickets Chirp" (**螻蟈鳴** Lóu Guō Míng), an evening melody that fills the air with anticipation of summer.

二候

# 蚯蚓出

## Earthworms Emerge

The second omen is "Earthworms Emerge" (蚯蚓出 Qiū Yǐn Chū), as they respond to the warming earth, signaling readiness for growth.

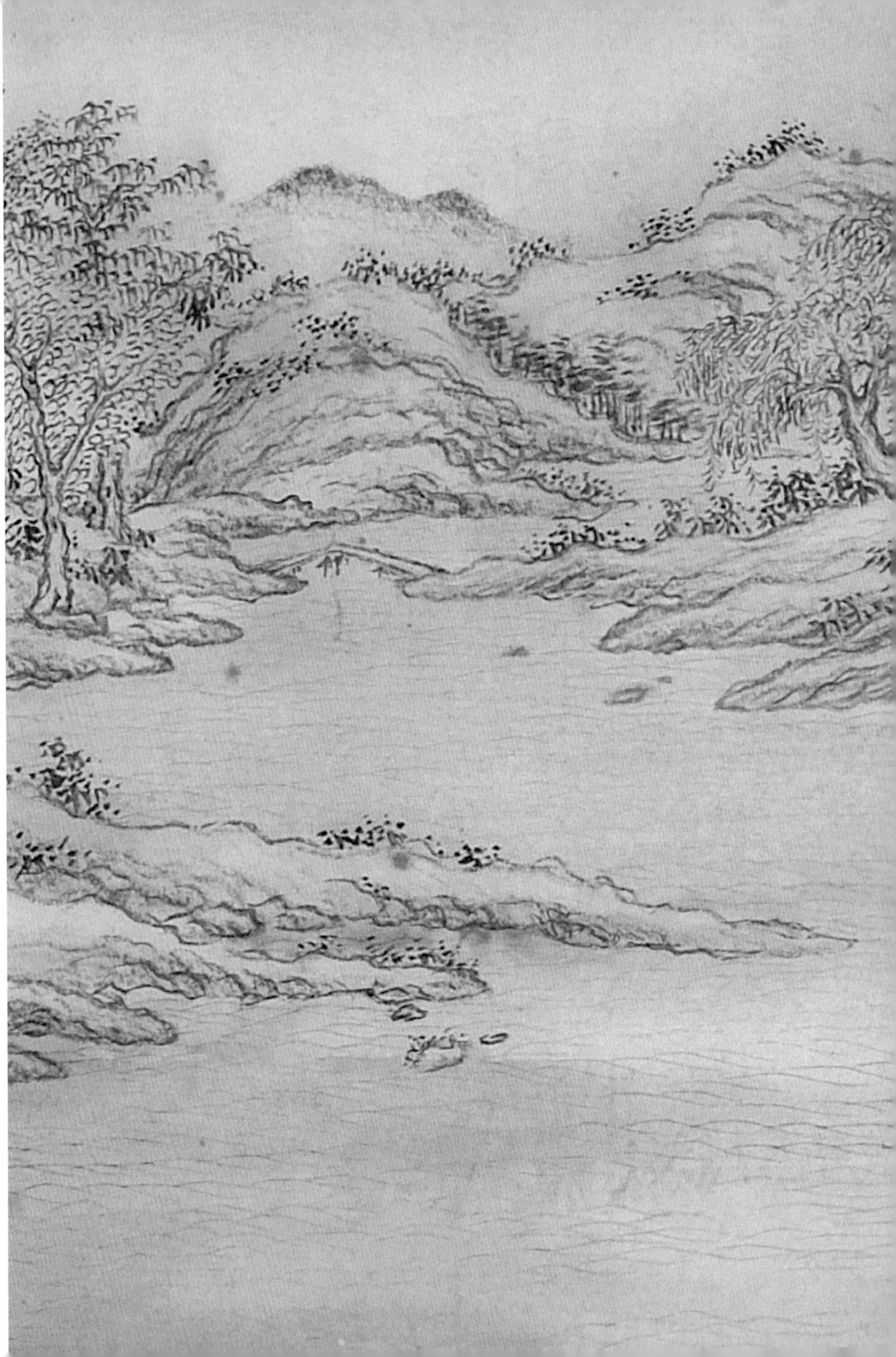

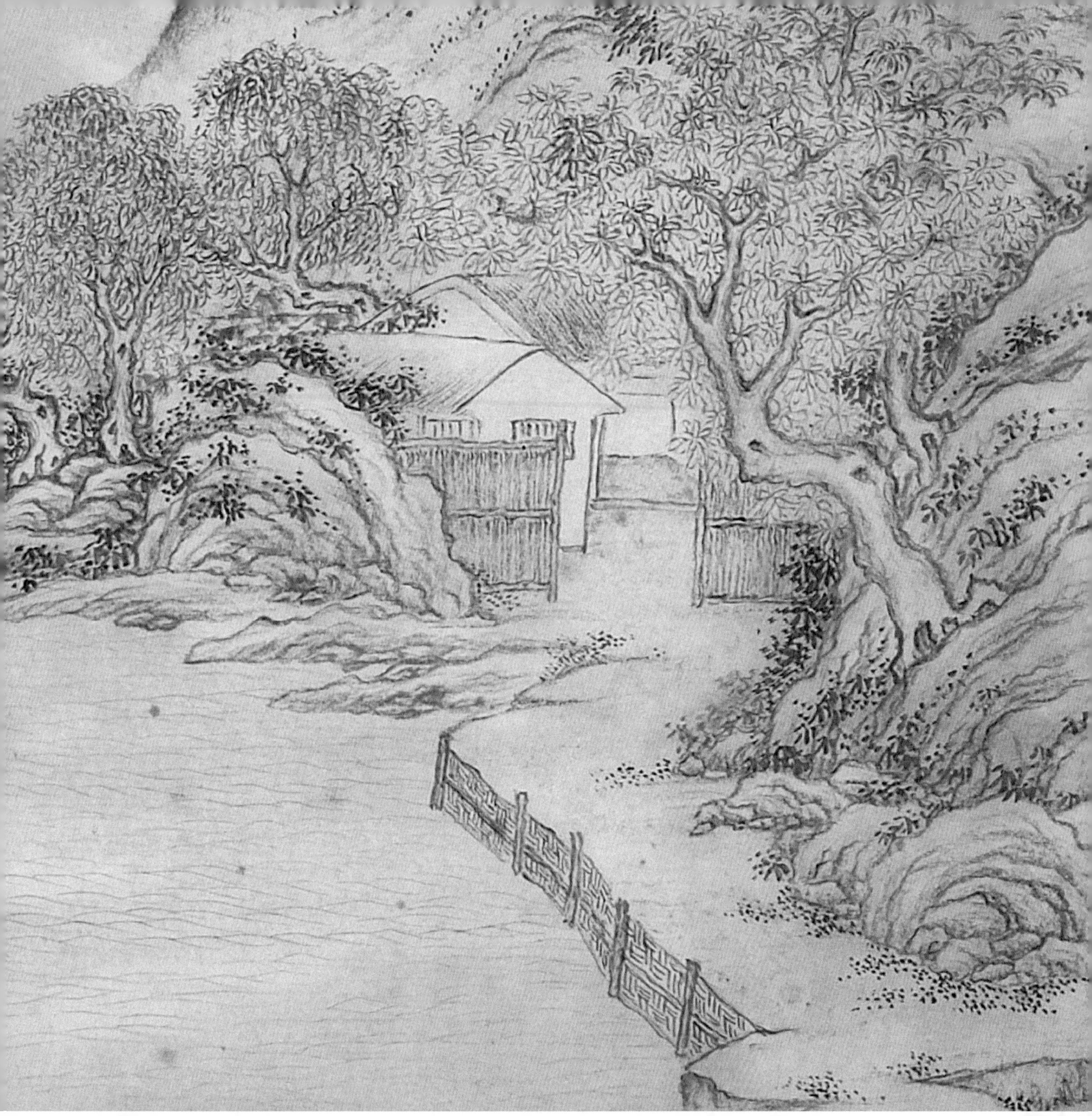

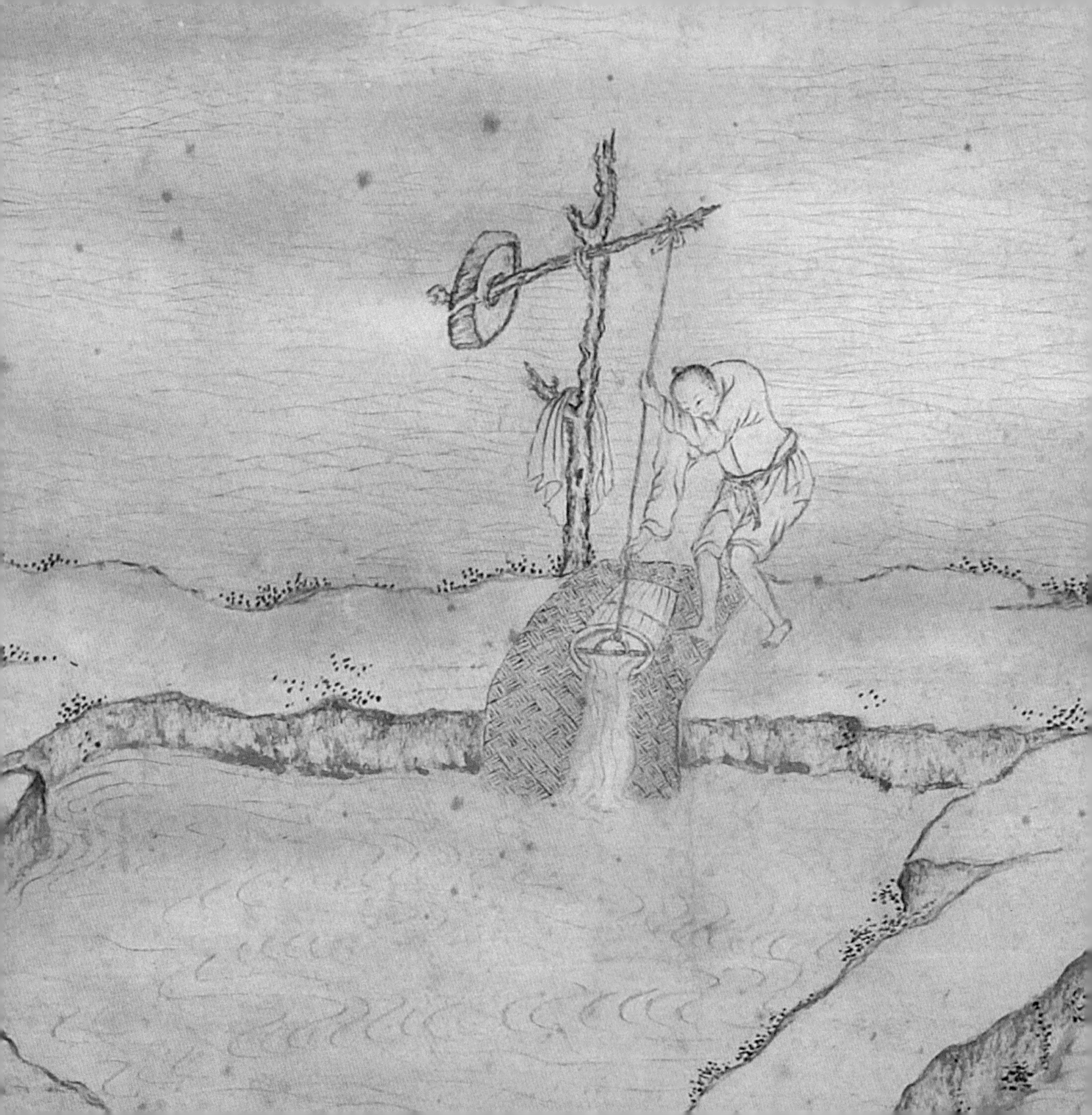

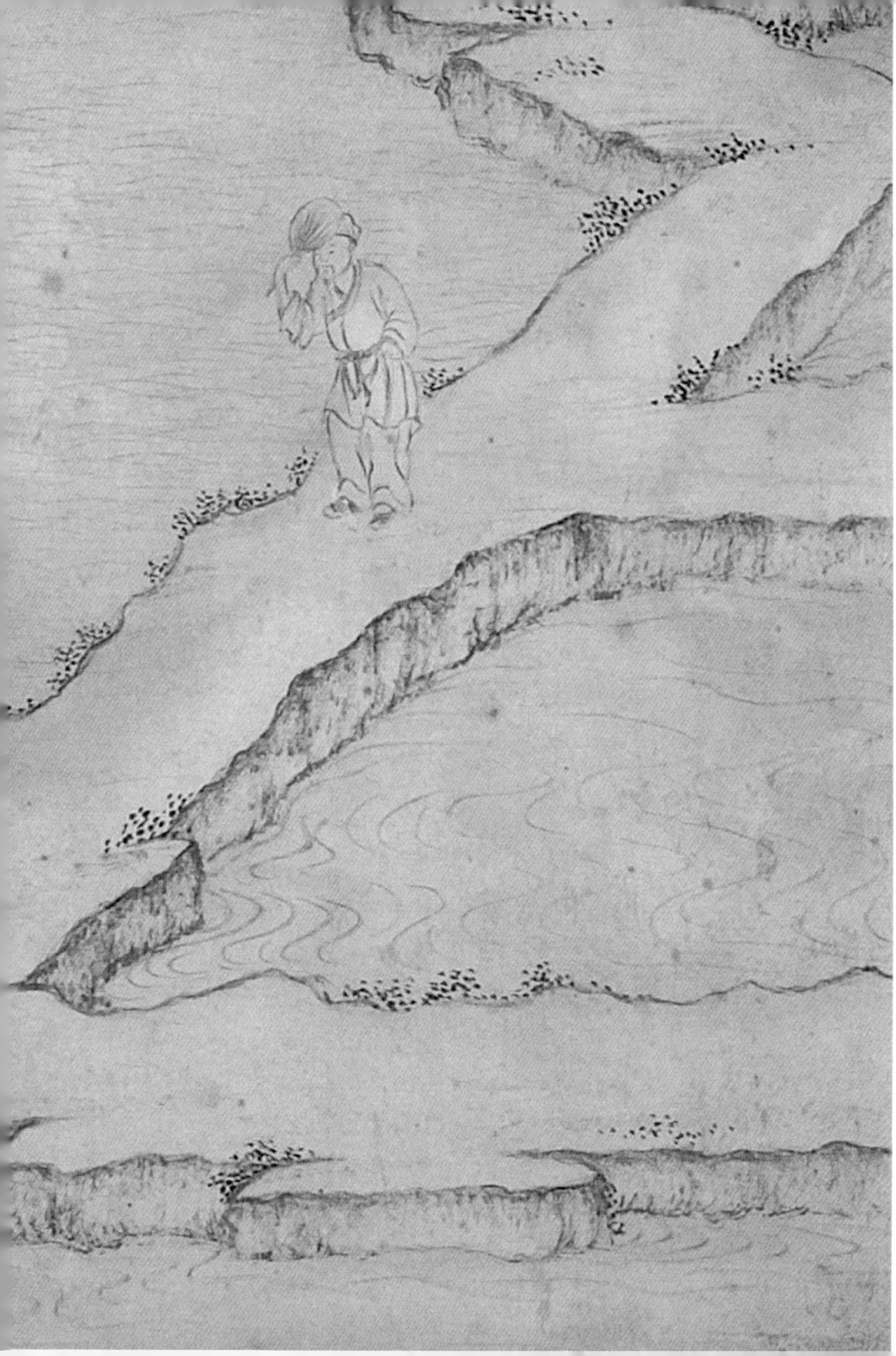

三候

# 王瓜生

## King Melon Grows

The third interval, “King Melon Grows” (王瓜生 Wáng Guā Shēng), describes the rapid growth of a specific medicinal vine in North China, known for its climbing nature and red fruits in summer. This period reflects the thriving vitality and subtleness of yin energy turning to yang energy in nature.

# 滿

Mǎn

Xiǎo

小

# LESSER FULLNESS OF GRAIN

May 21 (± 1)

# 小滿

## Lesser Fullness of Grain

Xiǎomǎn : Meaning "Lesser Fullness," implies a time when growth is abundant but not yet at its peak. It symbolizes the moment just before the culmination of summer's vitality. For agricultural communities, especially in northern China, it's a critical phase where wheat grains are plump but not fully ripe, and in the south, it's the time when rice begins to head. The term reflects a delicate balance, as the peak of Yang energy is almost approaching.

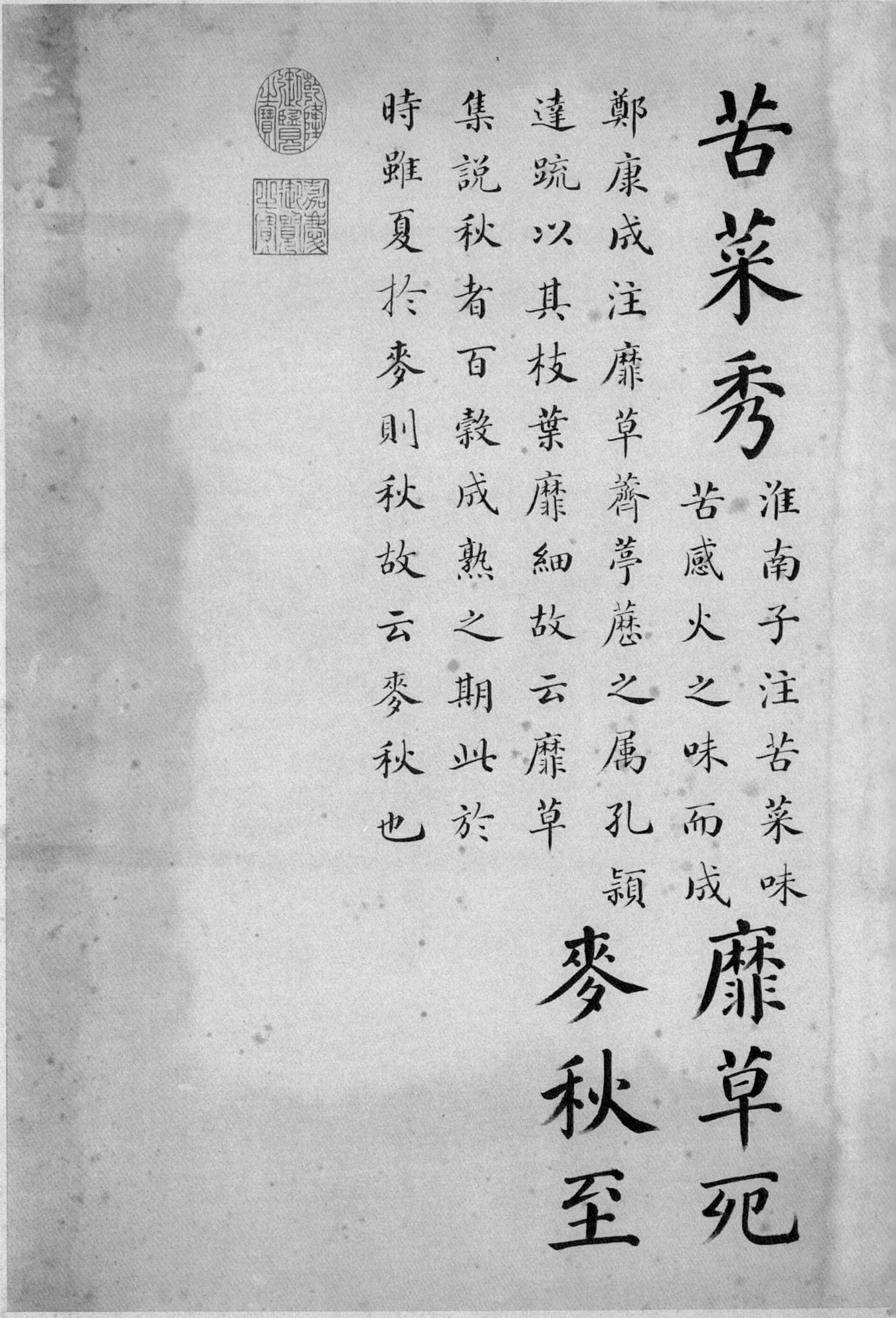

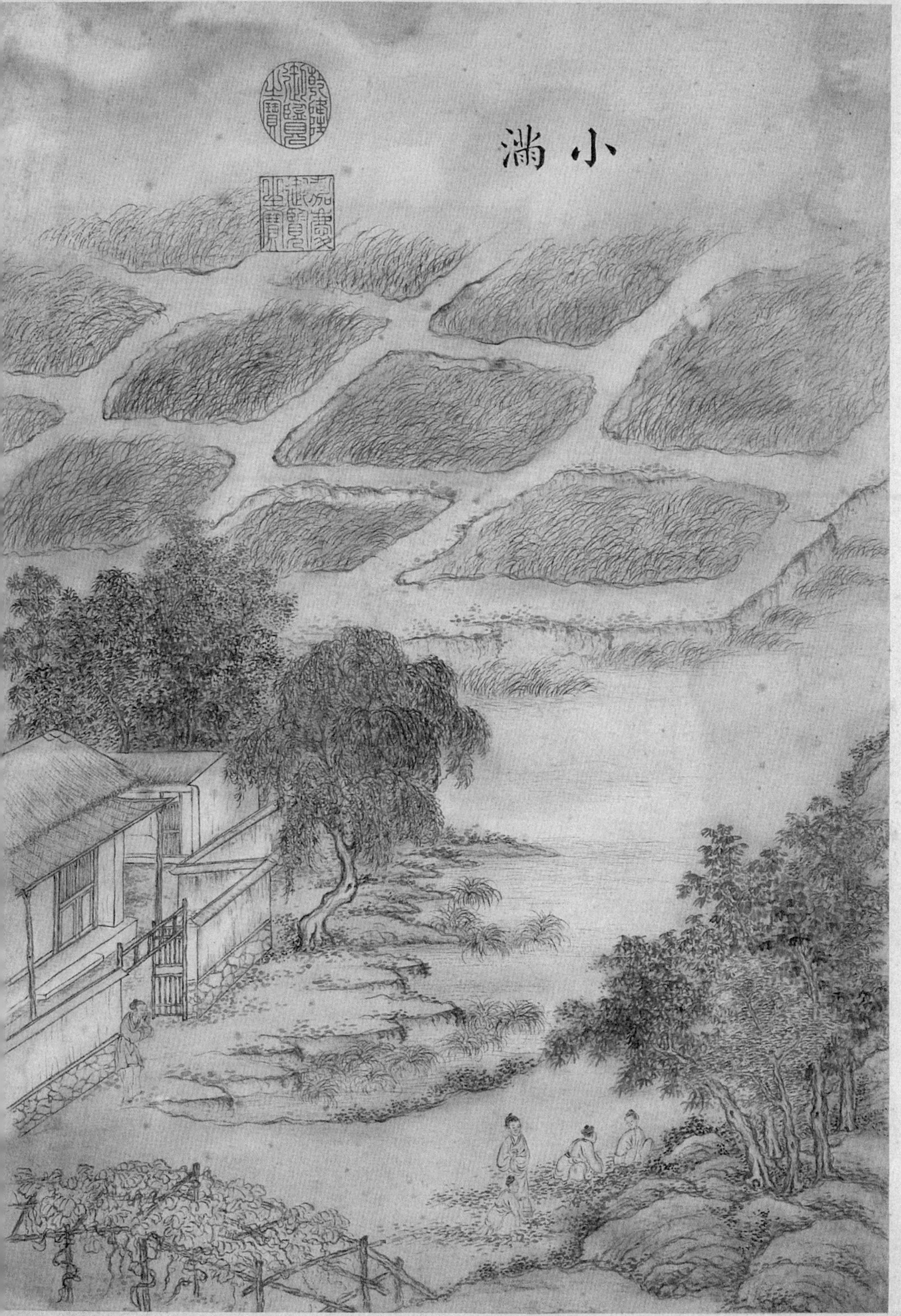


初候
## 苦菜秀
## Bitter Herbs in Bloom

二候
## 靡草死
## Delicate Grasses Wilt

三候
## 麥秋至
## Wheat Harvest Begins

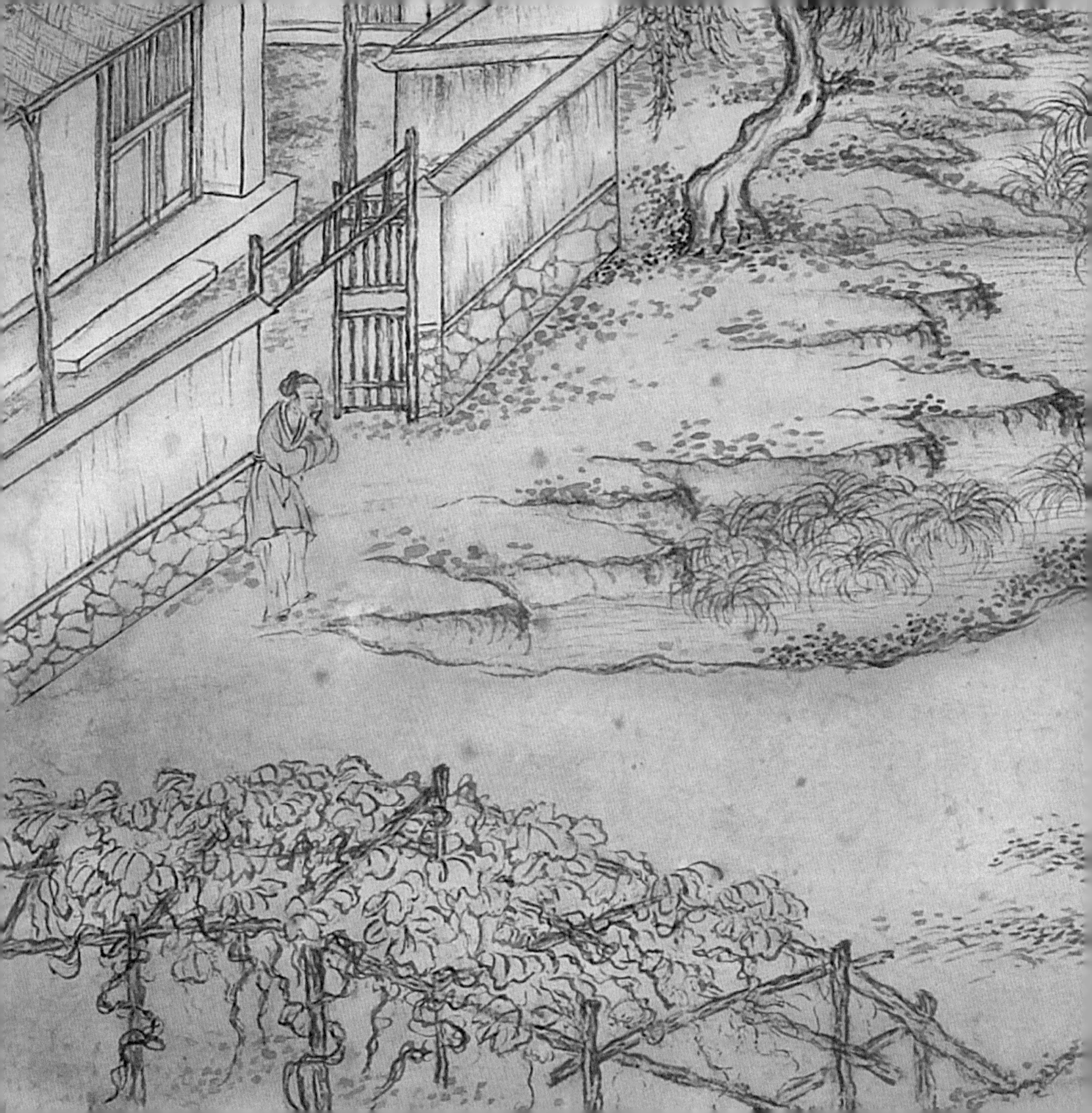

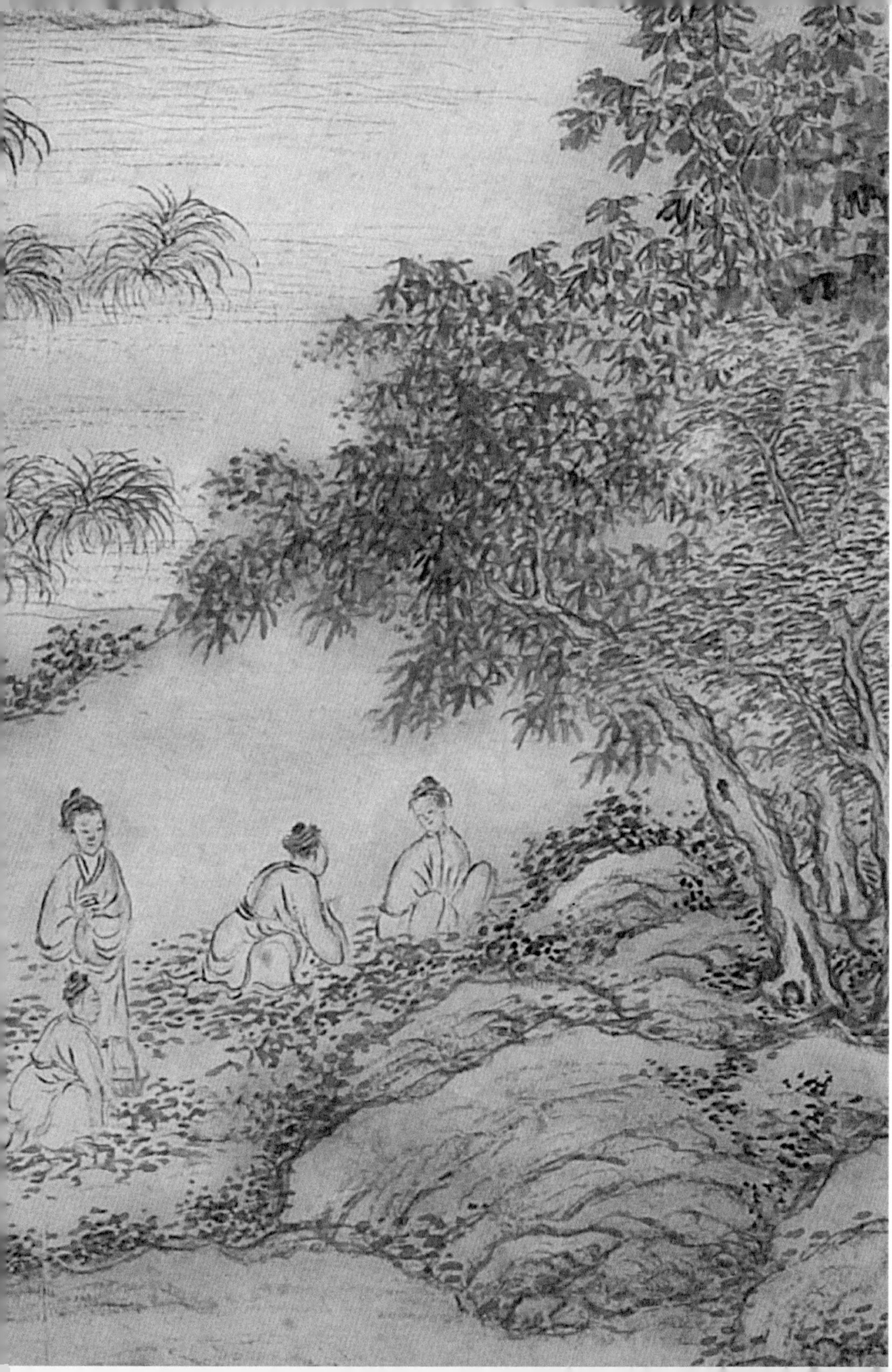

初候

# 苦菜秀

## Bitter Herbs in Bloom

The first hint of Xiǎomǎn is "Bitter Herbs in Bloom" (苦菜秀 Kǔ Cài Xiù), signaling the readiness of certain herbs and plants for harvest, embodying the fullness of spring.

二候

# 靡草死

## Delicate Grasses Wilt

The second sign is “Delicate Grasses Wilt” (靡草死 Mí Cǎo Sǐ), under the potent summer sun, reminding us of the cycle of growth and retreat.

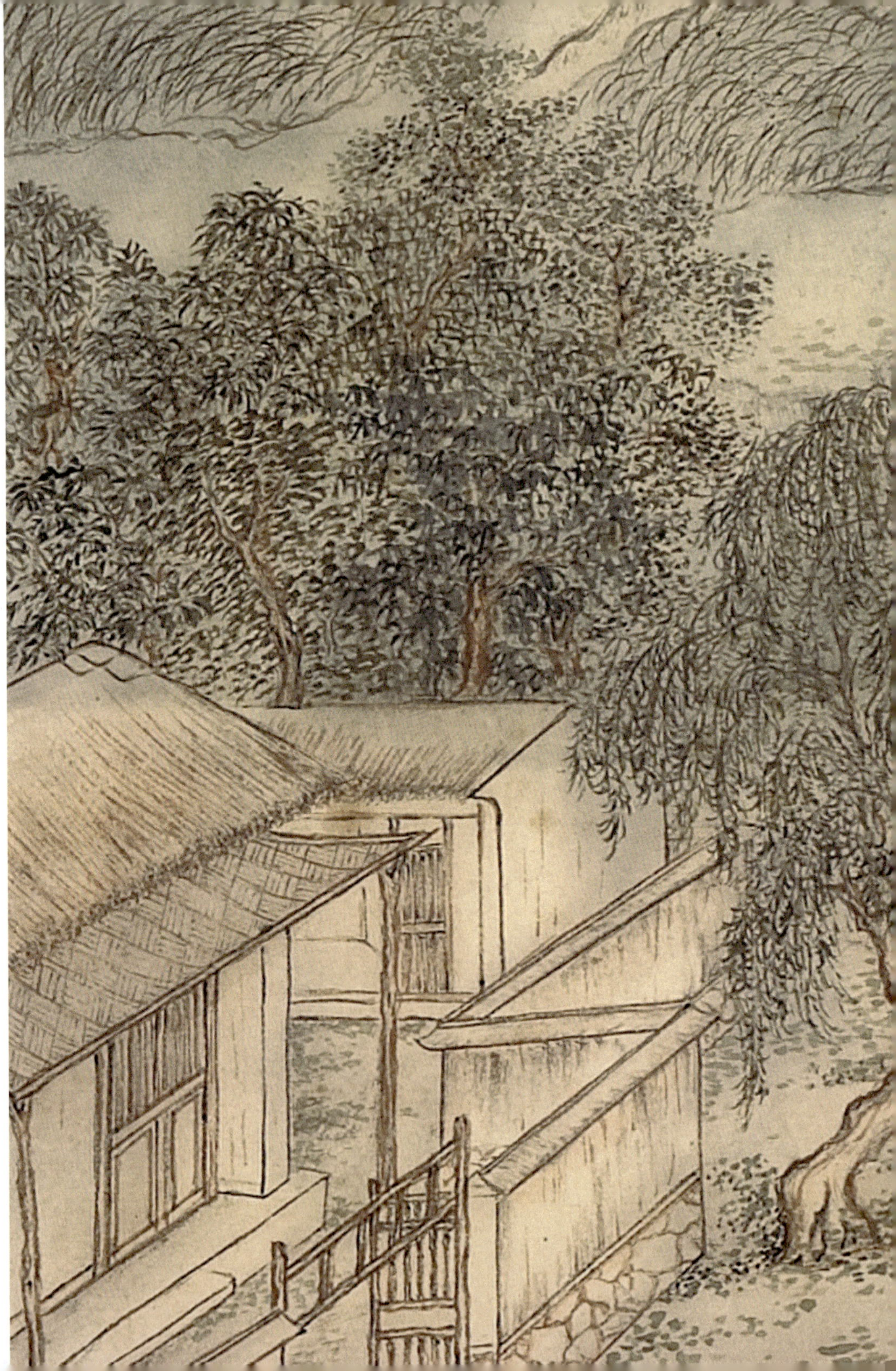

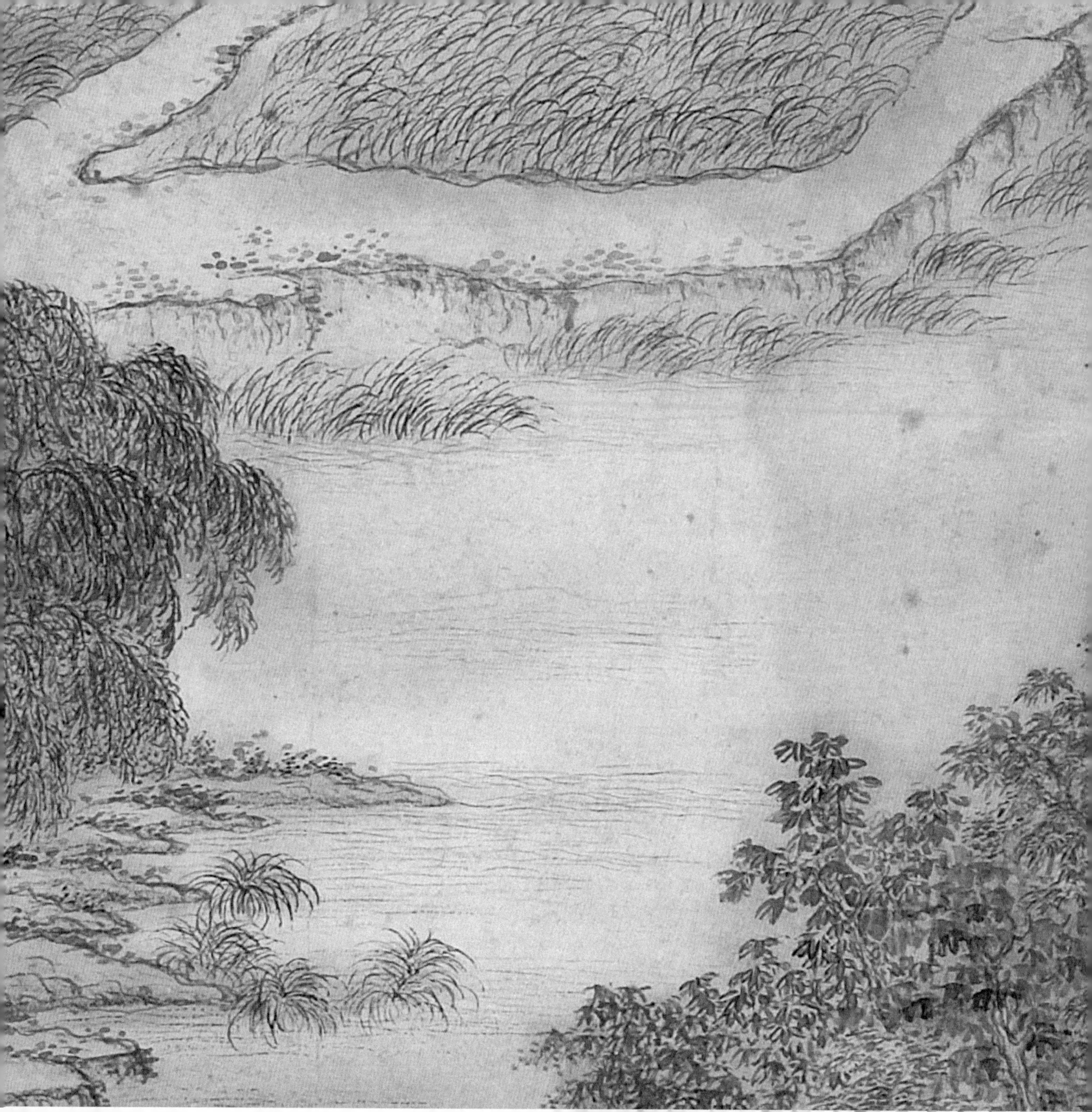

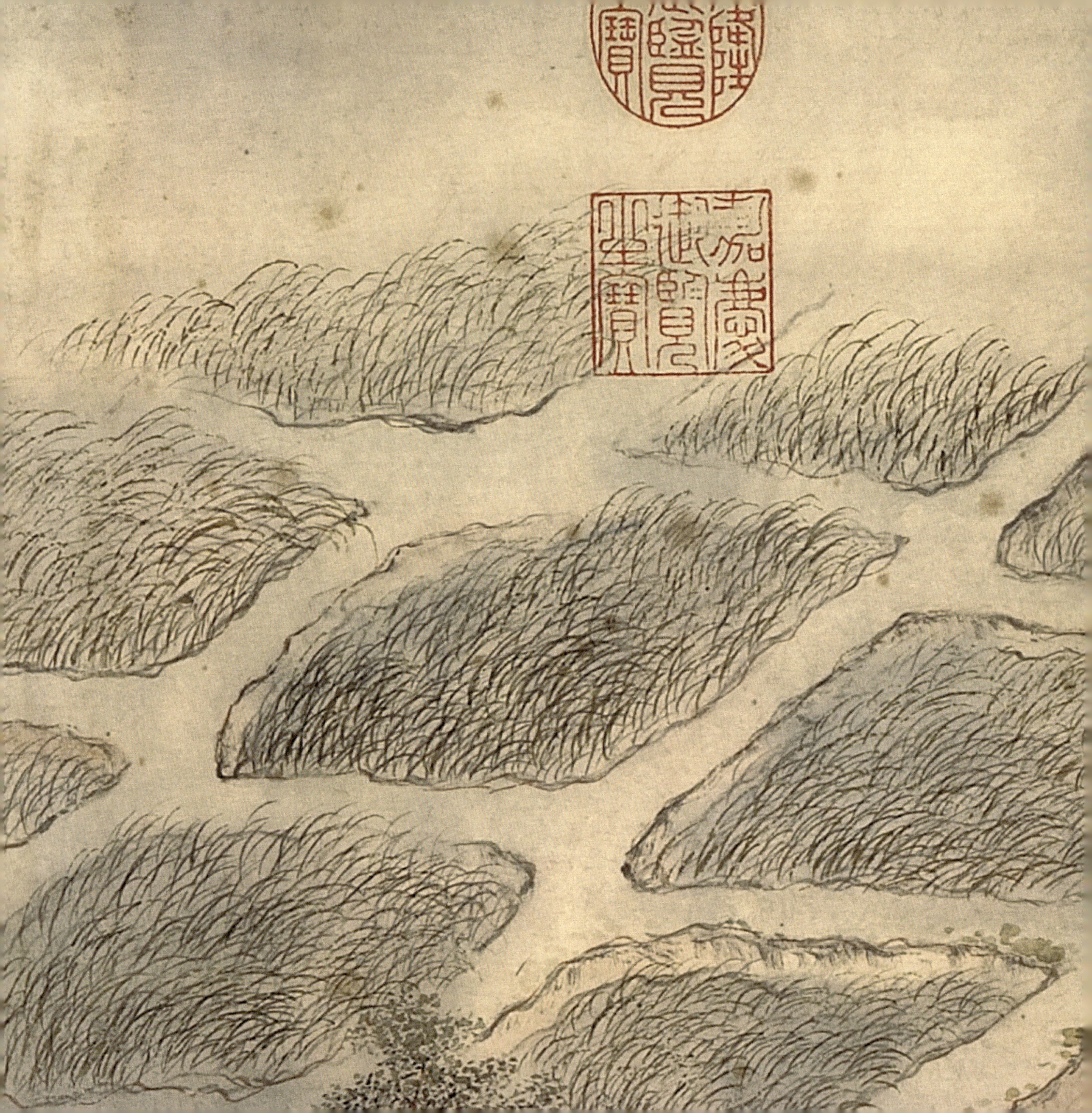

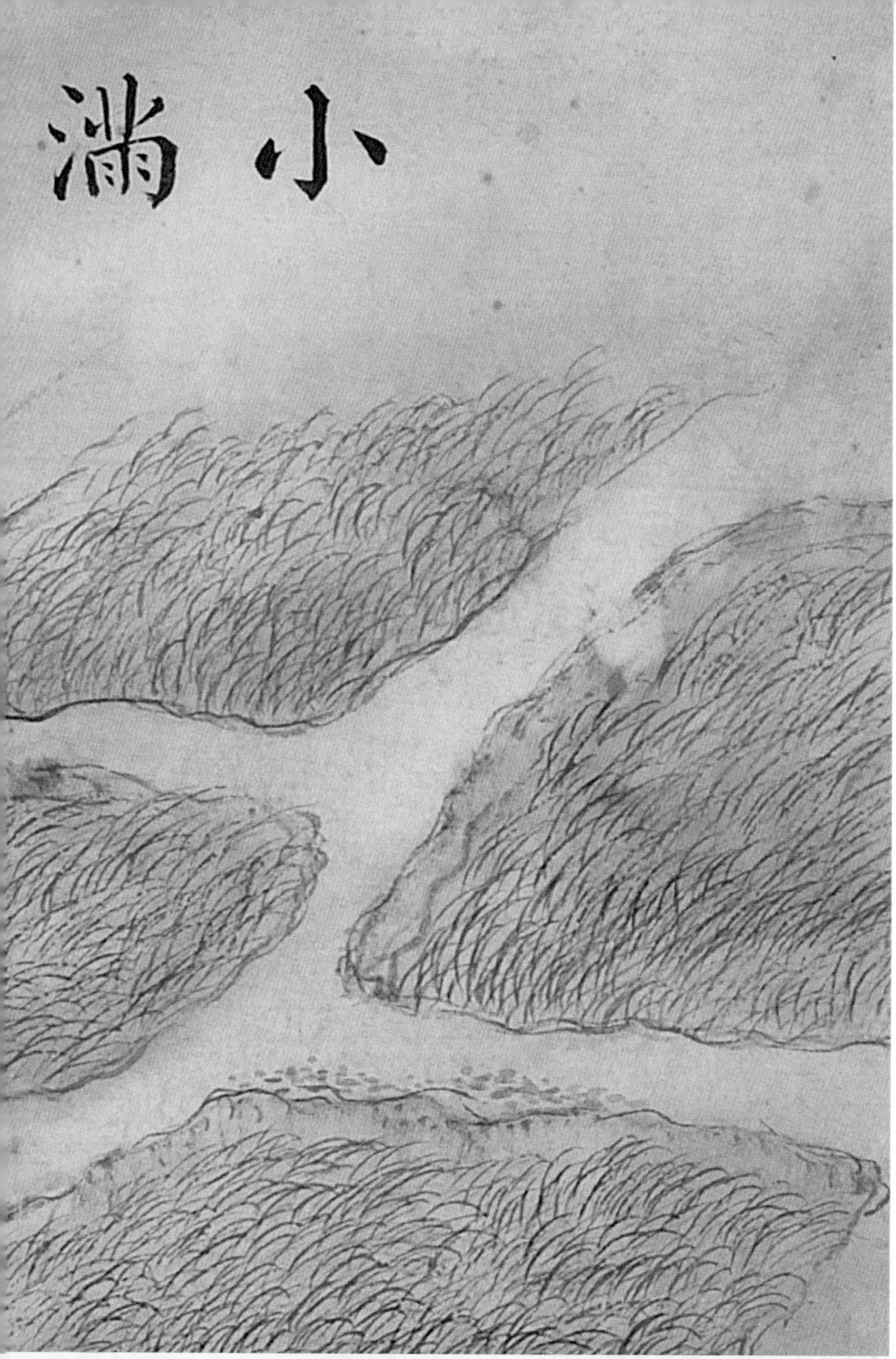


三候

# 麥秋至

## Wheat Harvest Begins

The third indication, “Wheat Harvest Begins” (麥秋至 Mài Qiū Zhì), reflects the moment when wheat grains, previously filled but not yet matured, reach full ripeness. This period signifies the start of the wheat harvest, a crucial time in the agricultural calendar particularly noted for its early arrival compared to other grains. It’s a moment of culmination and reward, as the golden wheat fields are ready to be reaped, marking a period of abundance and the transition into summer’s bounty.

種

Zhòng

Máng 芒

# GRAIN IN BEARD

June 6 (± 1)

# 芒種

## Grain in Beard

Mángzhōng : It is closely associated with the start of the plum rain season, especially along the Yangtze River, bringing continuous rain essential for the growth of grains. This period is indicative of a time for sowing certain crops and is traditionally seen as a precursor to a fruitful harvest. It represents a period where the harmony of wind and rain nourishes the fields, and agricultural activities are focused on leveraging the favorable weather conditions.

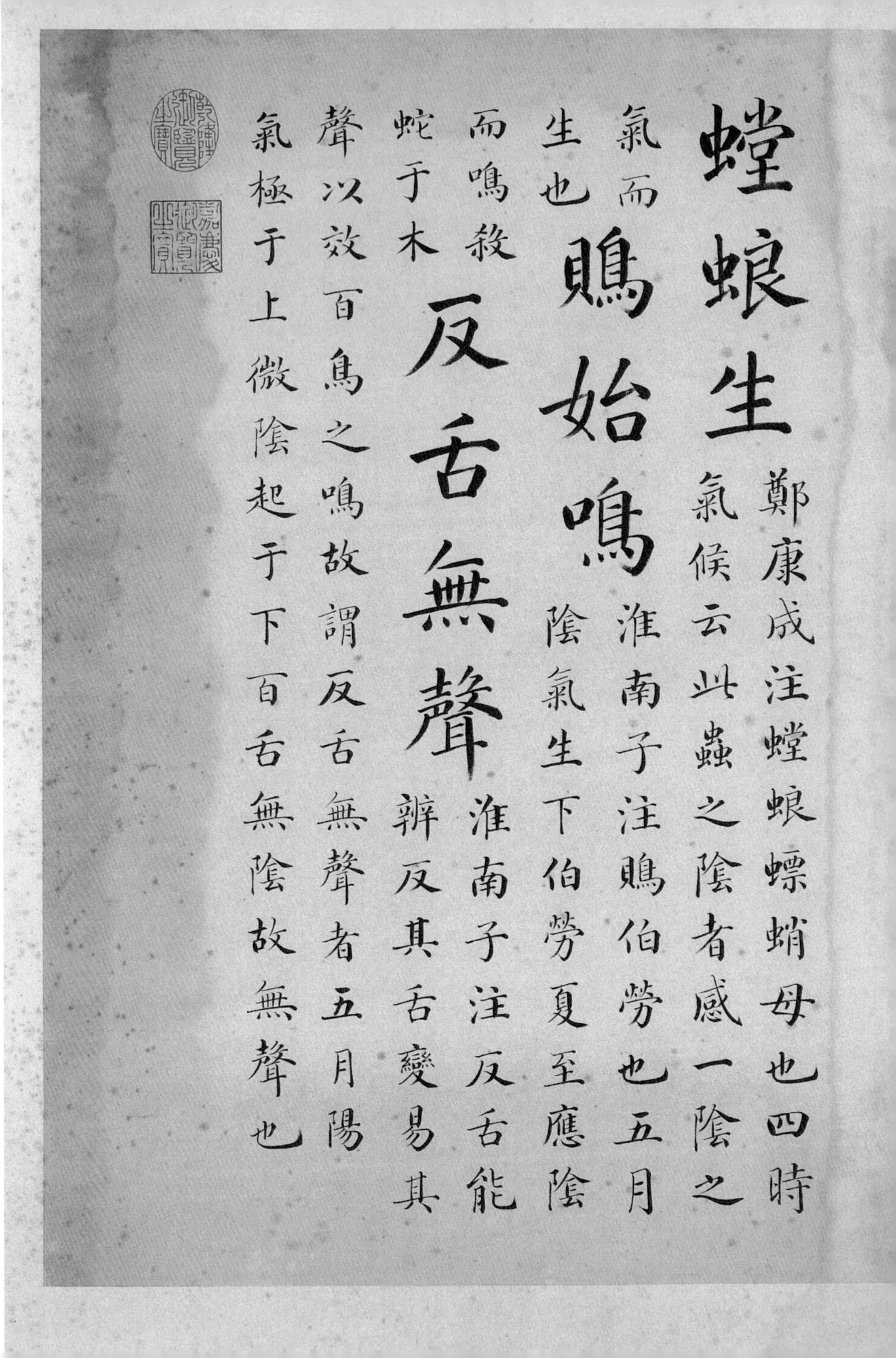

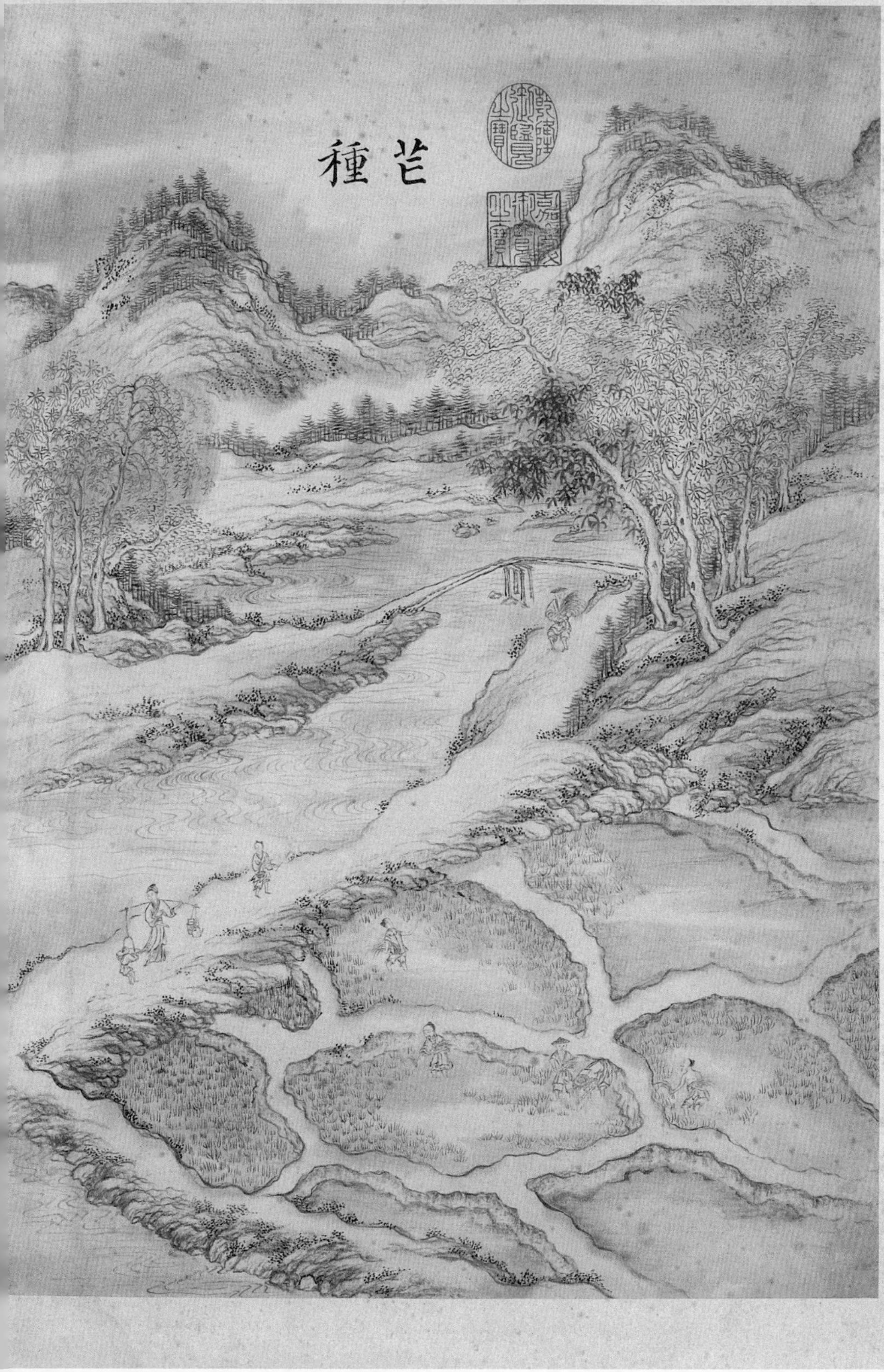


初候

## 螳螂生

# Praying Mantises Emerge

二候

## 鵙始鳴

# Shrikes Start Calling

三候

## 反舌無聲

# The Rook's Silent Period

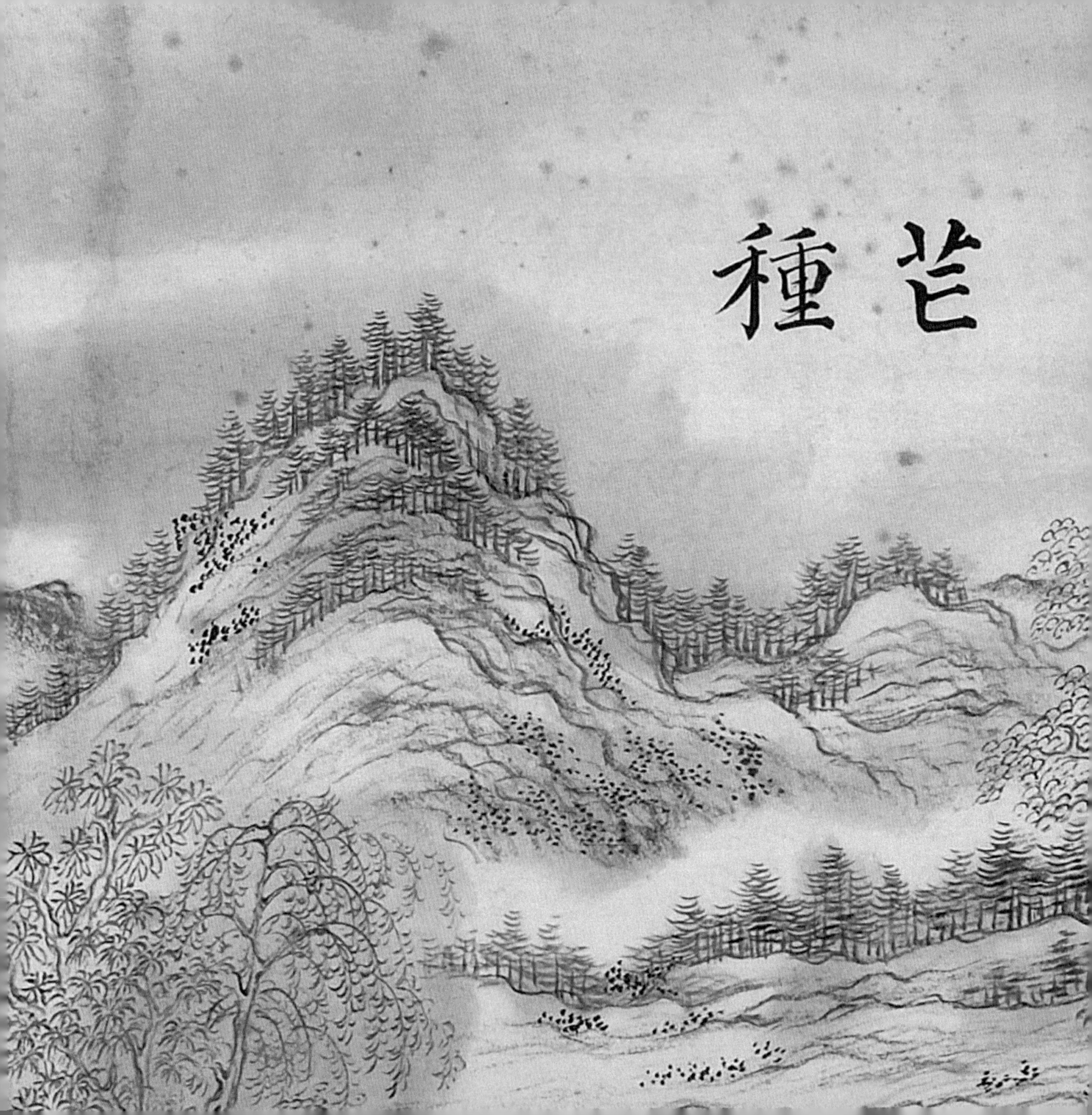
芒種

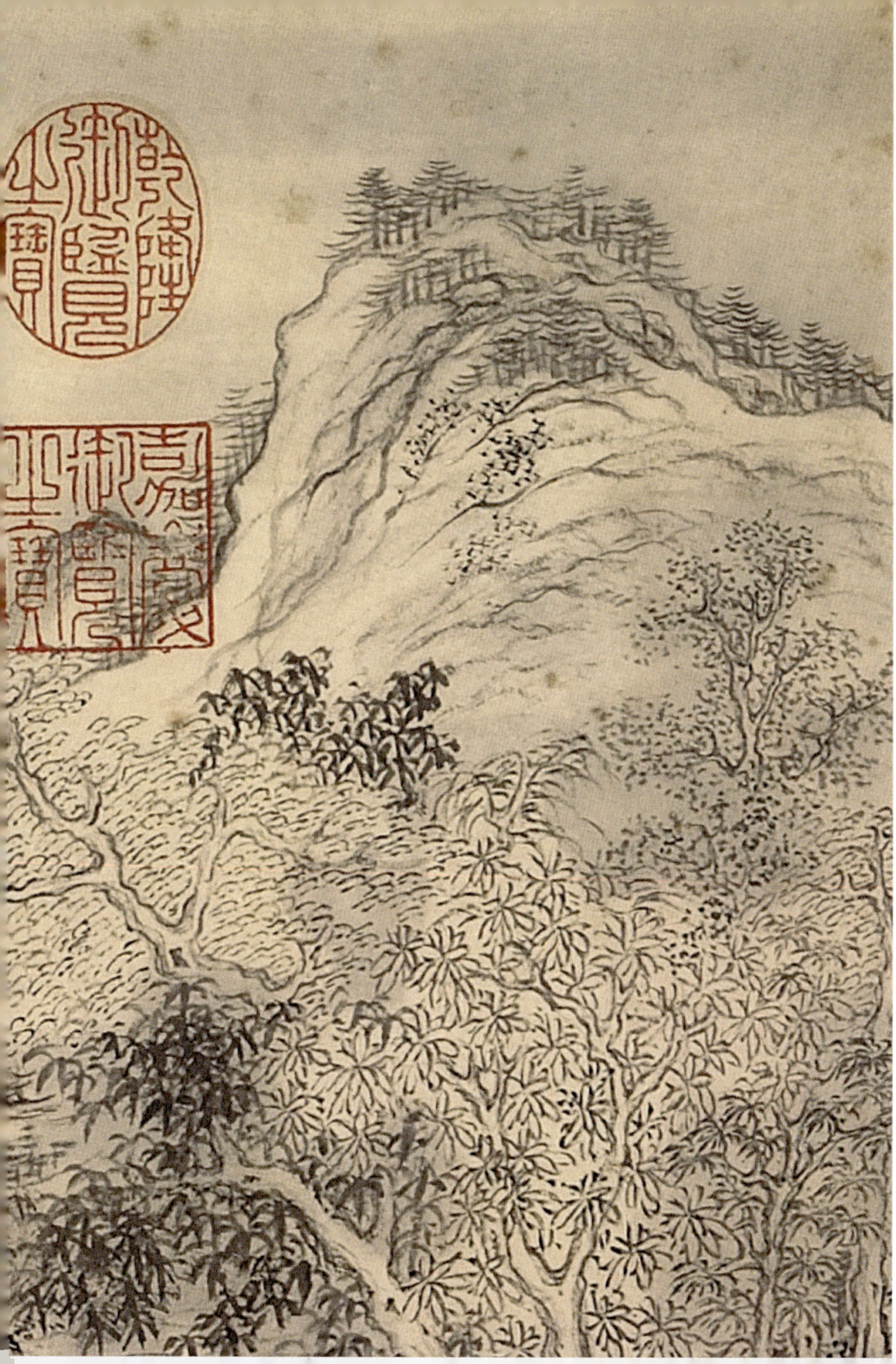

初候

# 螳蜋生

## Praying Mantises Emerge

The first beacon of Mángzhōng is "Praying Mantises Emerge" (螳蜋生 Táng Láng Shēng), as these intriguing creatures hatch, signaling the deepening of summer.

二候

# 鶪始鳴

## Shrikes Start Calling

The second cue is “Shrikes Start Calling” (鶪始鳴 Jú Shǐ Míng), their distinct calls echoing the rising temperatures and vibrant life.

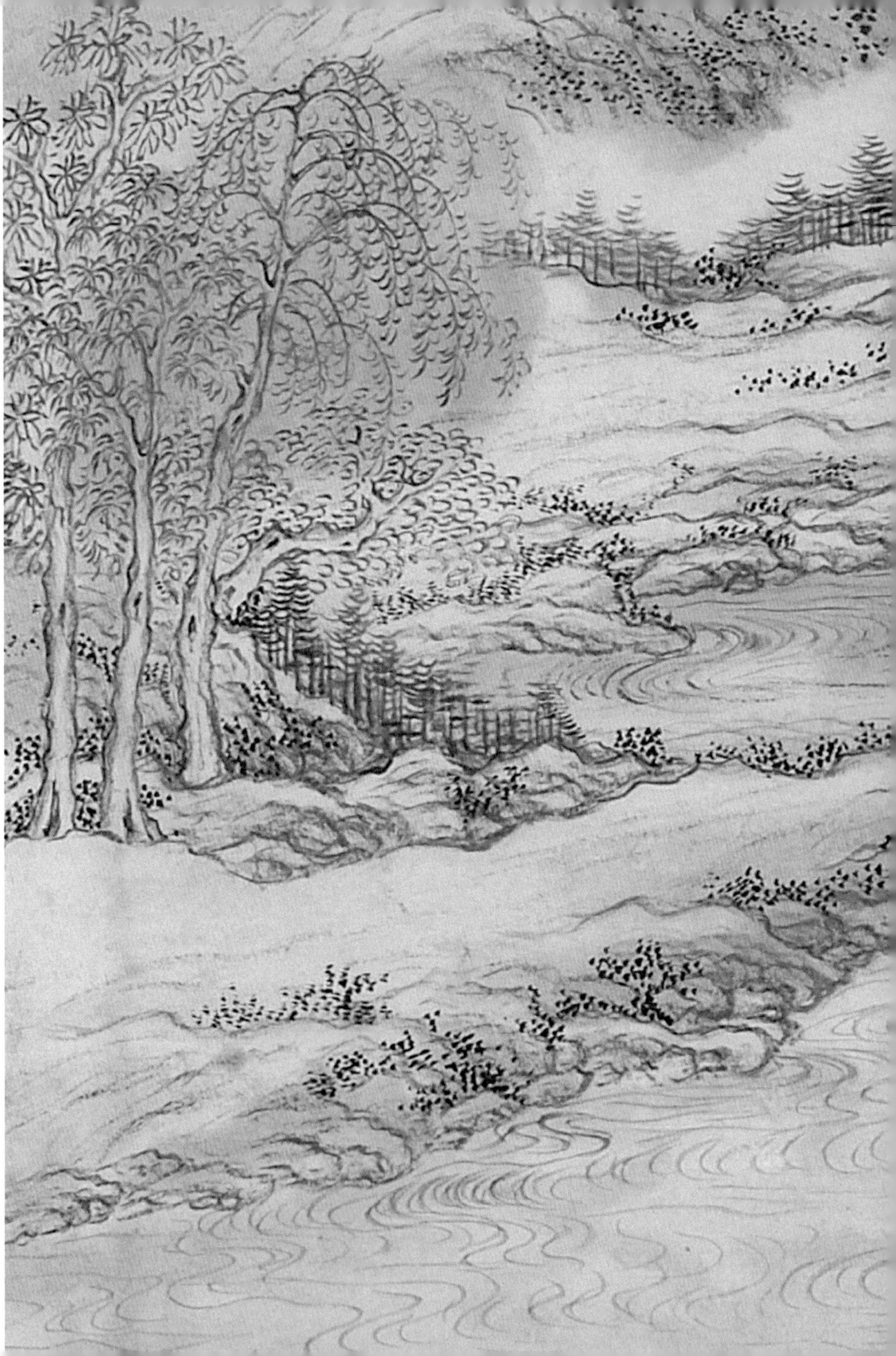

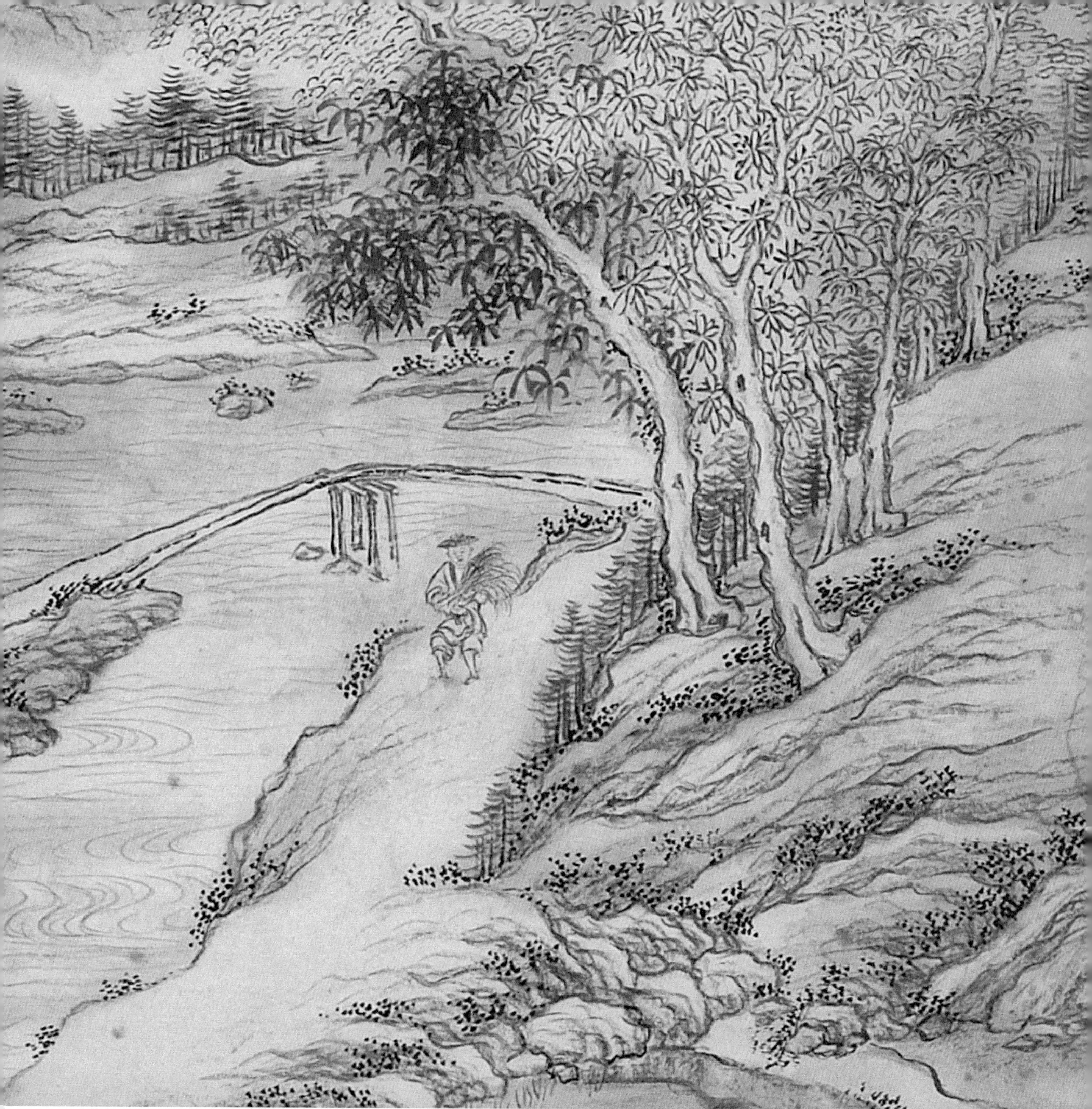

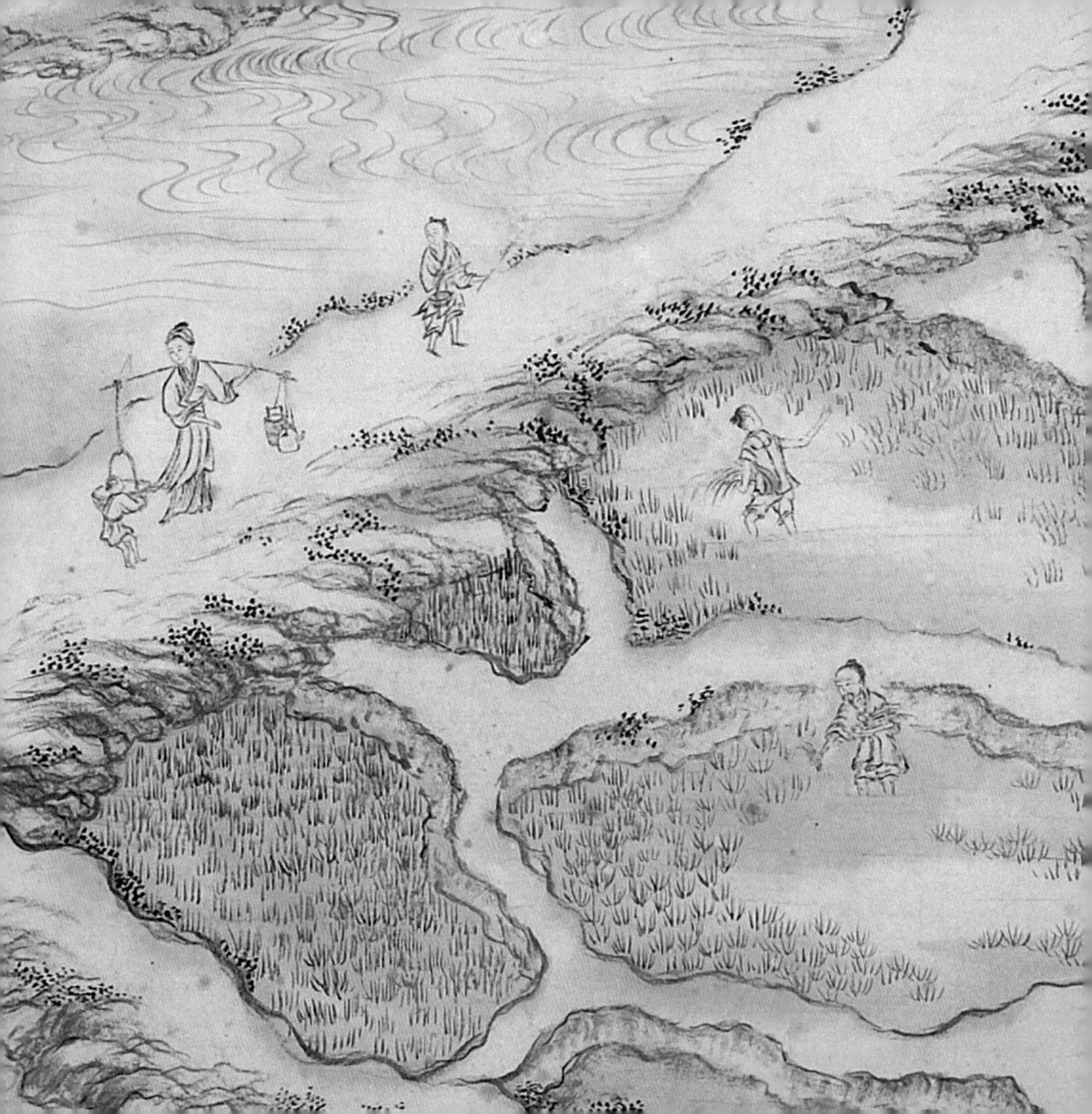

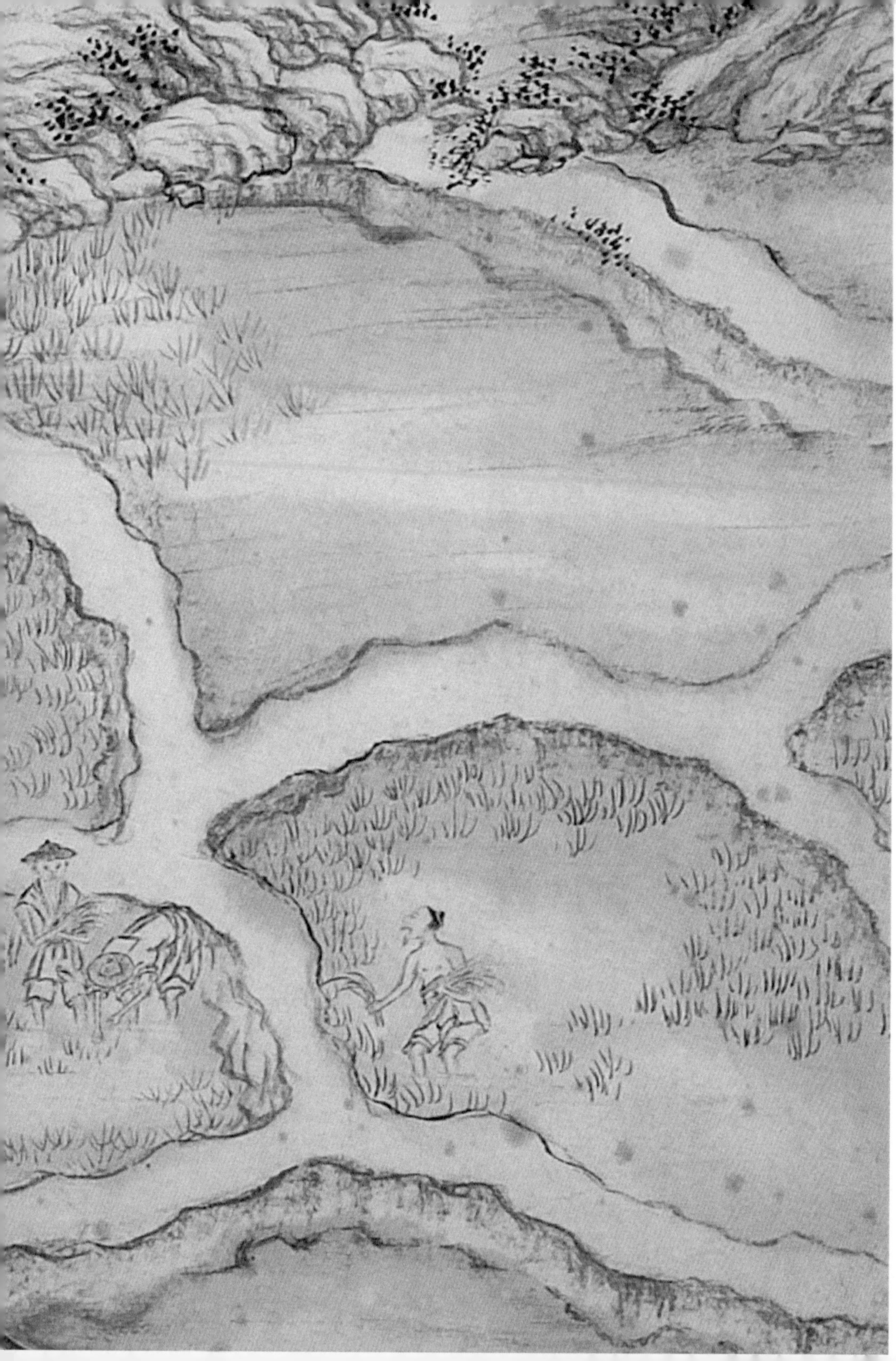

三候

# 反舌無聲

## The Rook’s Silent Period

The third harbinger is “The Rook’s Silent Period” (反舌無聲 Fǎn Shé Wú Shēng), symbolizing a moment of quietude amidst the season’s abundance, perhaps reflecting a deeper shift in nature.

# 至

Zhì

Xià

夏

# THE SUMMER SOLSTICE

June 21 (± 1)

# 夏至

## The Summer Solstice

Xiàzhì : The Summer Solstice, is marked as the day with the longest daylight hours, known historically as "the extreme of day length." It stands as the apex of Yang energy, with the longest day signifying the height of summer's power. However, it also marks the beginning of Yin's resurgence, as from this day forward, daylight gradually recedes. It is a period of both peak and transition, where the most Yang state begins to give way to the growing Yin, mirrored in the natural world by the changing behaviors of plants and animals.

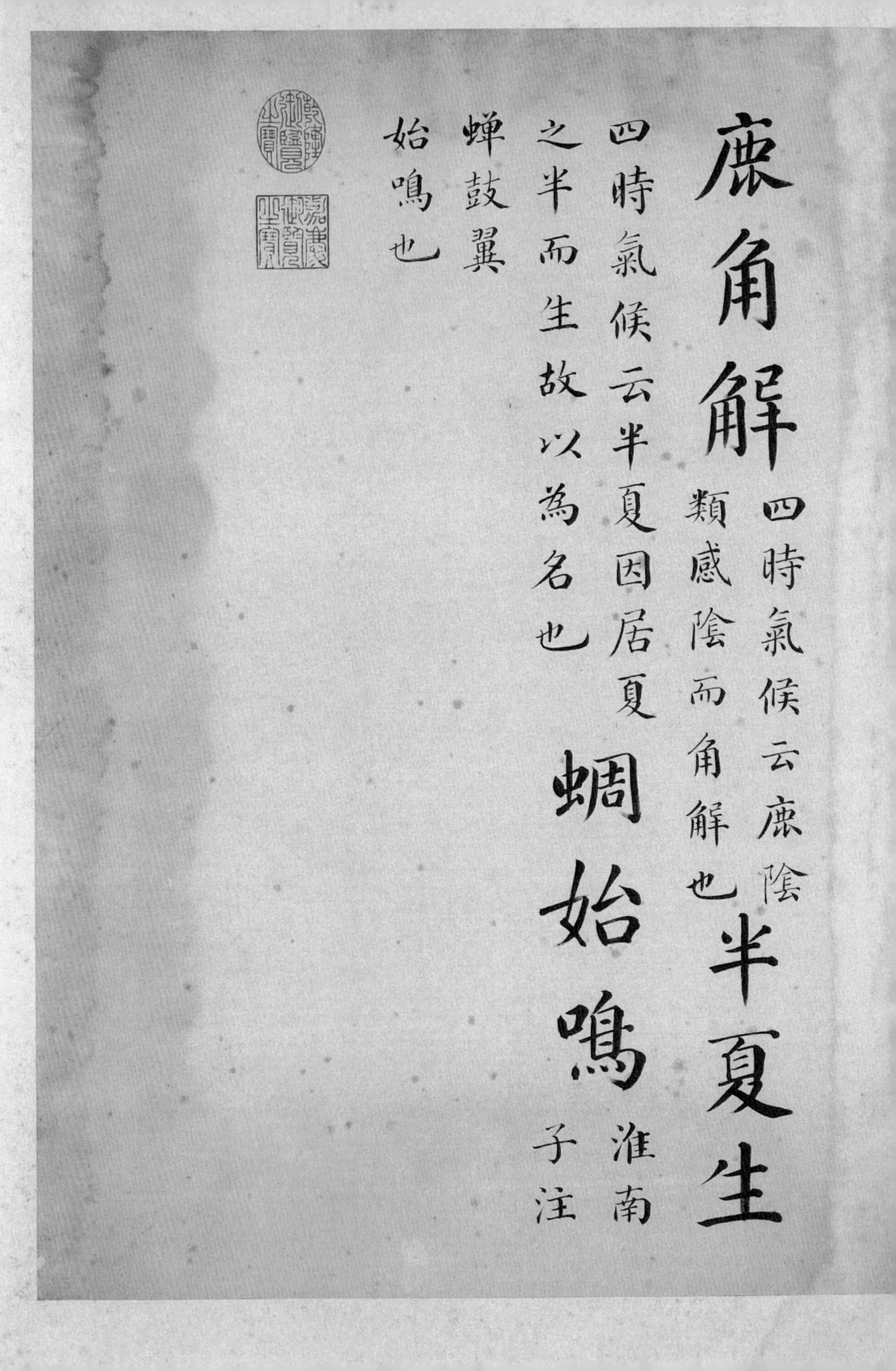

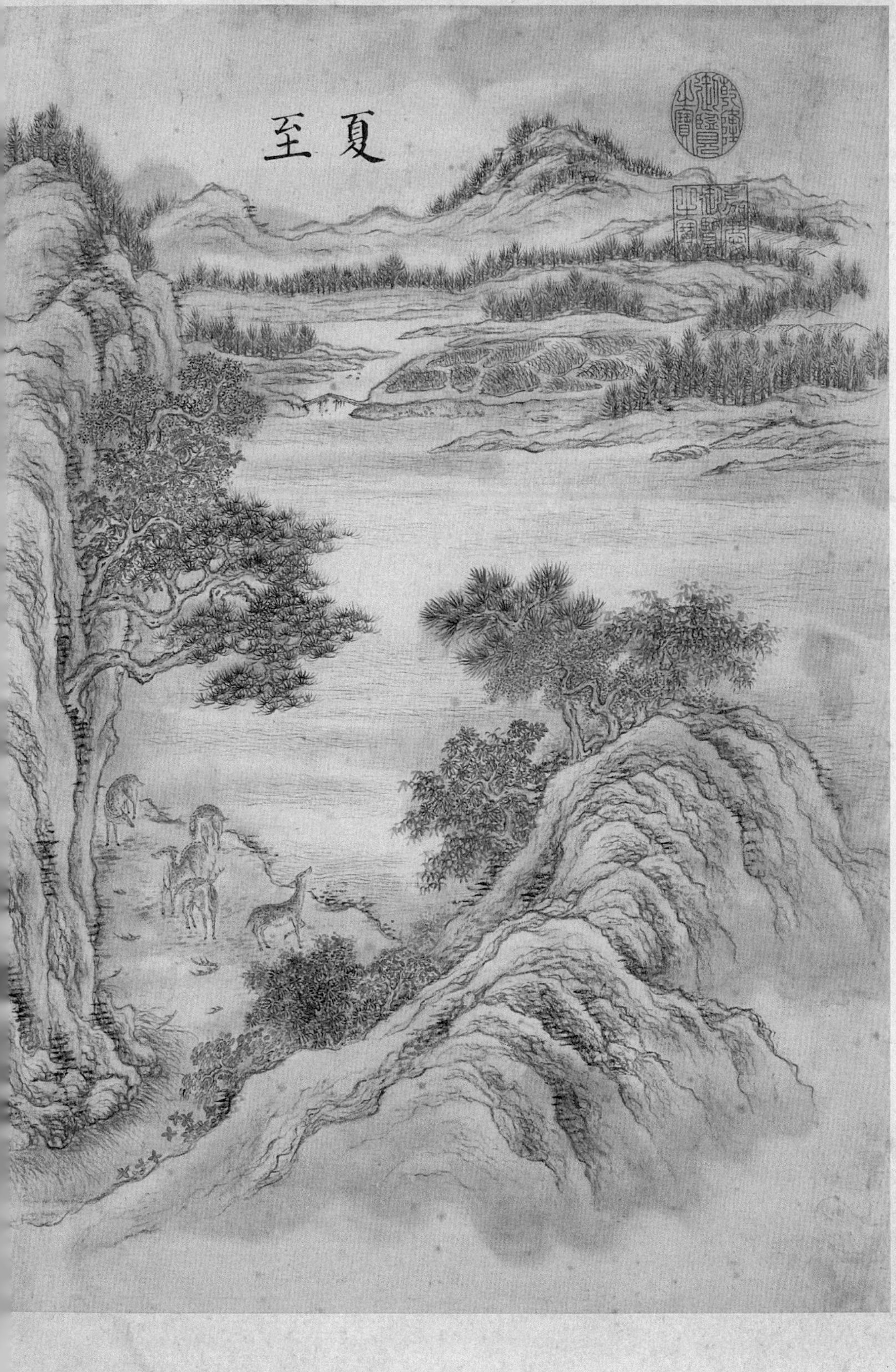


初候

## 鹿角解
## Deer Shed Antlers

二候

## 蜩始鳴
## Cicadas Begin to Sing

三候

## 半夏生
## Pinellia Ternata Grows

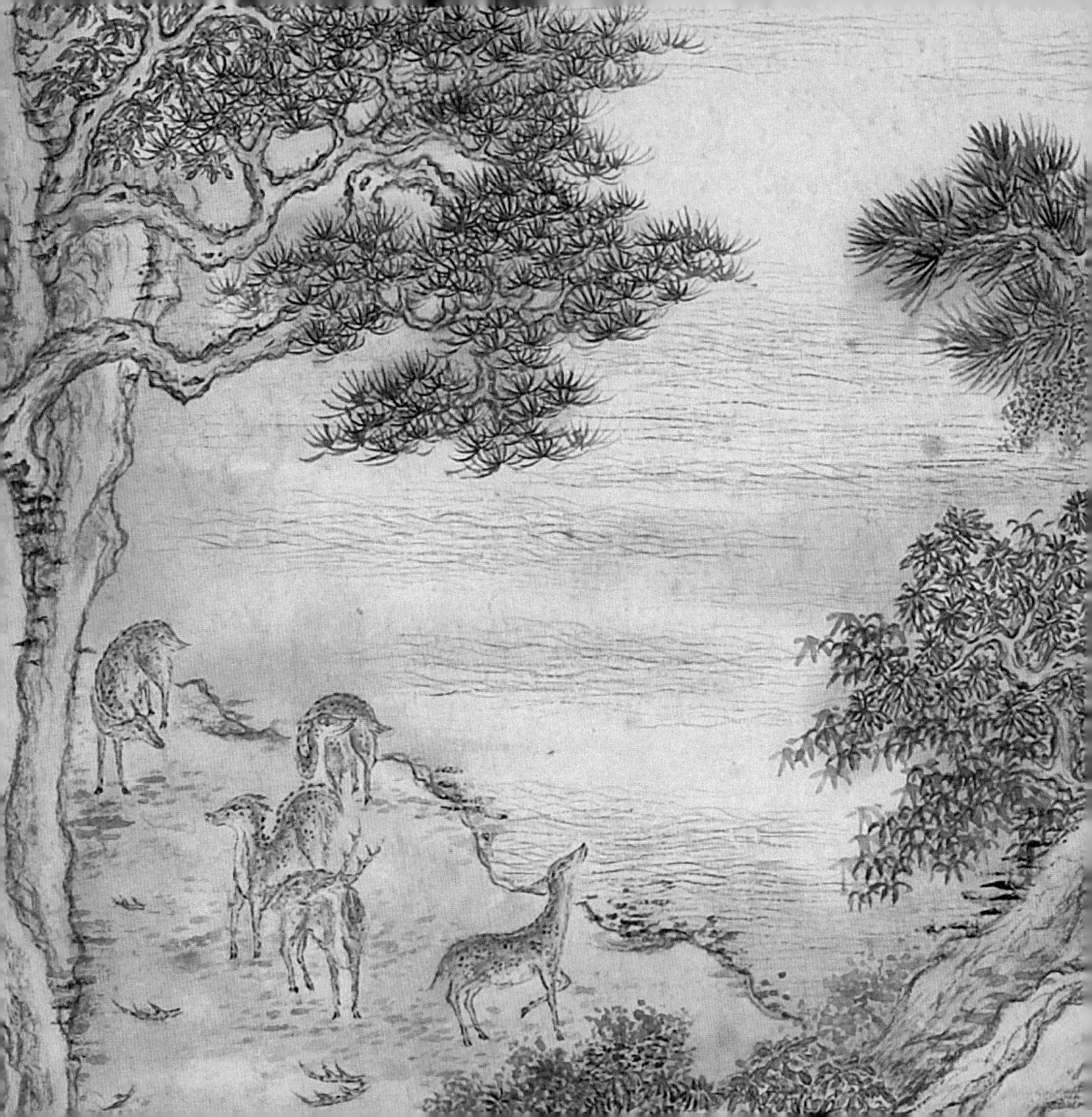

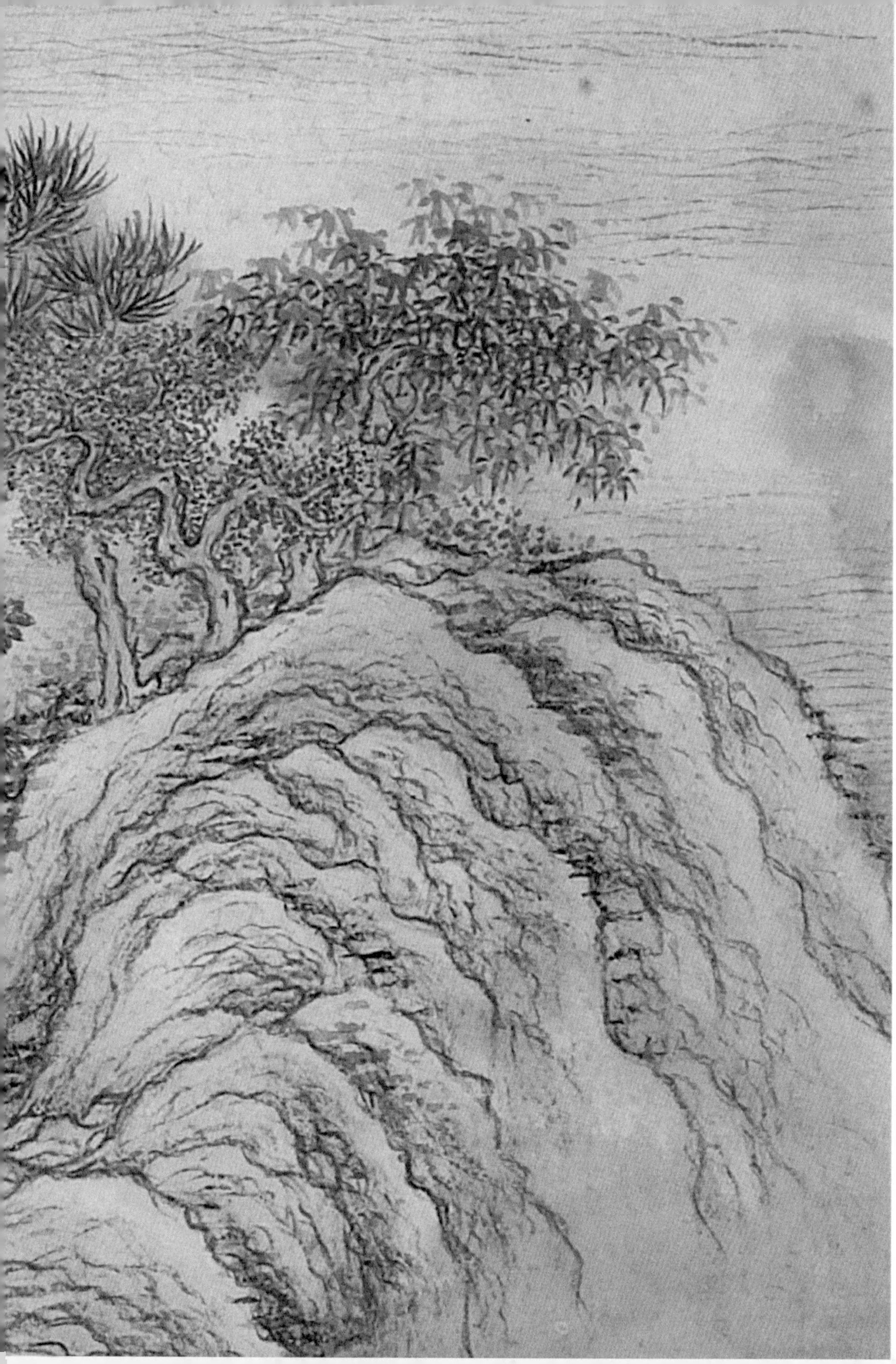

初候

# 鹿角解

## Deer Shed Antlers

The initial marker of Xiàzhì is "Deer Shed Antlers" (鹿角解 Lù Jiǎo Jiě), an emblematic change reflecting the peak and subsequent ebbing of Yang energy.

二候

# 蜩始鳴

## Cicadas Begin to Sing

The second harbinger is “Cicadas Begin to Sing” (蜩始鳴 Tiáo Shǐ Míng), their chorus becoming the soundtrack of deep summer.

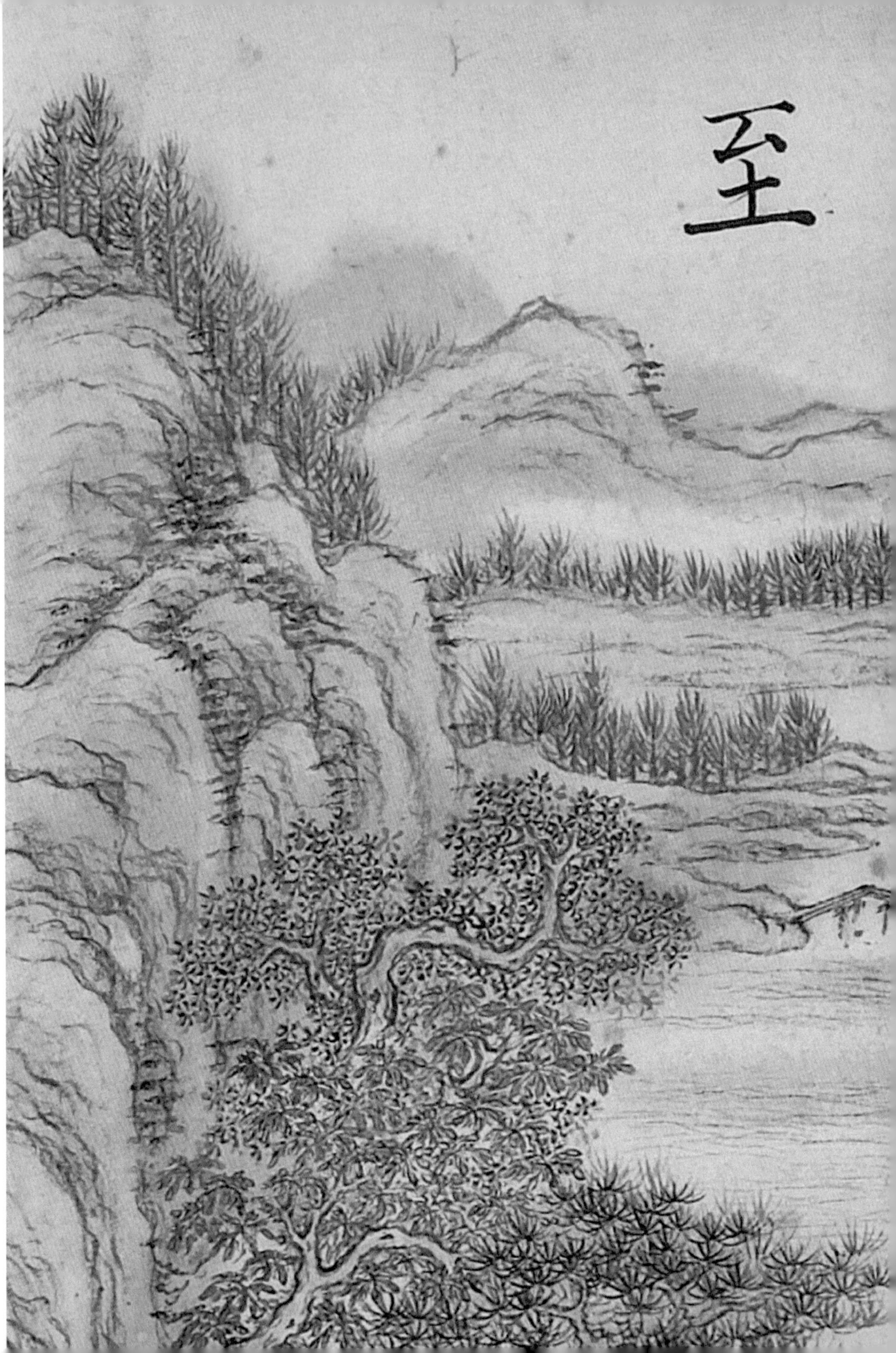

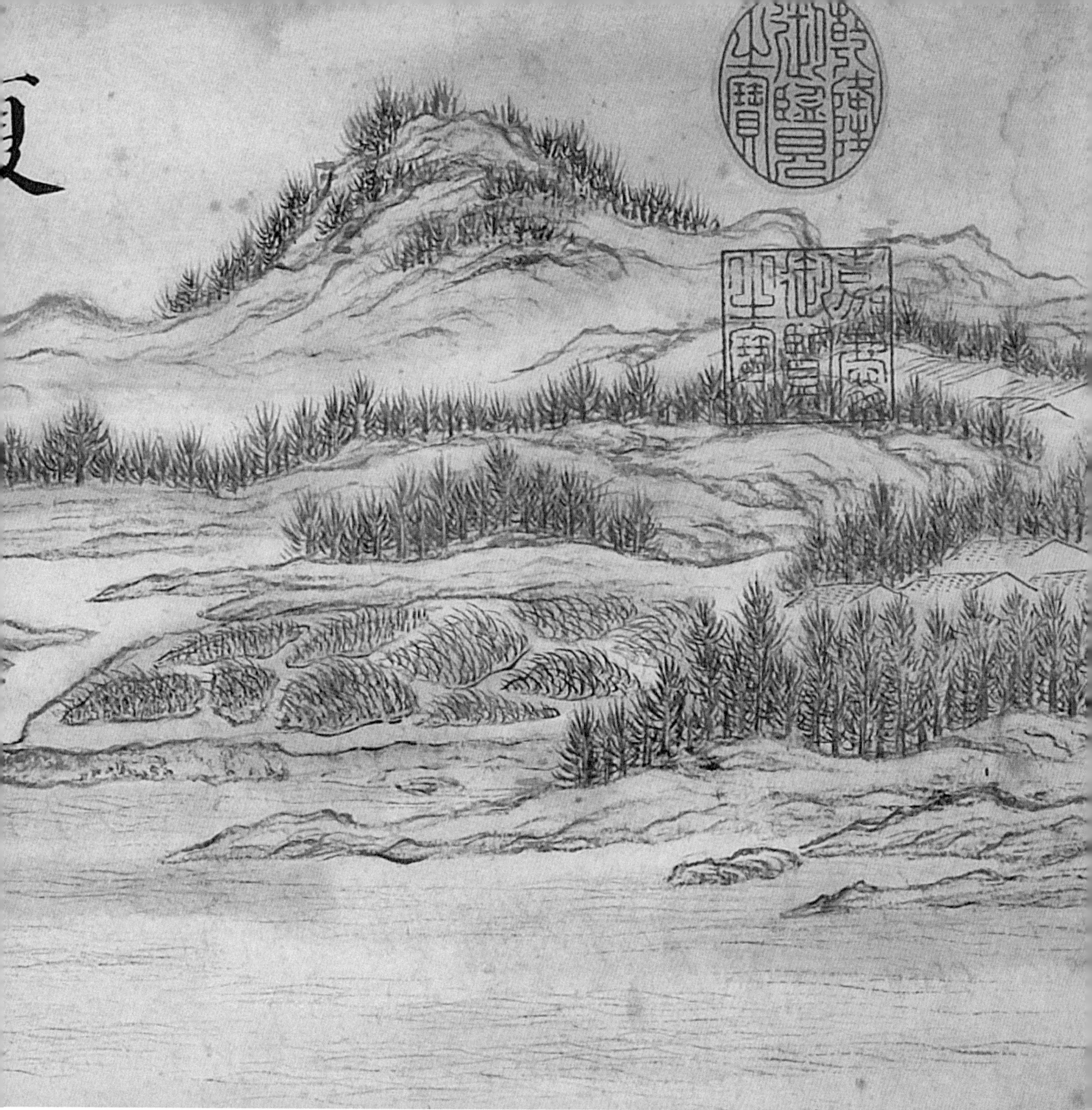
乾隆御覽之寶
嘉慶御覽之寶

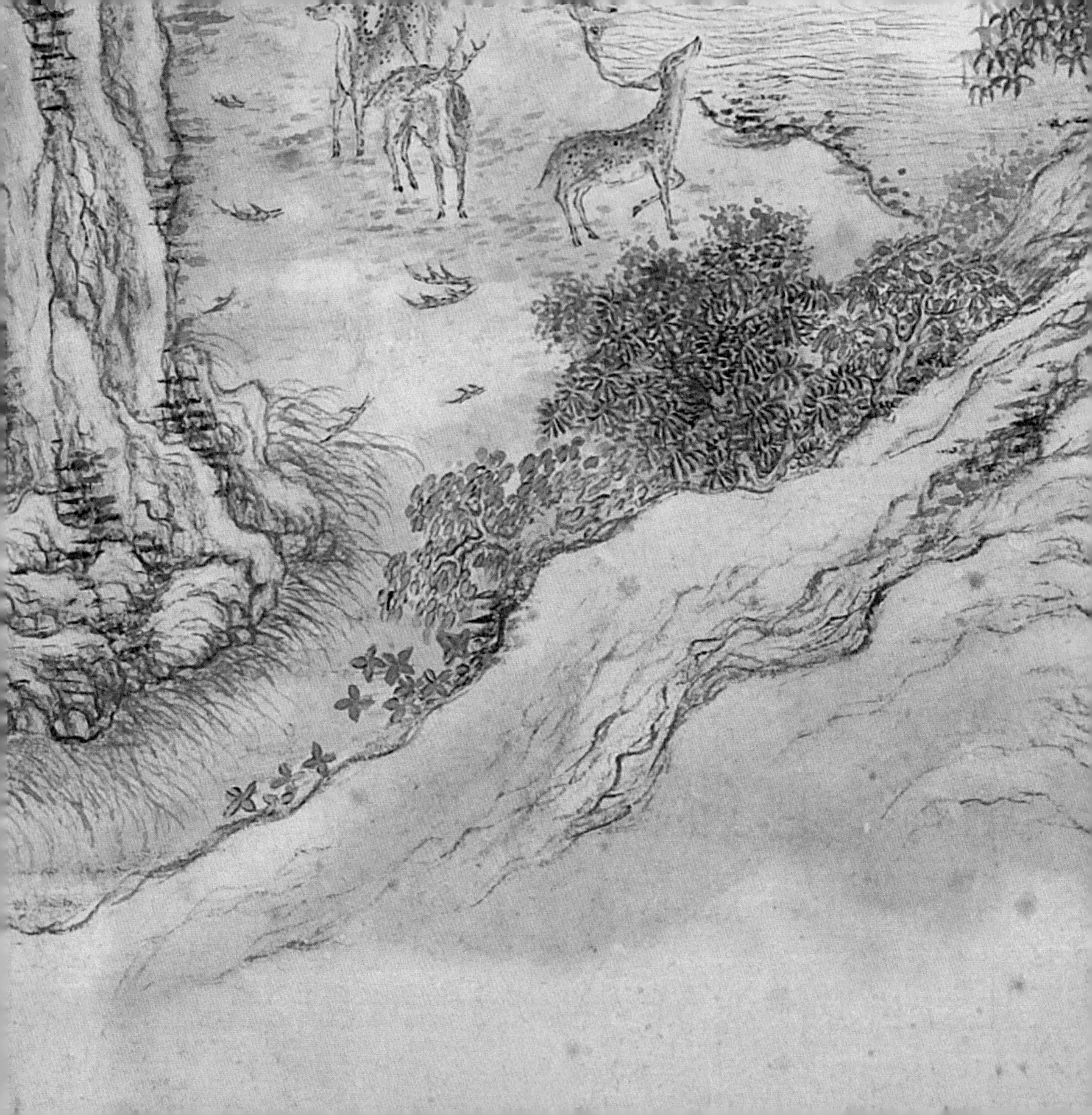

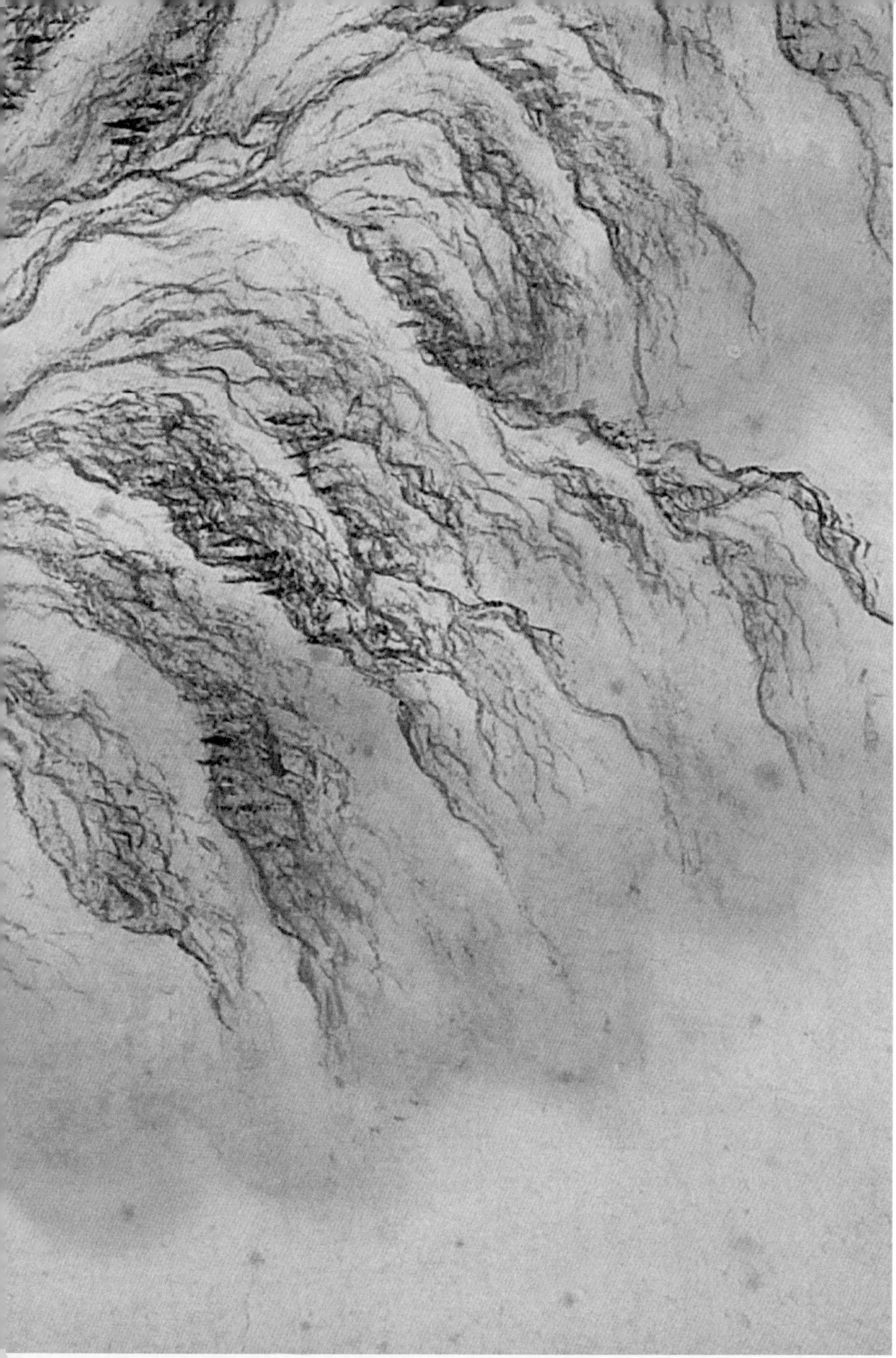

三候

# 半夏生

## Pinellia Ternata Grows

The third sign, "Pinellia Ternata Grows" (半夏生 Bàn Xià Shēng), refers to the growth of "Bànxià," a wild medicinal herb. The term "Bànxià" literally means "half summer," derived from its characteristic growth period during the middle of the summer season.

# 暑

Shǔ

# Xiǎo

## LESSER HEAT

July 7 (± 1)

# 小暑

## Lesser Heat

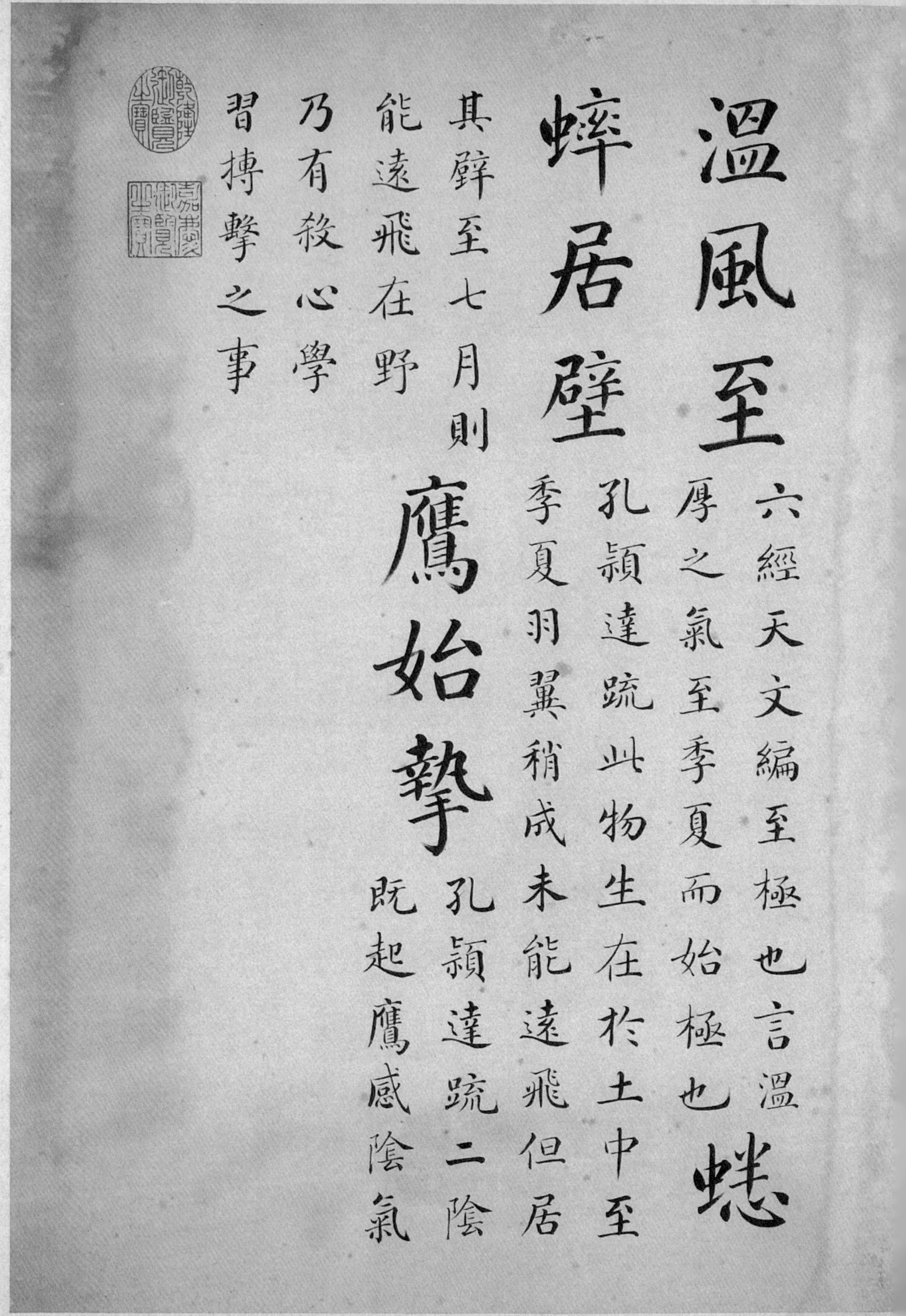


Xiǎoshǔ : Signifing the onset of the hot phase of summer, where the warmth starts to intensify but has not yet reached its extreme. It's a period when the latent heat within the earth begins to manifest more noticeably, signaling the approach of the hotter days ahead. Xiǎoshǔ is a prelude to the peak heat, marking a time of gradual increase in temperature and preparing for the more intense heat of Dàshǔ.

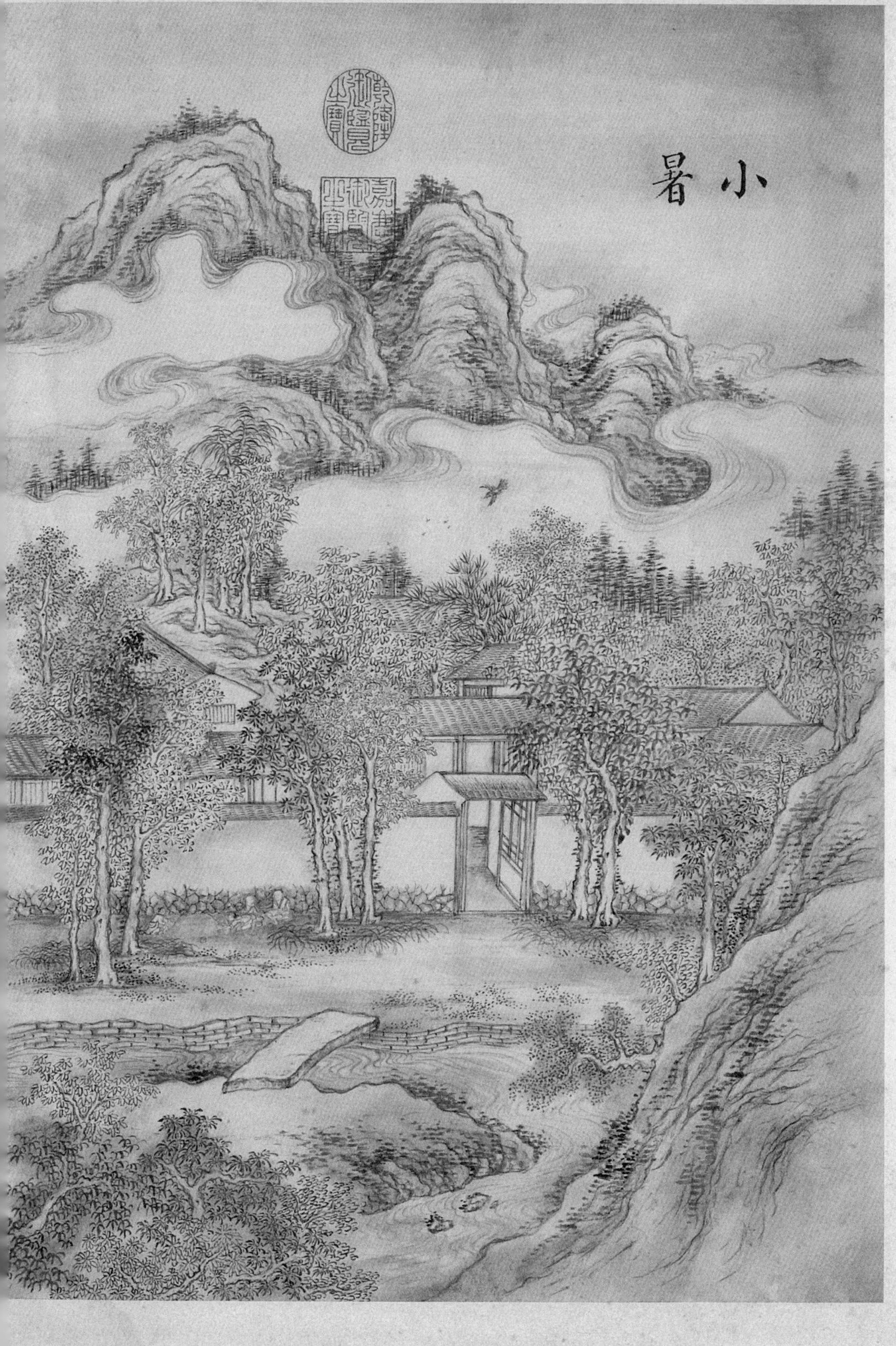


初候

# 溫風至

# Warm Winds Arrive

二候

# 蟋蟀居壁

# Crickets Take to the Hearth

三候

# 鷹始摯

# Eagles Begin Fierceness

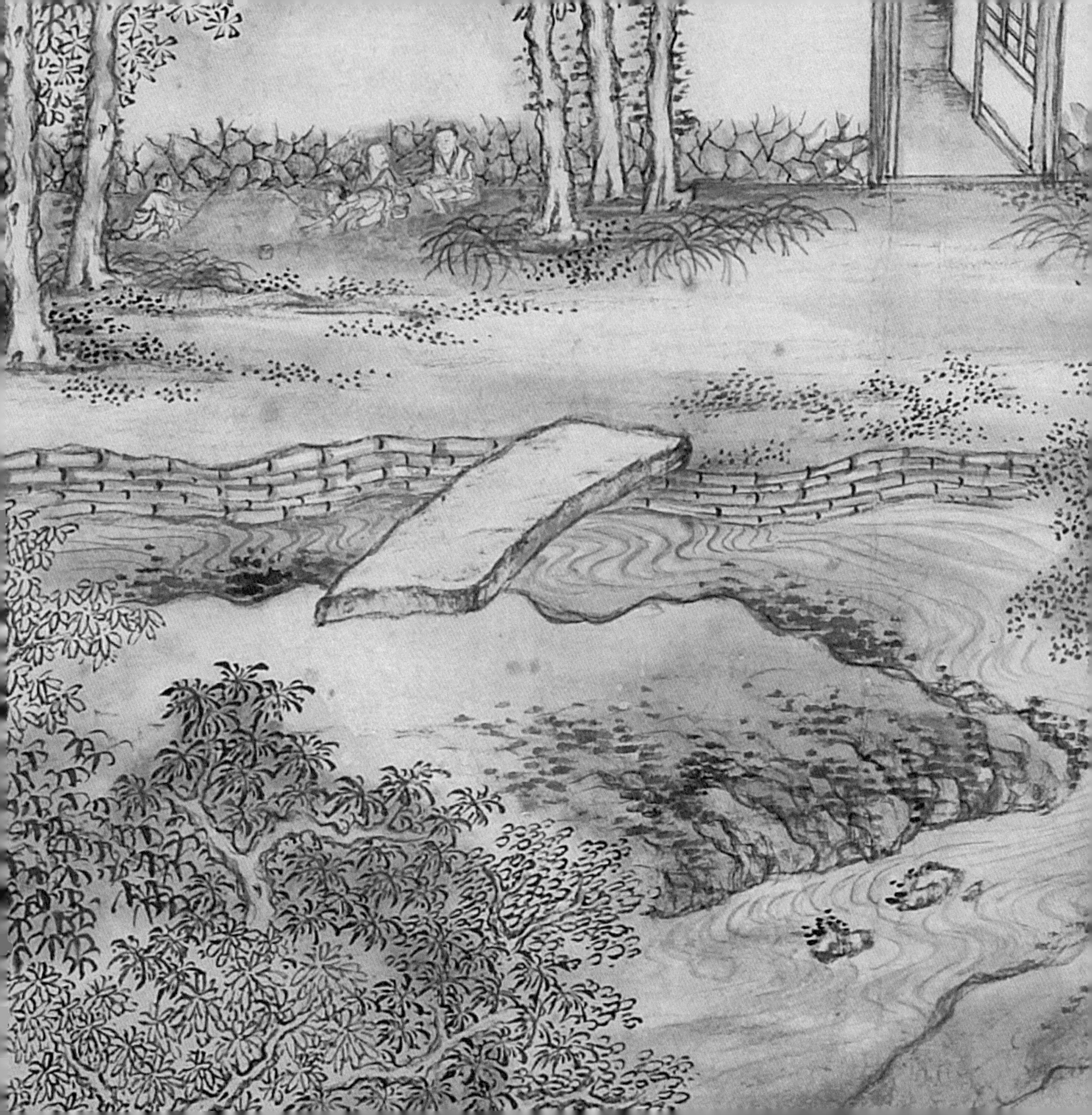

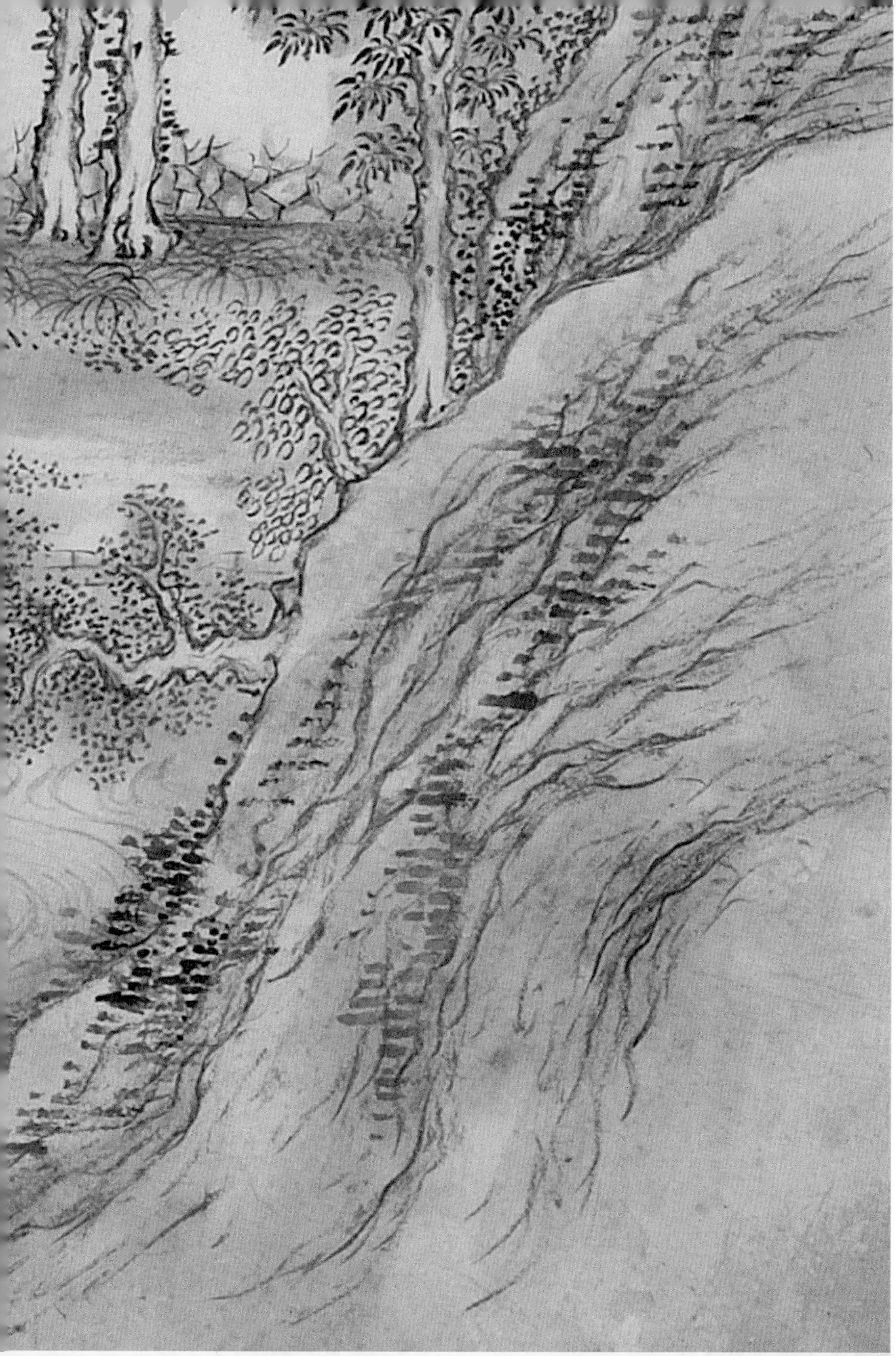

初候

# 溫風至

## Warm Winds Arrive

The first whisper of Xiǎoshǔ is “Warm Winds Arrive” (溫風至 Wēn Fēng Zhì), caressing the land and heralding the onset of heat.

二候

# 蟋蟀居壁

## Crickets Take to the Hearth

The second inkling is “Crickets Take to the Hearth” (蟋蟀居壁 Xī Shuài Jū Bì), as they move closer to human habitation, seeking comfort in the warmth.

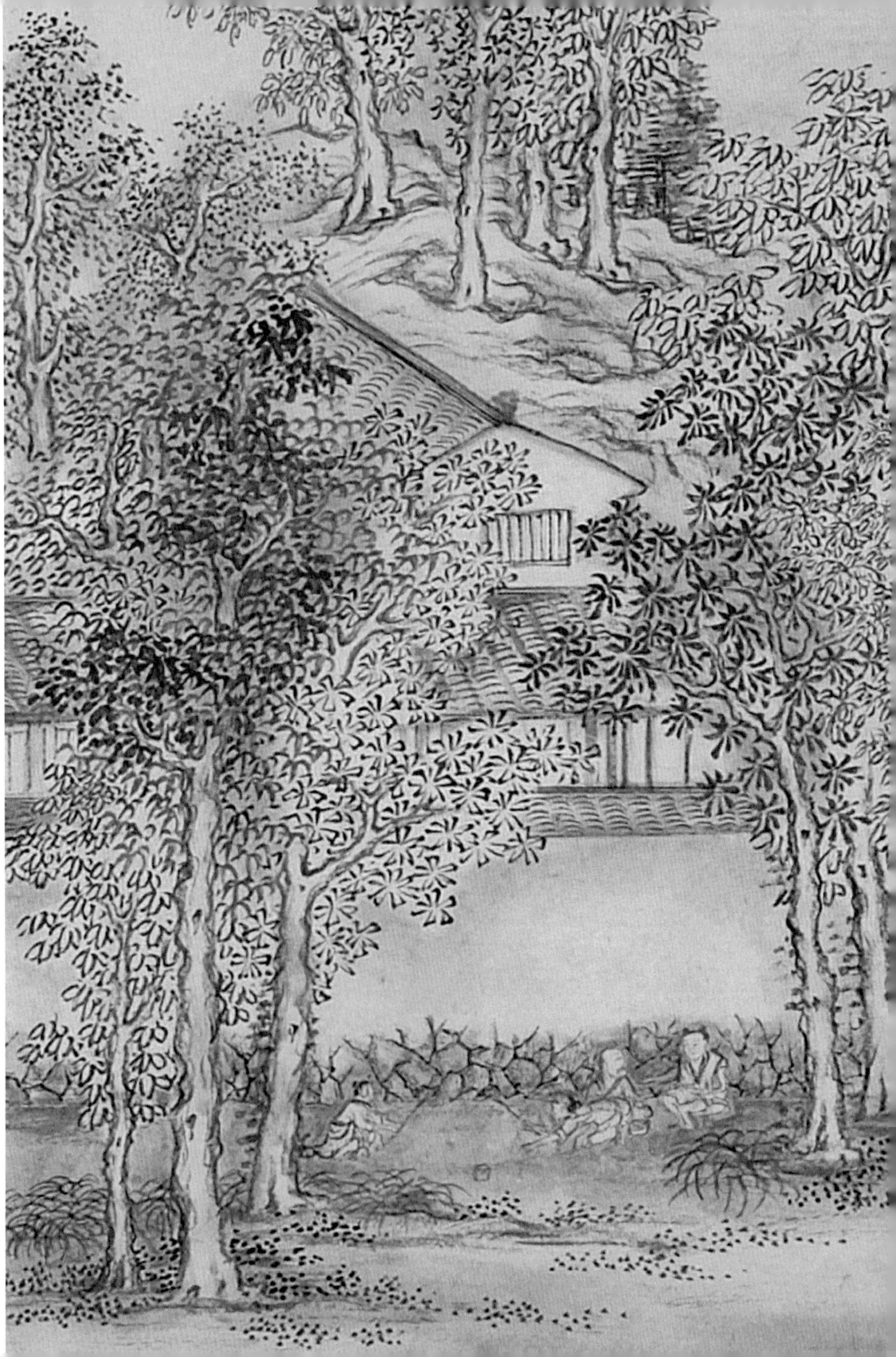

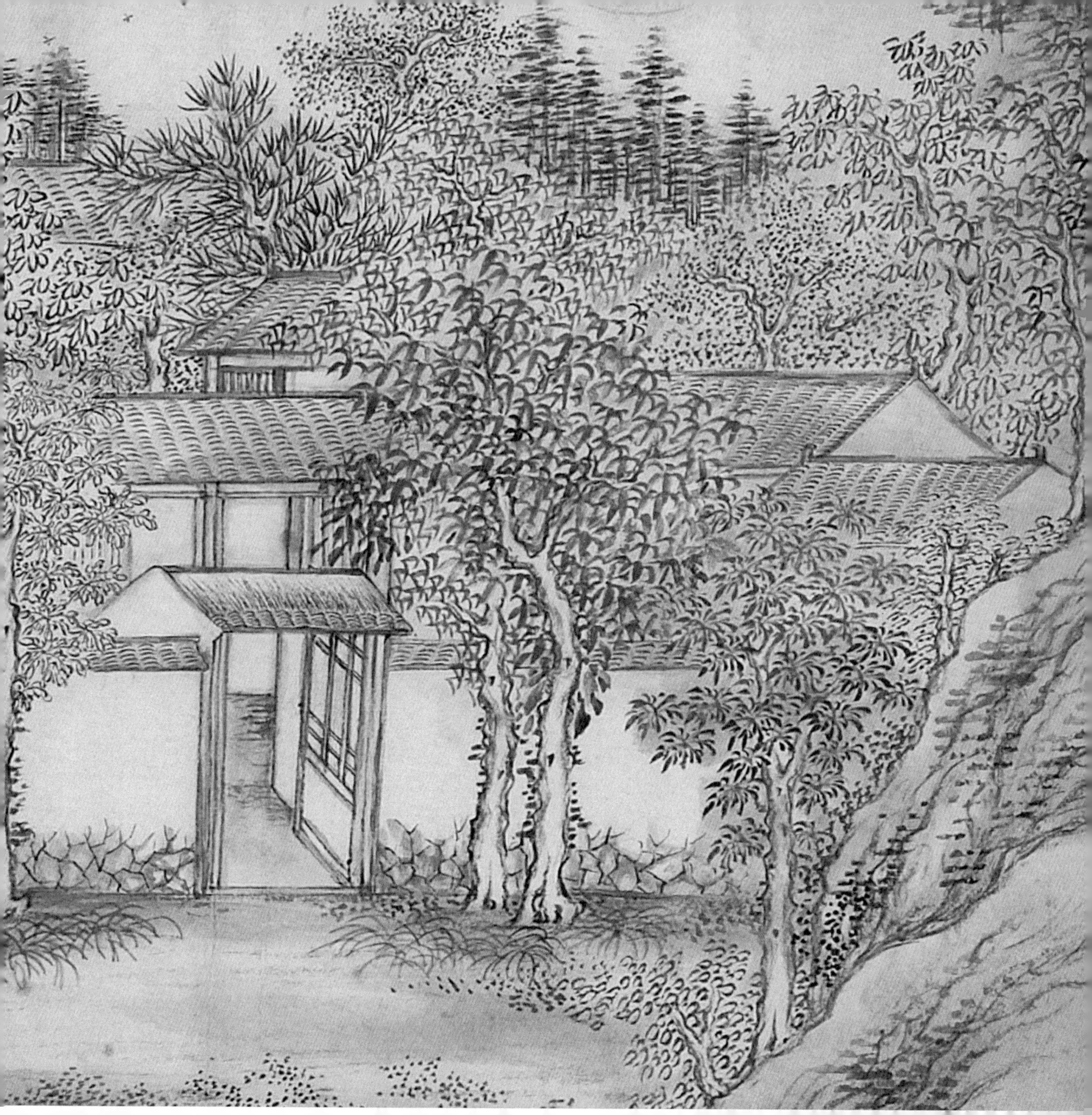

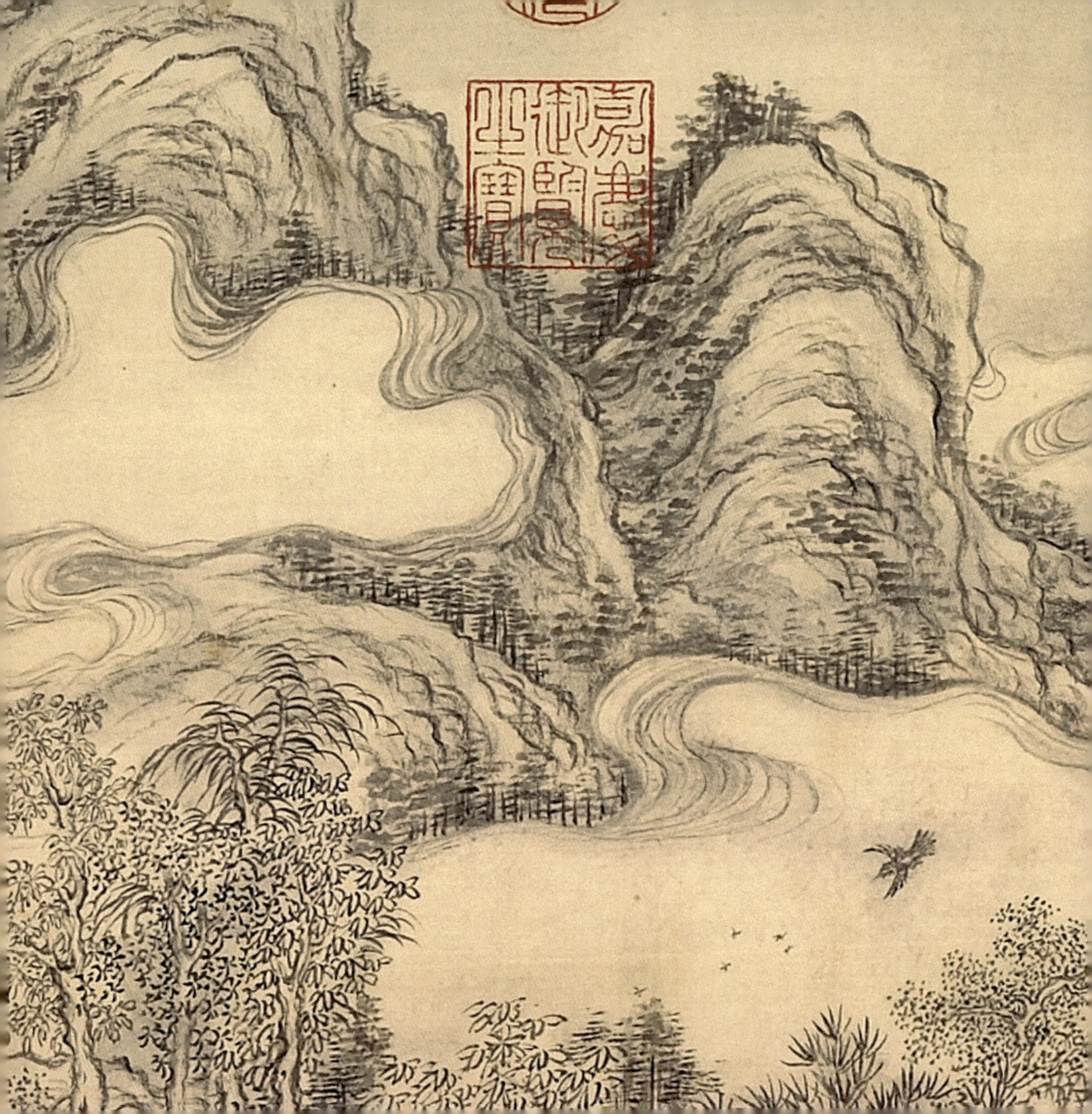
嘉慶御覽之寶

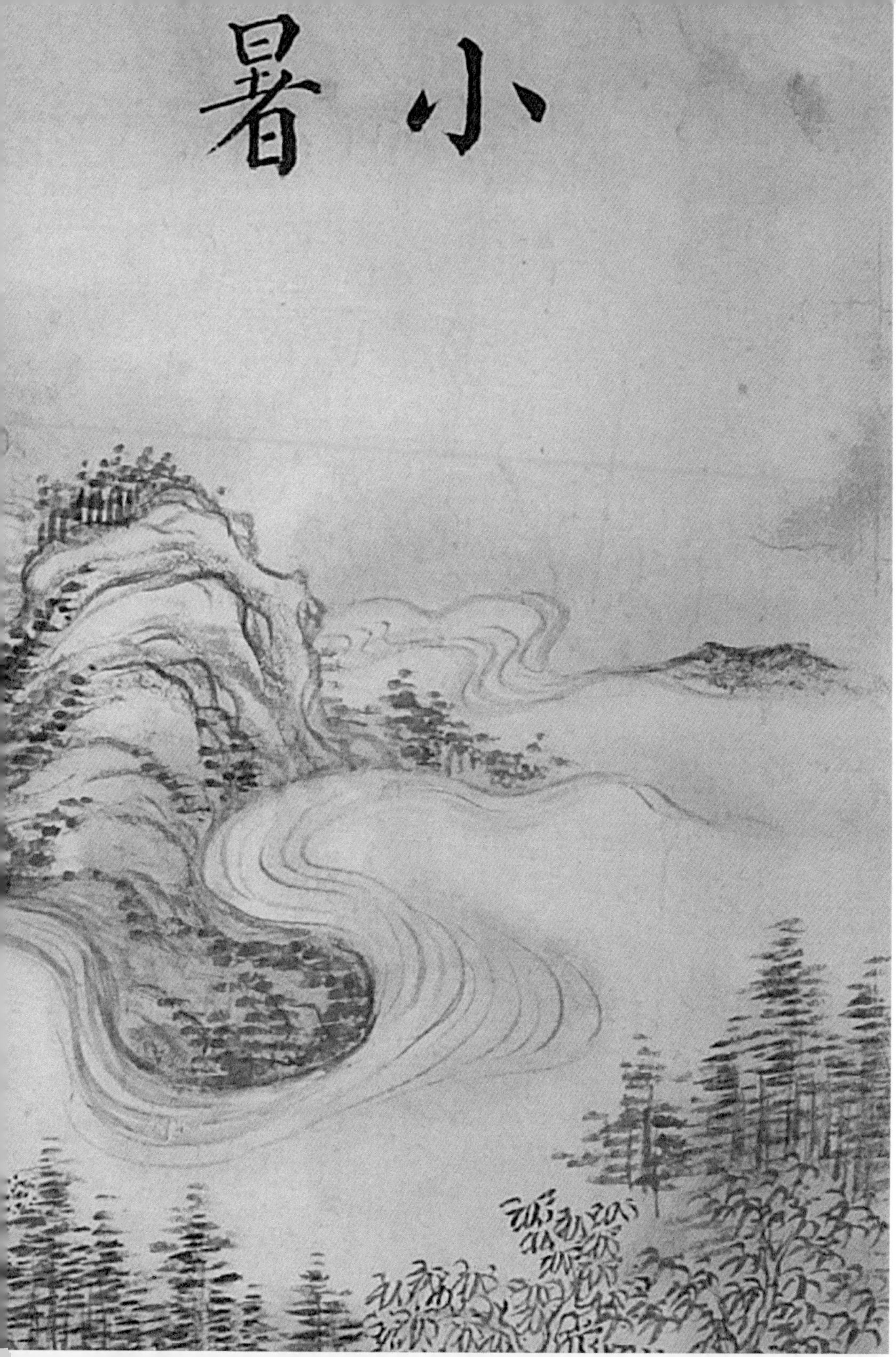


三候

# 鹰始挚

## Eagles Begin Fierceness

The third sign, “Eagles Begin Fierceness” (鹰始挚 Yīng Shǐ Zhì), refers to the period when young eagles, led by the adults, leave the nest and start learning the vital skills of soaring and hunting.

# 暑

Shǔ

Dà

大

# GREATER HEAT

July 23 (± 1)

# 大暑

## Greater Heat

Dàshǔ: Representing the peak of the summer heat, known as the most intense period of warmth throughout the year. It's a time when the heat is most potent, often referred to as the "Dog Days" of summer. Following Dàshǔ, the term "fu" implies a period of hidden or latent heat, suggesting that during this time, people seek ways to shield themselves from the extreme temperatures, often seeking cooler environments or methods to escape the heat. It's a period characterized by seeking comfort and respite from the summer's intensity.

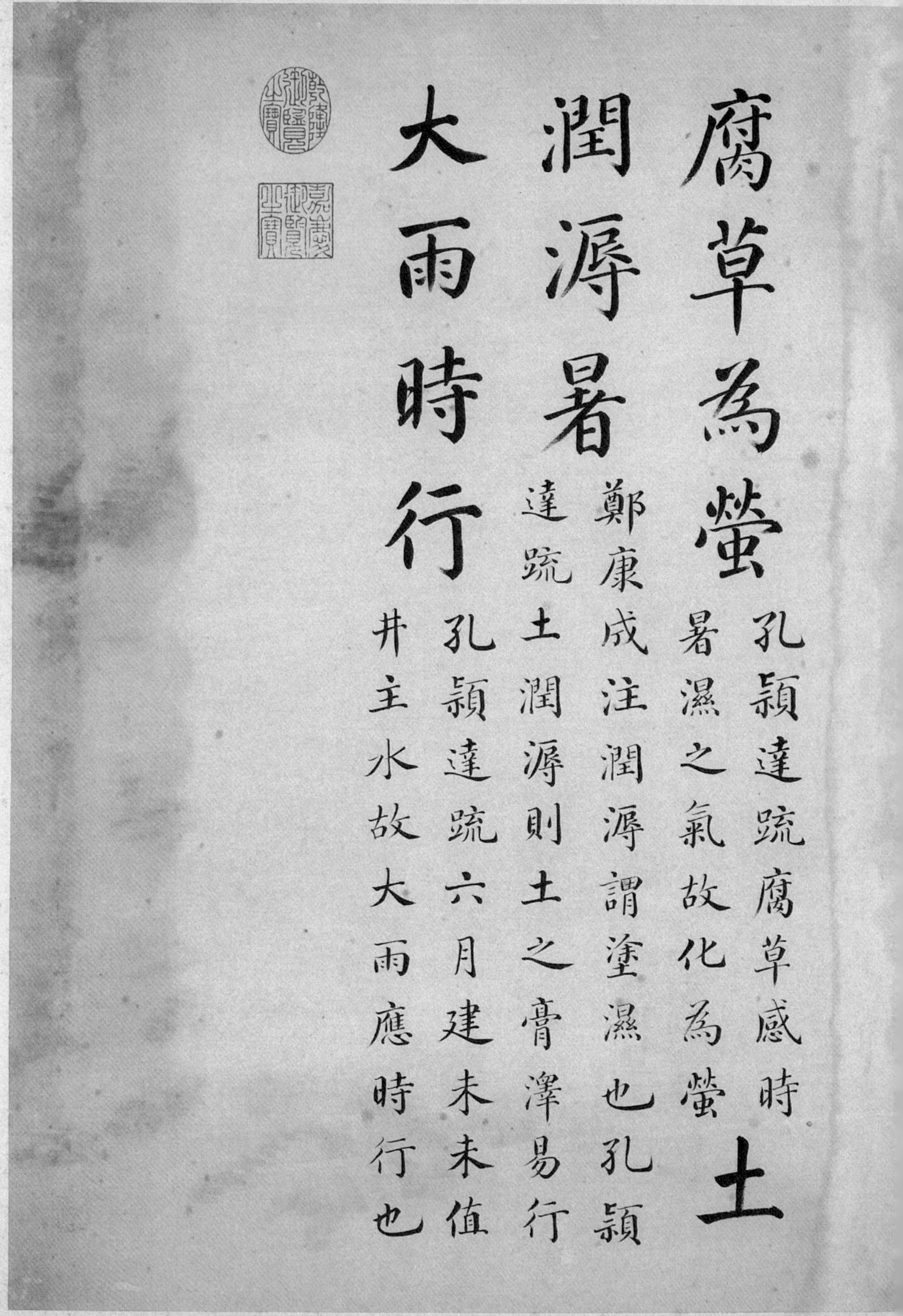
腐草為螢
孔穎達疏腐草感時暑濕之氣故化為螢
土潤溽暑
鄭康成注潤溽謂塗濕也孔穎達疏土潤溽則土之膏澤易行
大雨時行
孔穎達疏六月建未未值井主水故大雨應時行也

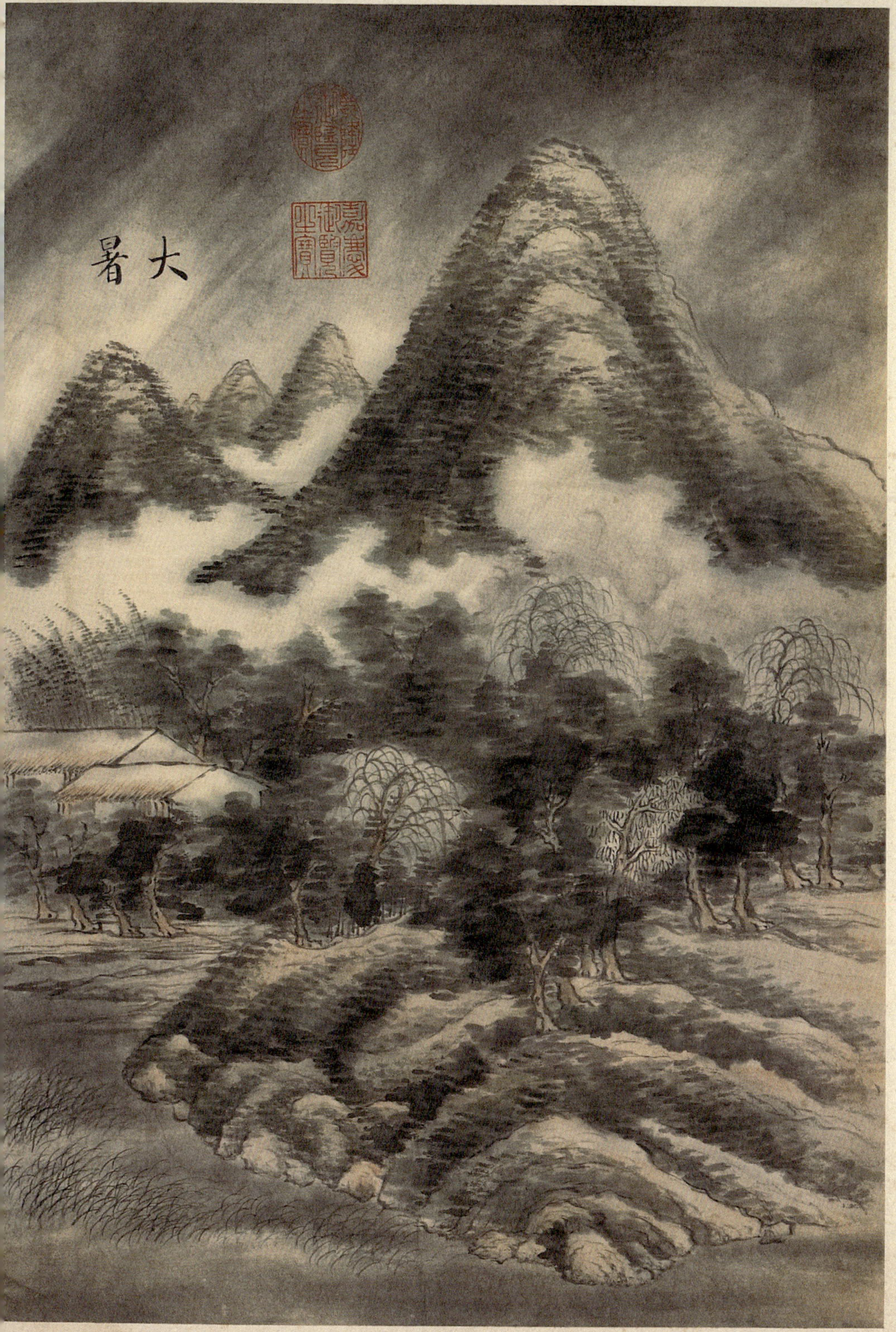


初候

## 腐草為螢
## Fireflies Born from Decaying Grass

二候

## 土潤溽暑
## Humid Heat Saturates the Earth

三候

## 大雨時行
## Frequent Heavy Rains

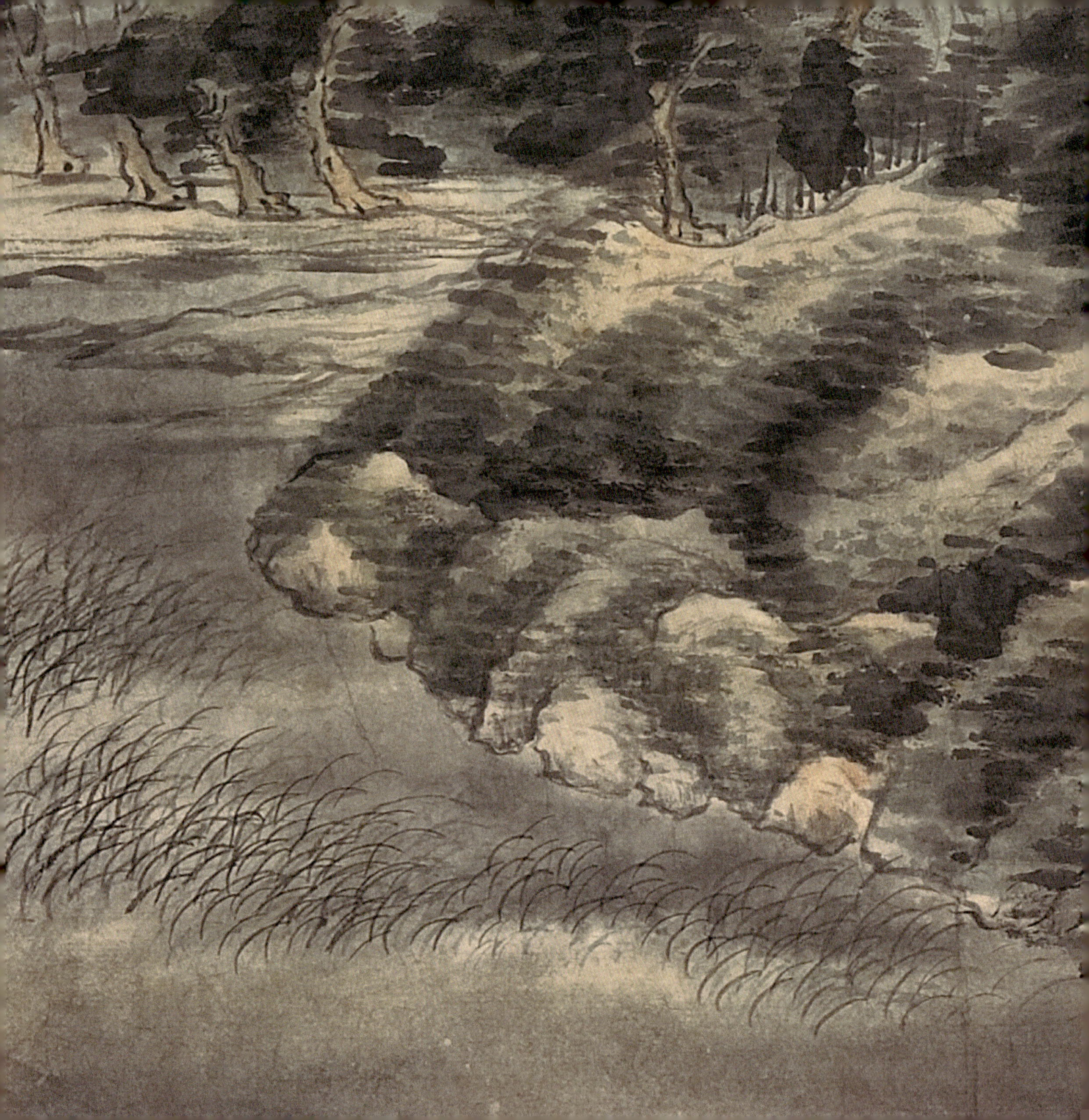

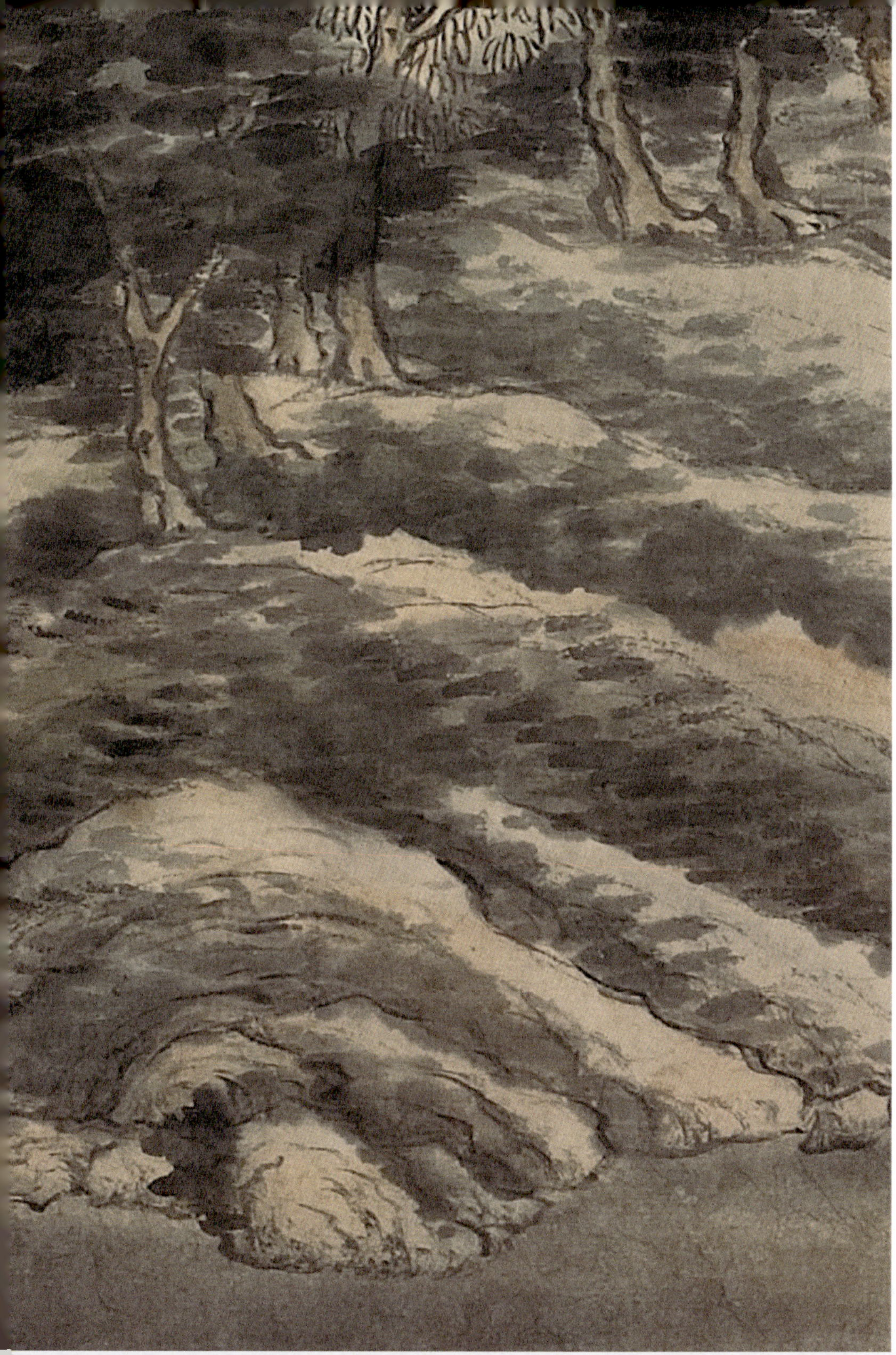

初候

# 腐草為螢

## Fireflies Born from Decaying Grass

The first indication is “Fireflies Born from Decaying Grass” (腐草為螢 Fǔ Cǎo Wéi Yíng).
This sign marks the appearance of fireflies, a captivating feature of summer evenings. The ancient belief was that fireflies emerged from decaying grass and leaves, though in reality, firefly eggs develop in the ground and amongst fallen leaves. As summer reaches its peak during Dàshǔ, these larvae complete their growth cycle, metamorphosing from pupae into the glowing adult fireflies that illuminate the night.

二候

# 土潤溽暑

## Humid Heat Saturates the Earth

The second signal is “Humid Heat Saturates the Earth” (土潤溽暑 Tǔ Rùn Rù Shǔ).

During this period, the earth becomes saturated with moisture, and the weather becomes unbearably hot and humid. The air feels heavy and oppressive. It’ s a period when the dampness of the soil and the sweltering heat combine to create a distinctly muggy and uncomfortable atmosphere.

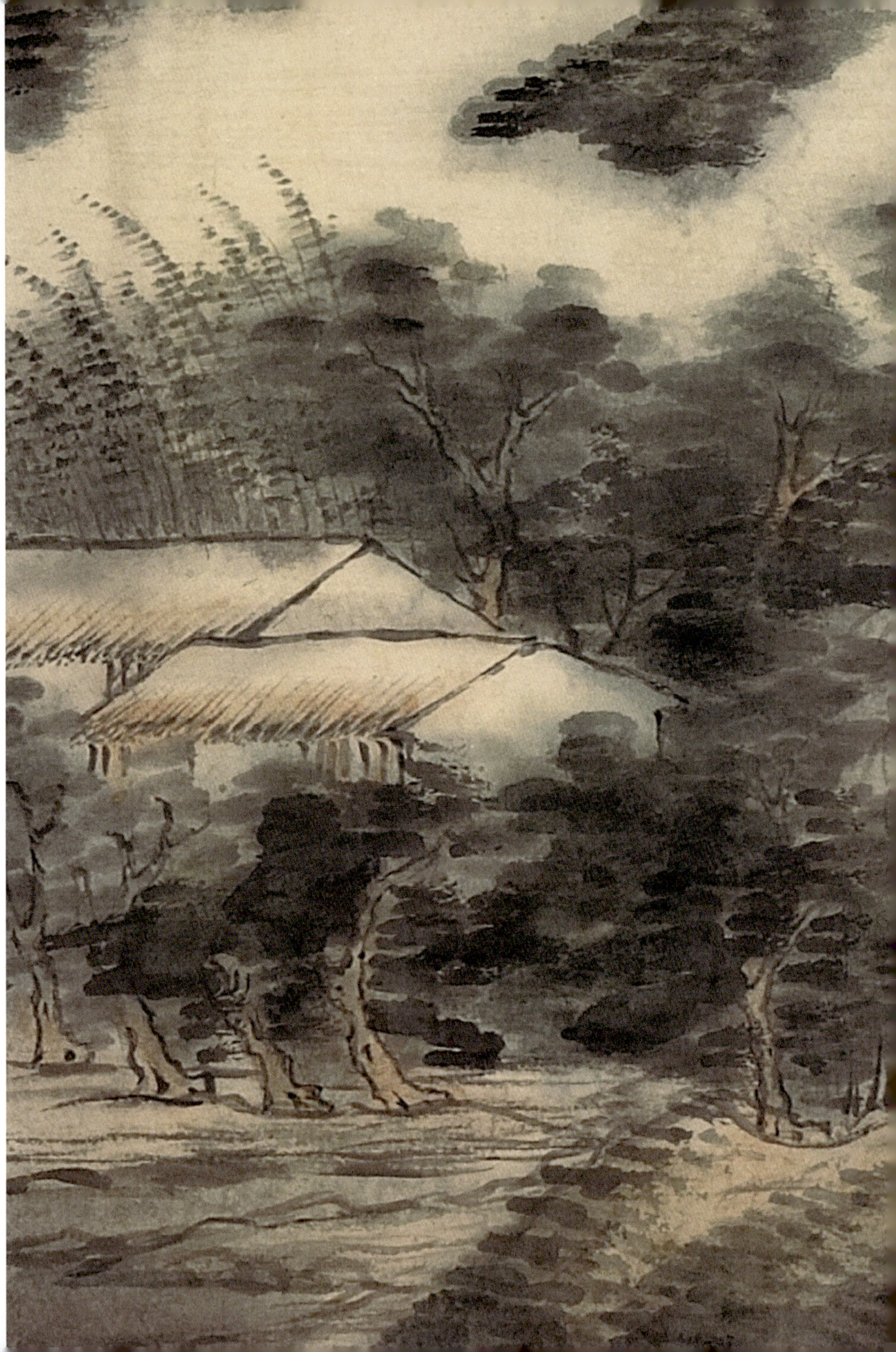

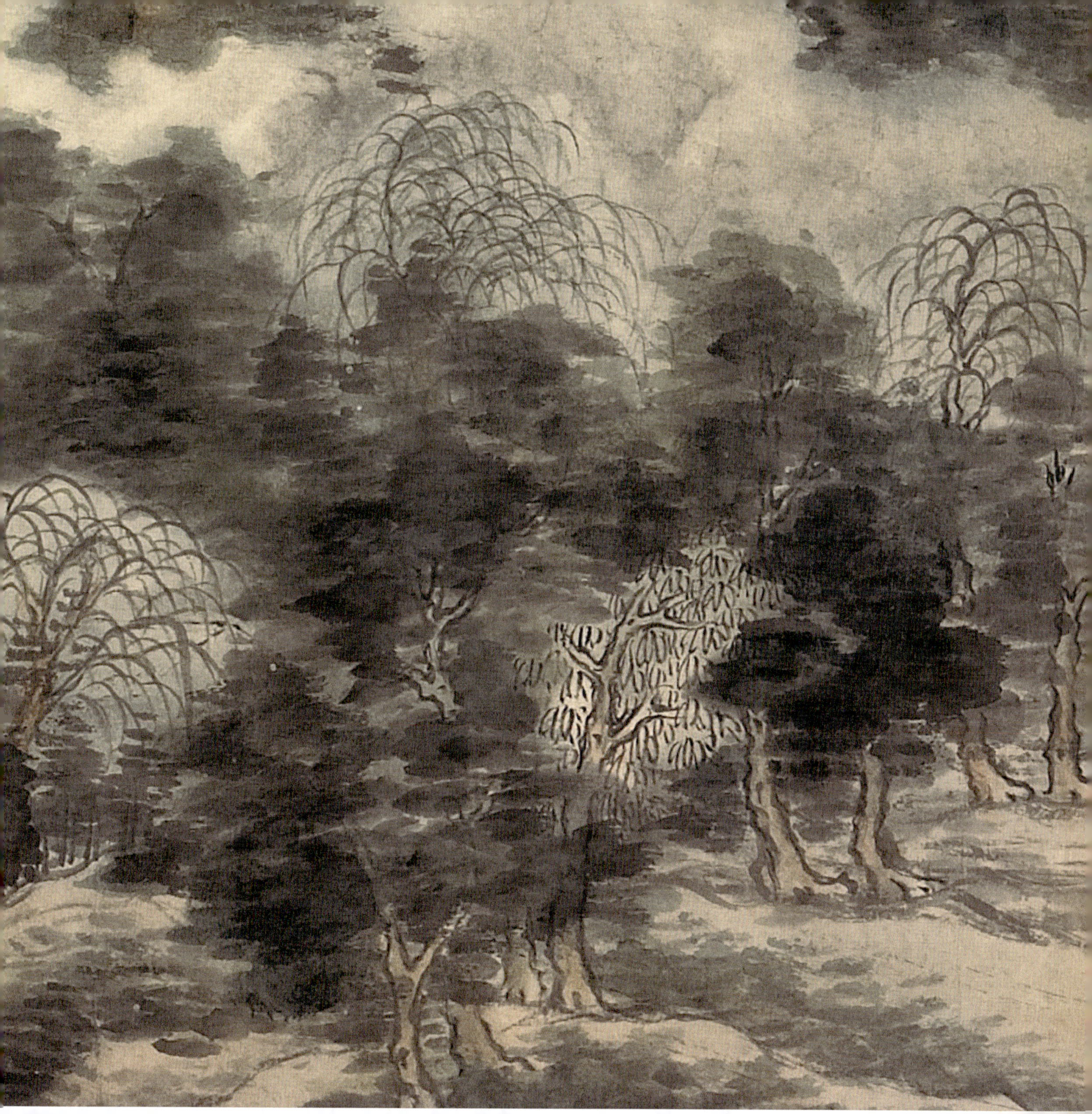

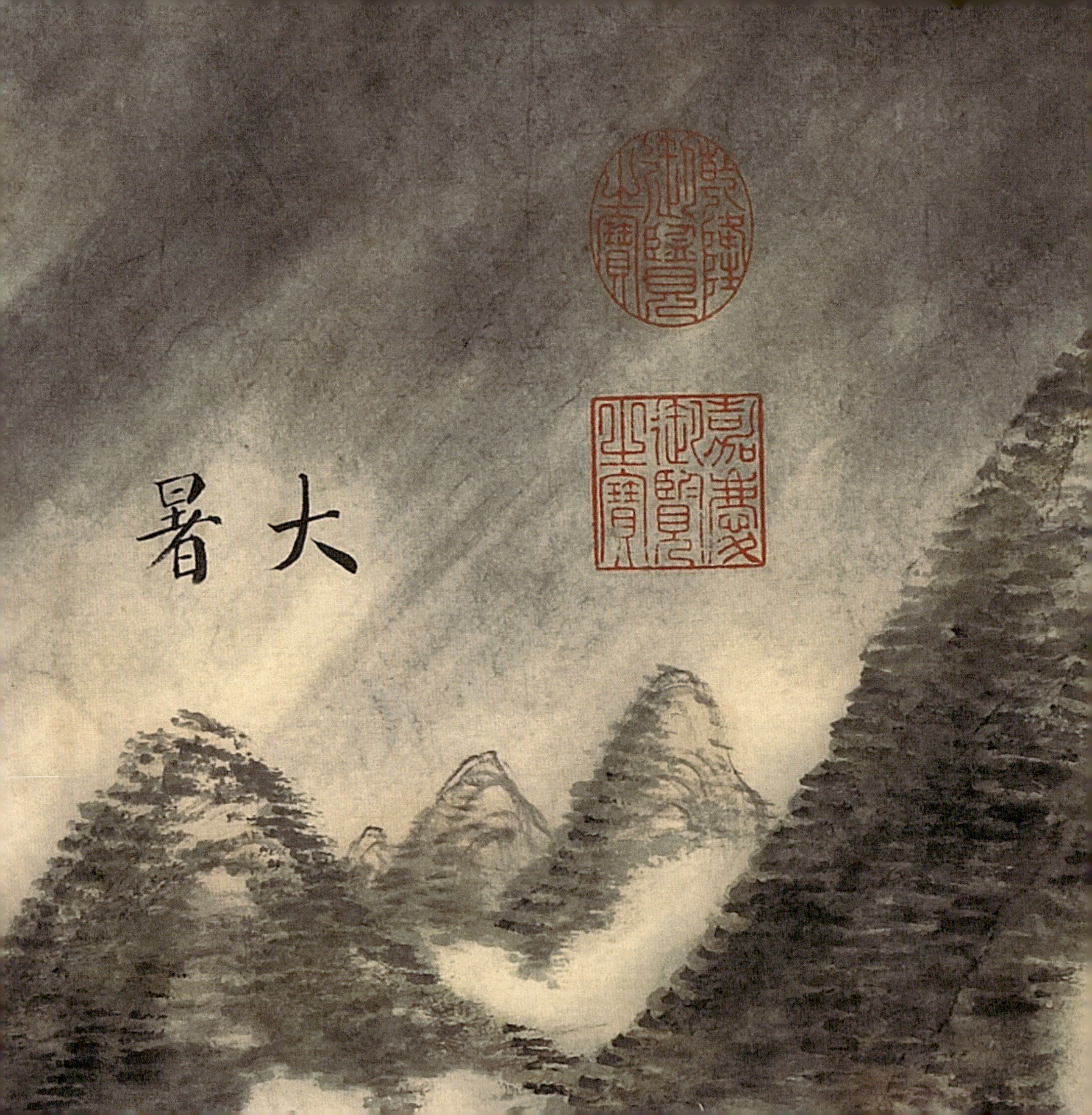
乾隆御覽之寶
嘉慶御覽之寶
大暑

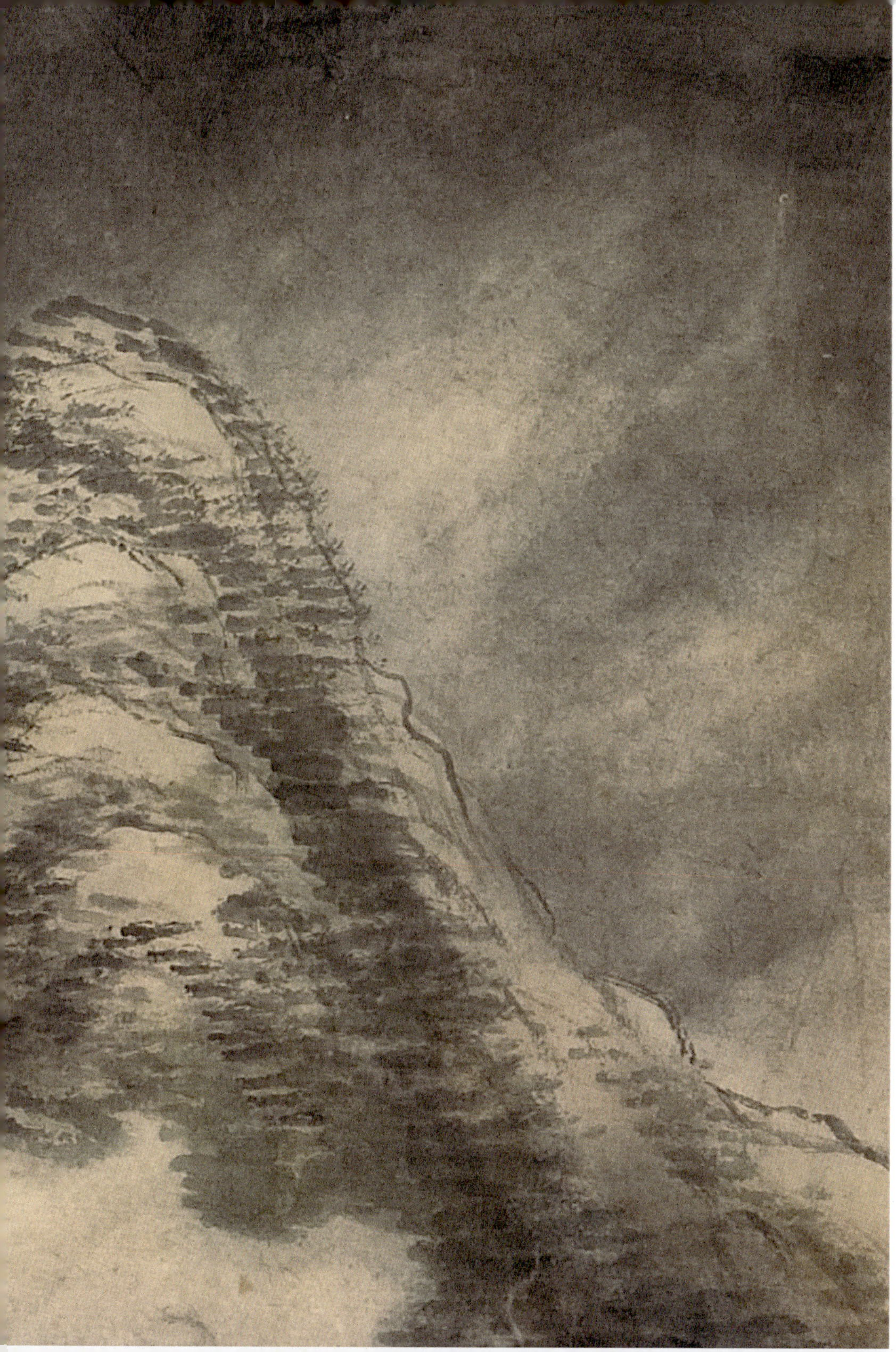

三候

# 大雨時行

## Frequent Heavy Rains

The third sign is “Frequent Heavy Rains” (大雨時行 Dà Yǔ Shí Xíng). The third sign is characterized by sudden and heavy downpours, often occurring in the afternoons after the accumulation of heat and humidity throughout the day. These torrential rains provide temporary relief from the sweltering heat, breaking the tension in the air and cooling down the earth. While the heavy showers may be brief, they are intense and dramatic, showcasing the dynamic and powerful aspects of nature during the peak of summer.

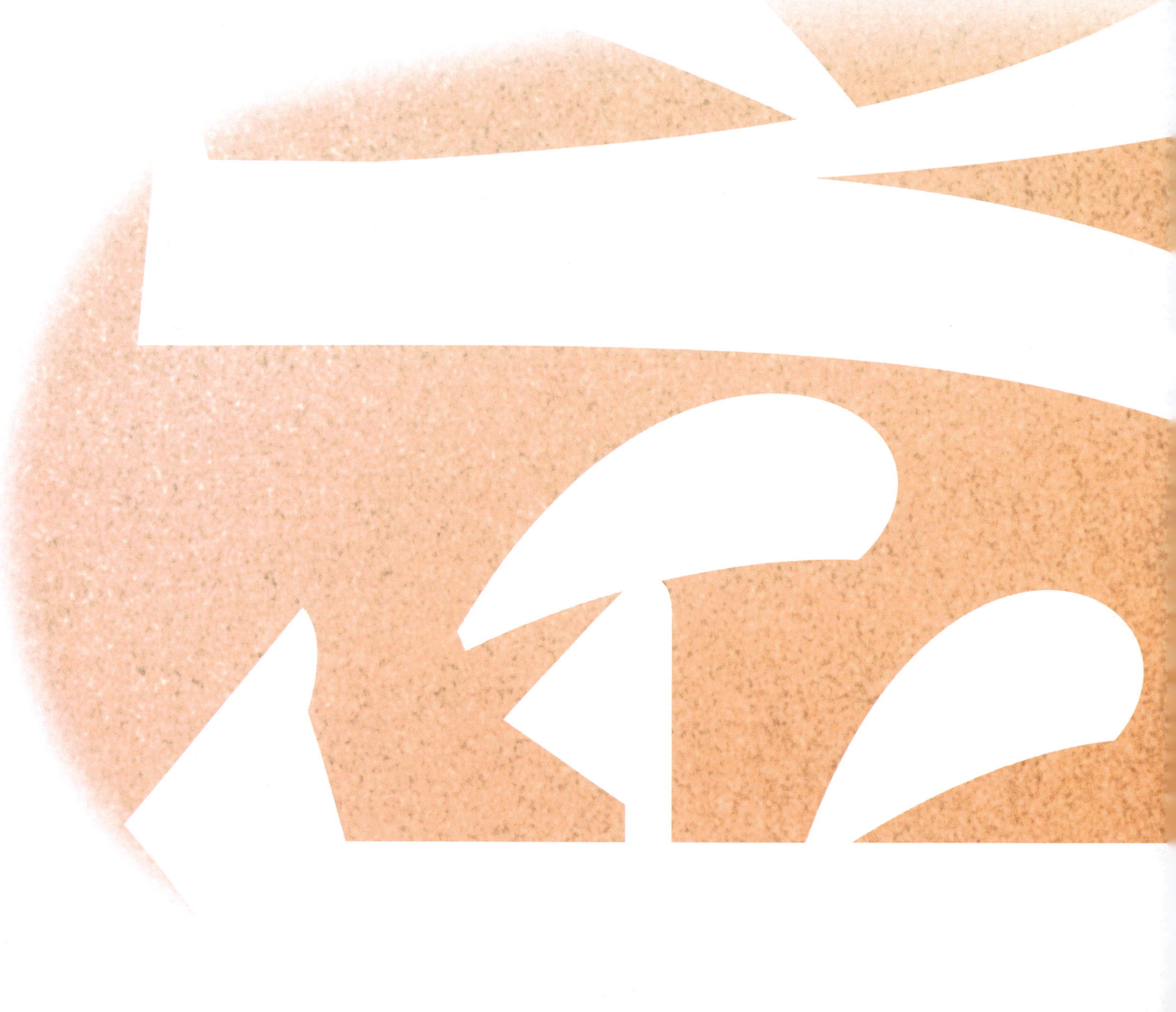

# AUTUMN

# 秋

Qiū

Lì

立

# BEGINNING OF AUTUMN

August 8 (± 1)

# 立秋

## Beginning of Autumn

Lìqiū : Heralding the beginning of autumn, representing a transition from the warmth of summer to the coolness of fall. It's a time when the crops mature and the harvest season approaches, symbolizing a shift in agricultural activities and the rhythm of life. It's a period of change, reflection, and preparation for the cooler days ahead.

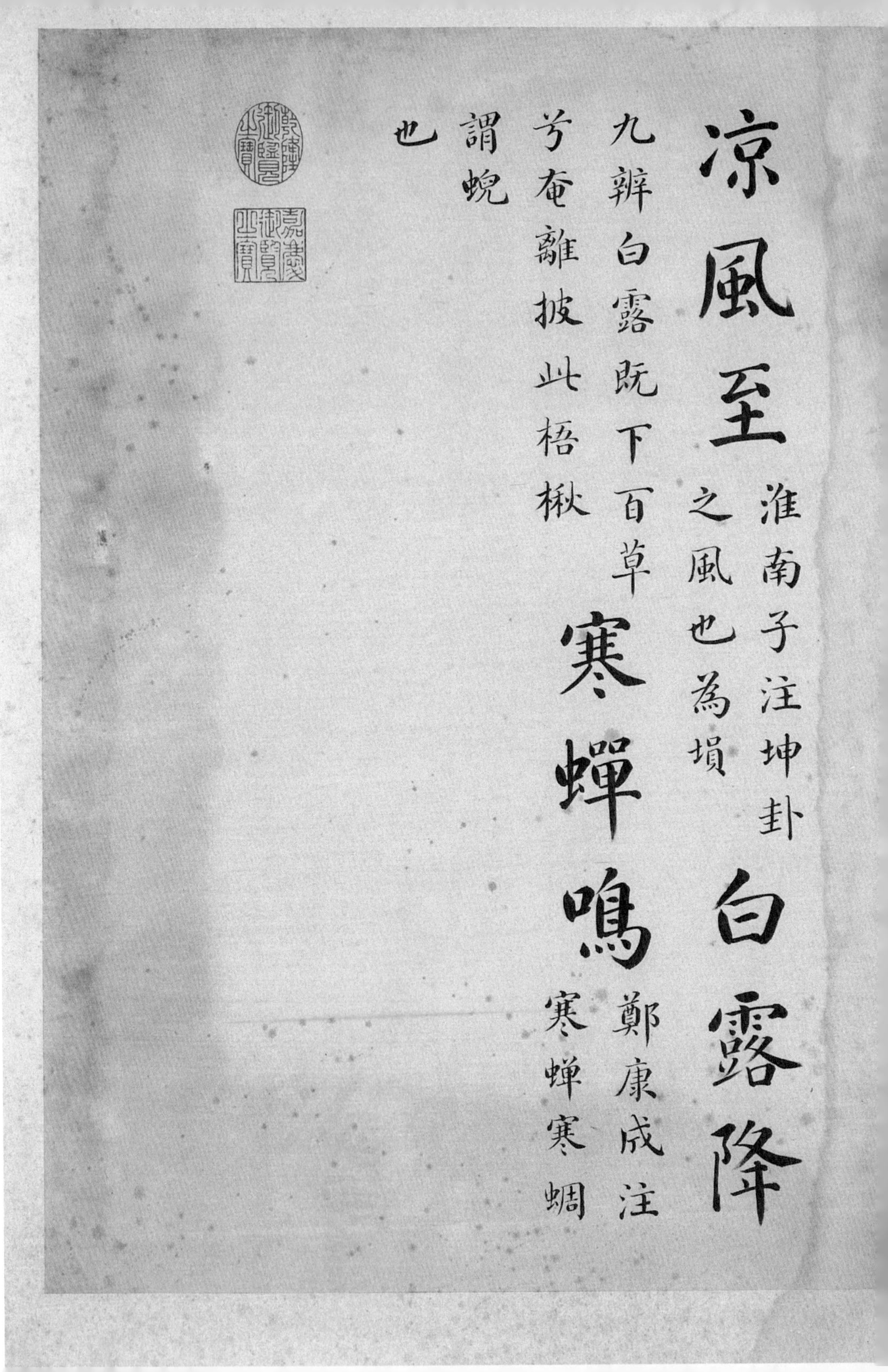
涼風至
淮南子注坤卦之風也爲埧
白露降
九辯白露既下百草兮奄離披此梧楸
寒蟬鳴
鄭康成注寒蟬寒蜩謂蜺也

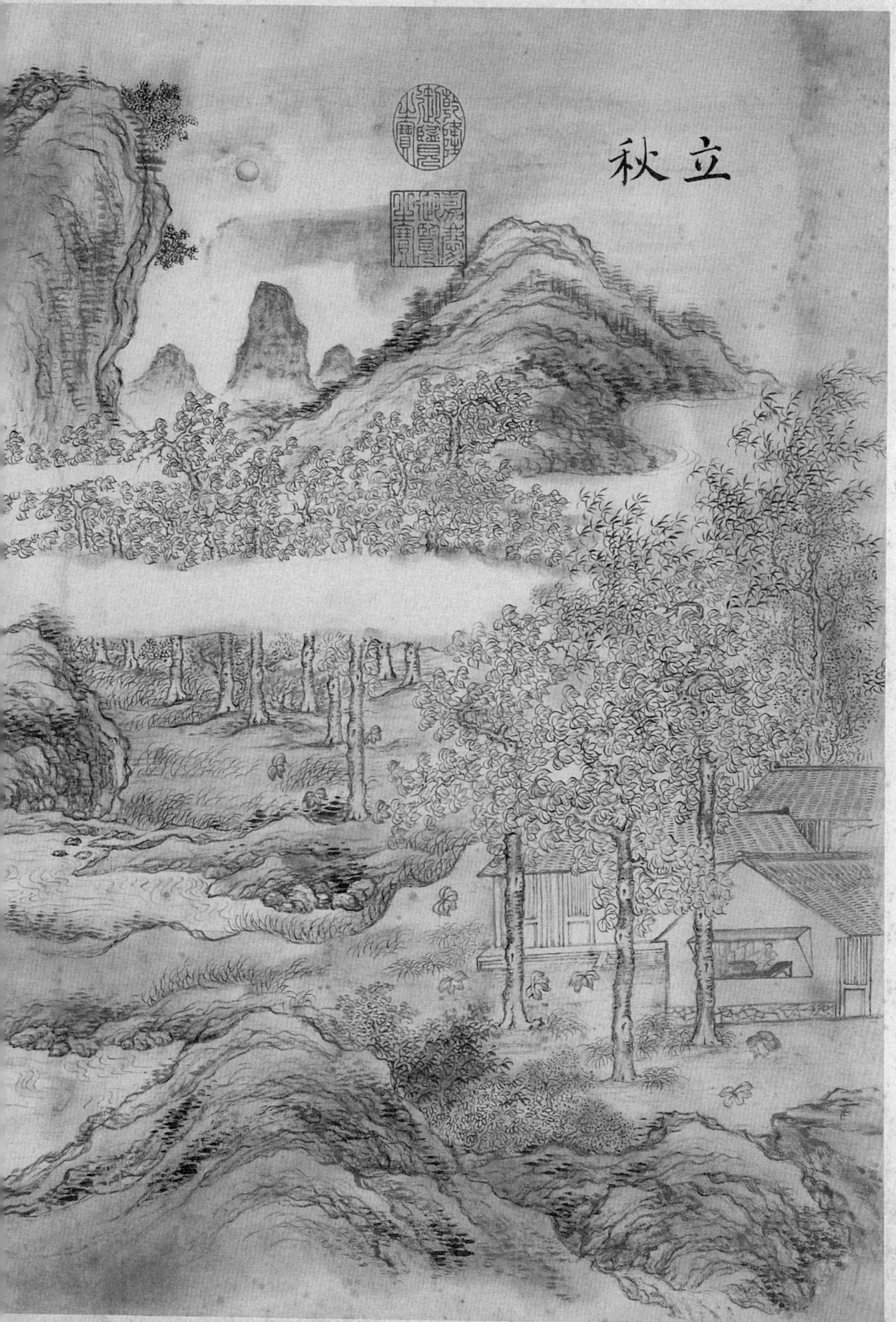


初候

**涼風至**

Cool Winds Commence

二候

**白露降**

White Dew Descends

三候

**寒蟬鳴**

Cicadas Sing of Coming Autumn

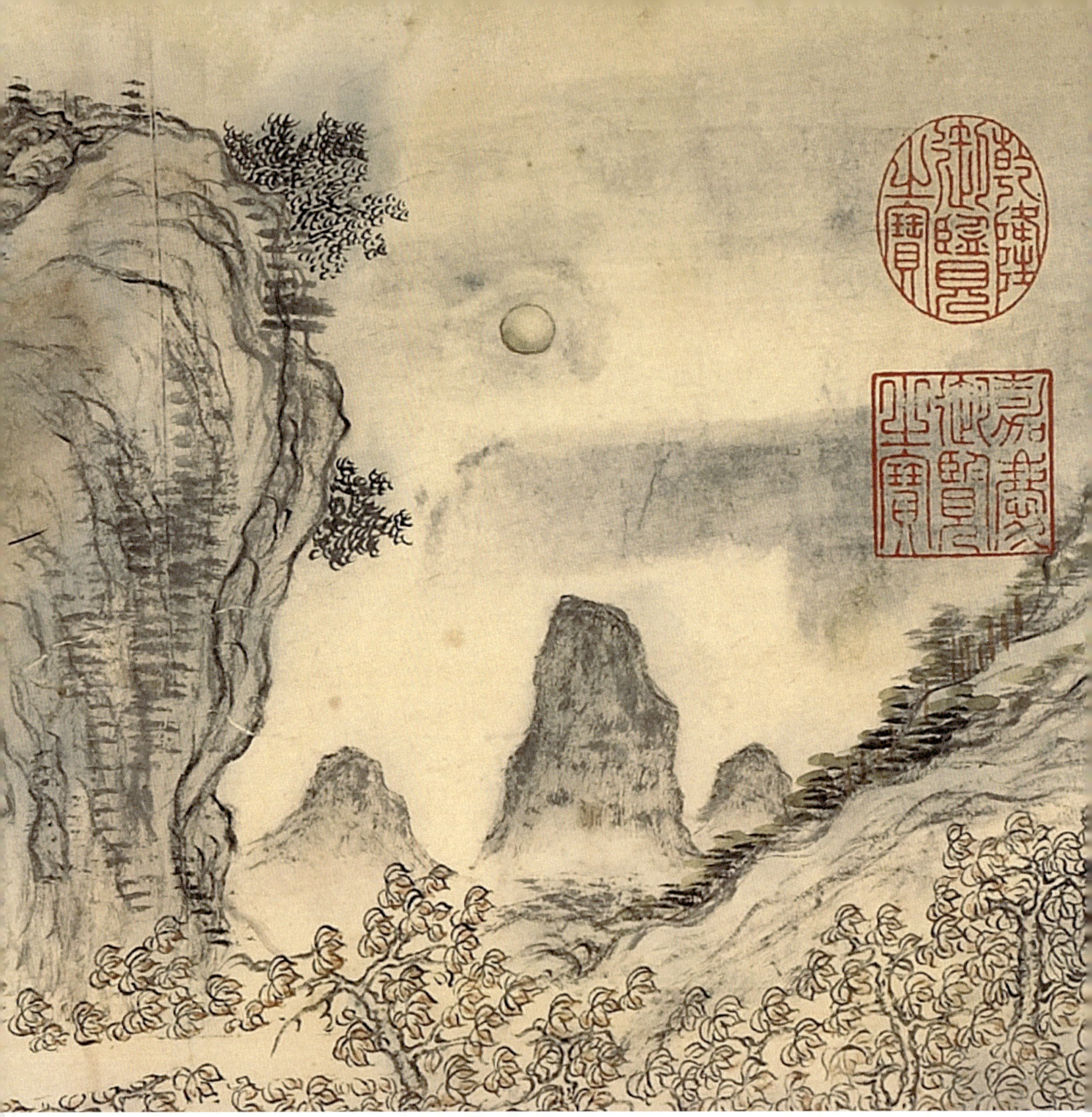
乾隆御覽之寶
嘉慶御覽之寶

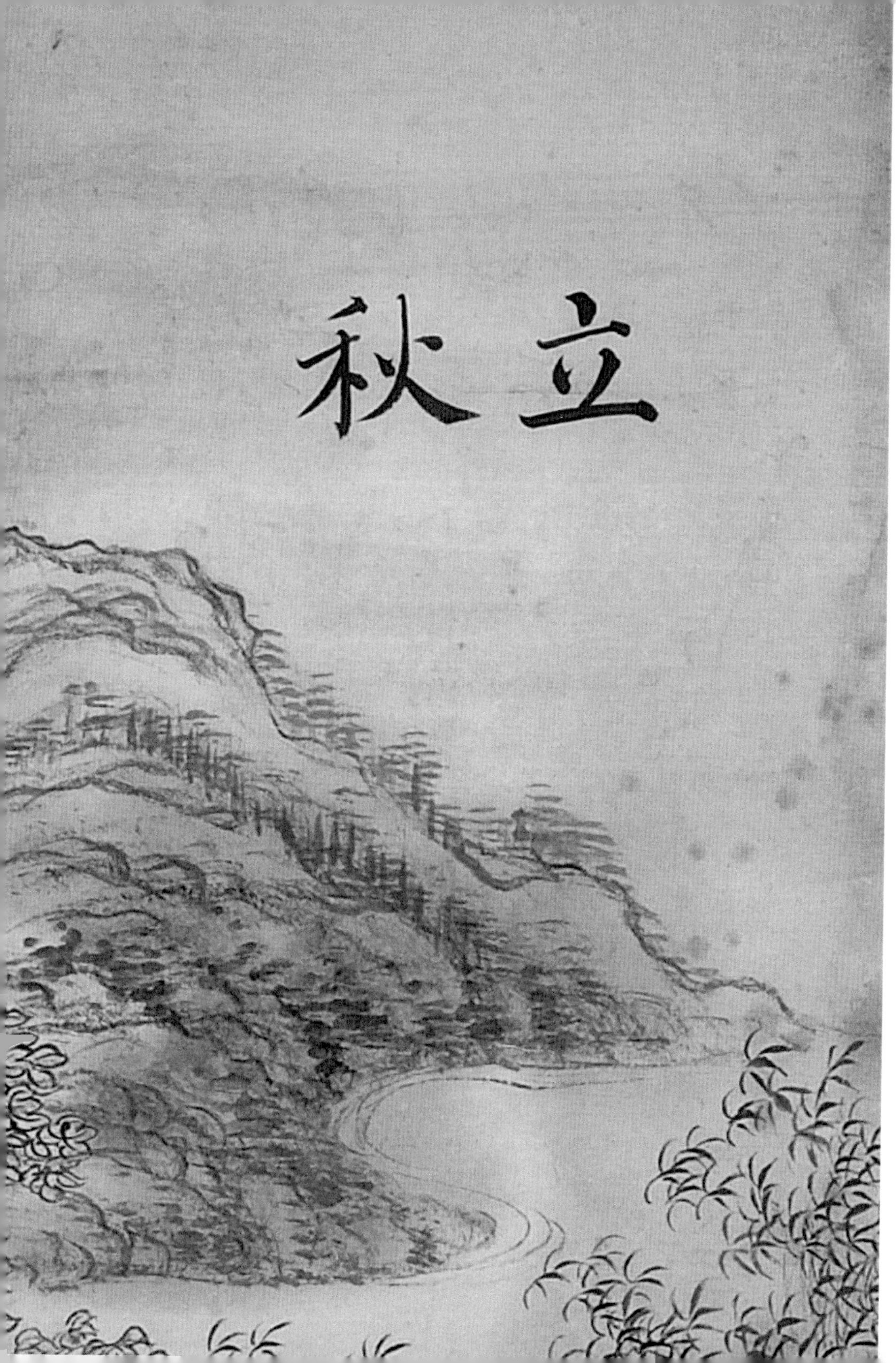


初候

# 涼風至
## Cool Winds Commence

The initial hint of Lìqiū is “Cool Winds Commence” (涼風至 Liáng Fēng Zhì), a much-welcomed change signaling the retreat of summer’s heat.

二候

# 白露降

## White Dew Descends

The second indication is "White Dew Descends" (白露降 Bái Lù Jiàng). After the commencement of Lìqiū, temperature differences between day and night grow more pronounced, leading to the formation of dew. This sign, "White Dew Forms," refers to the visible condensation of moisture in the early mornings, creating a layer of white dew that blankets the ground. It is not yet formed into beads but appears as a fine mist or frost-like layer, signaling the deepening of autumn and the subtle onset of cooler temperatures. This phenomenon symbolizes the gradual transition from the warmth of summer to the chill of fall.

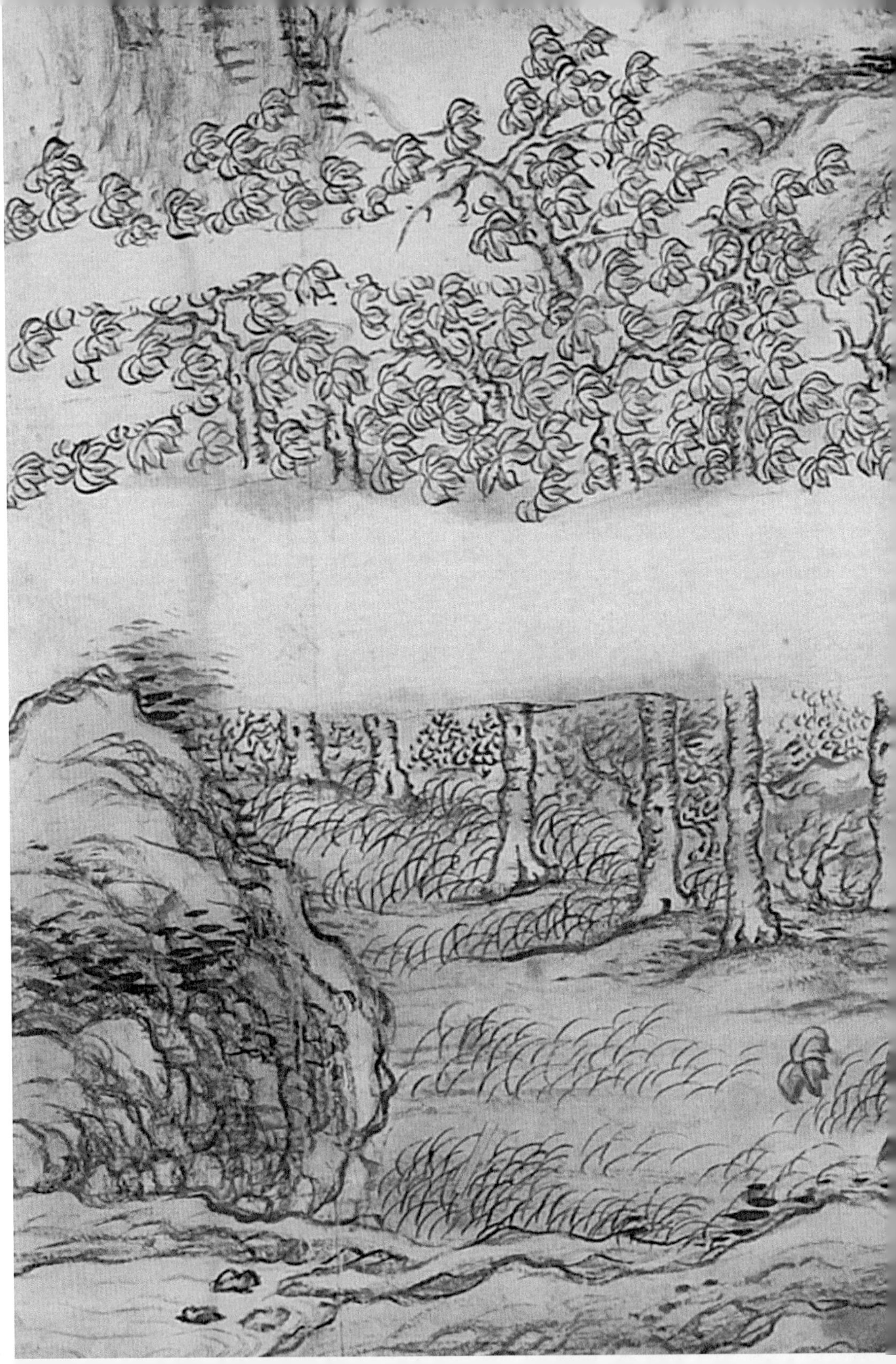

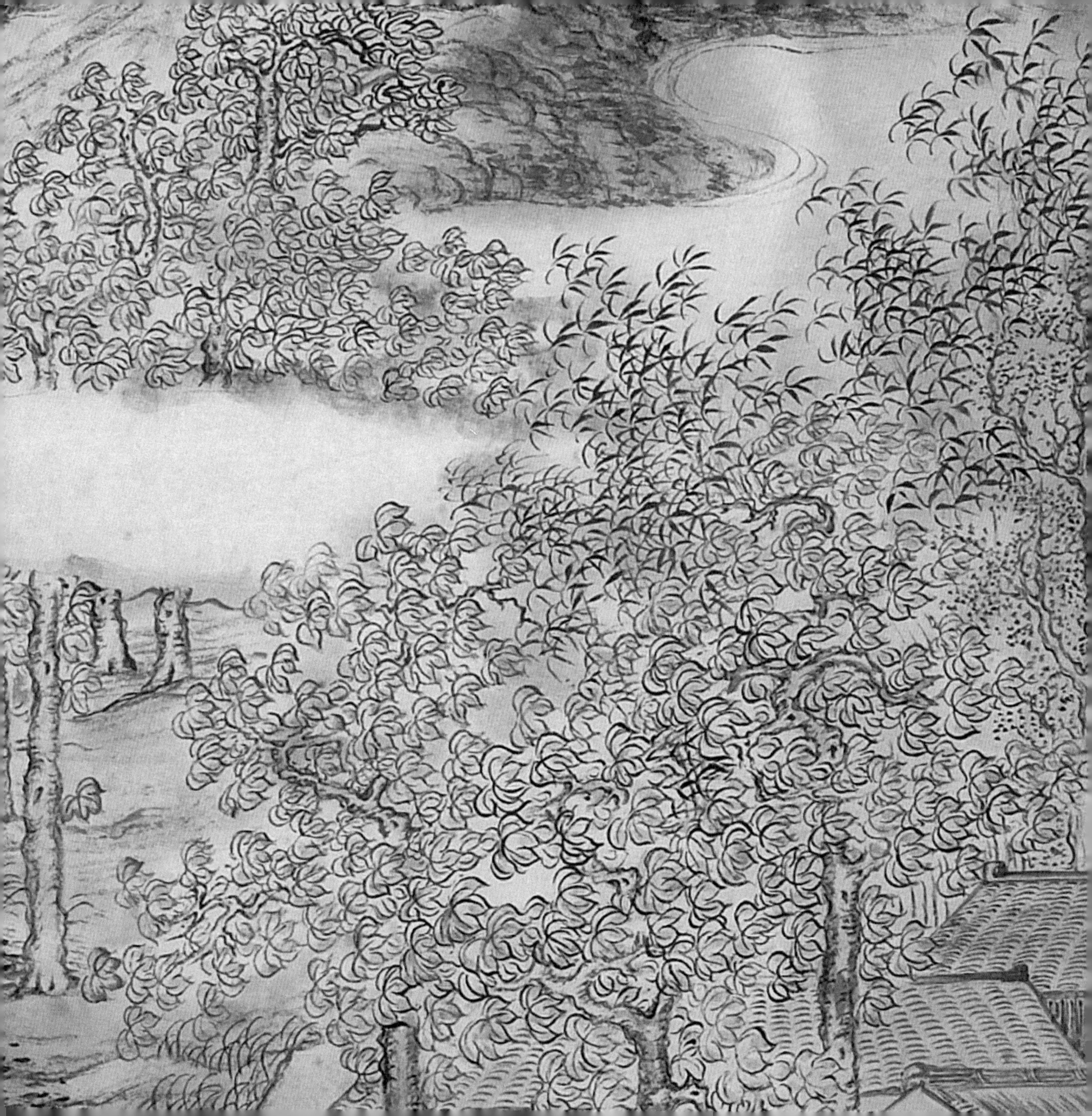

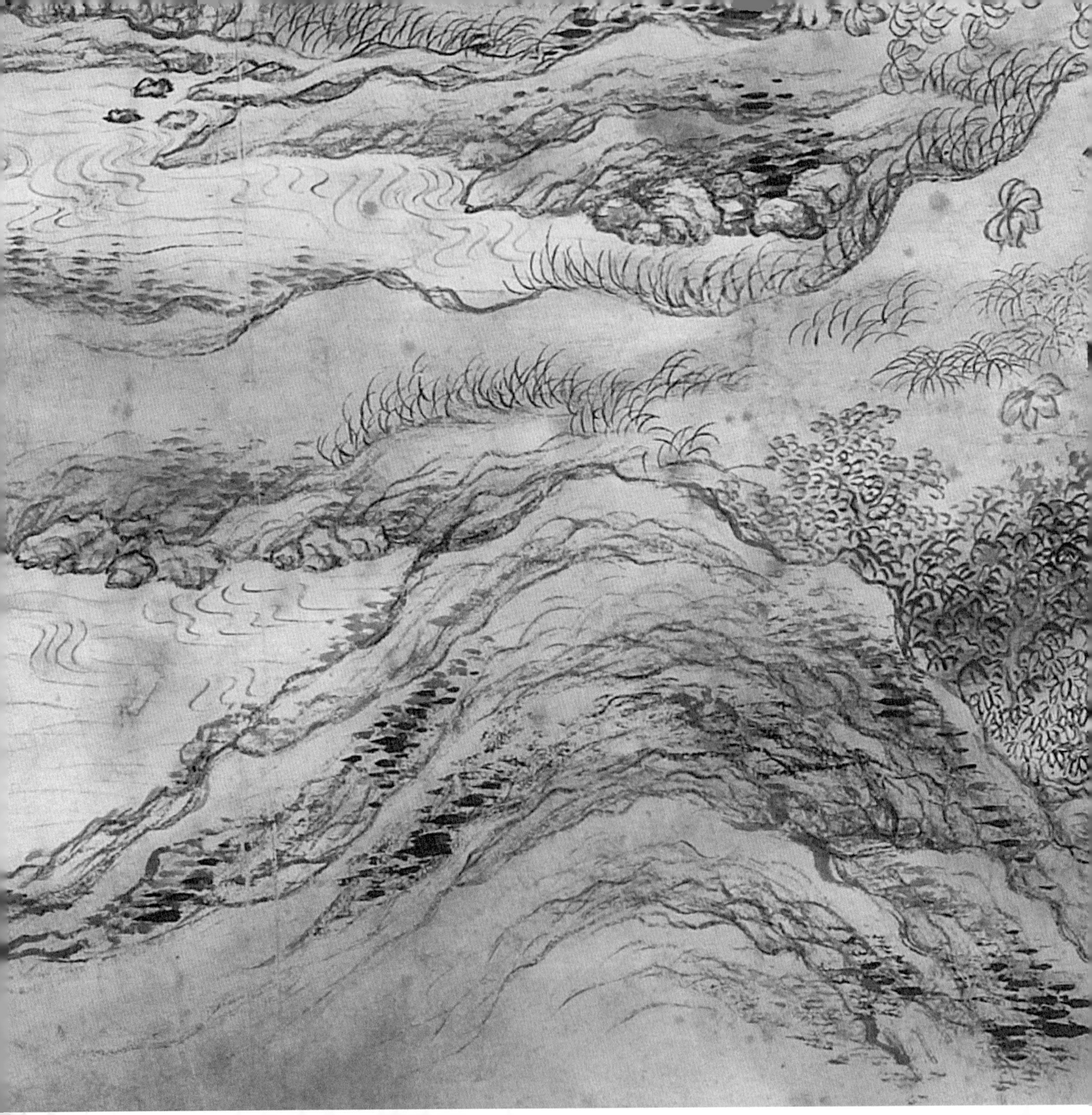

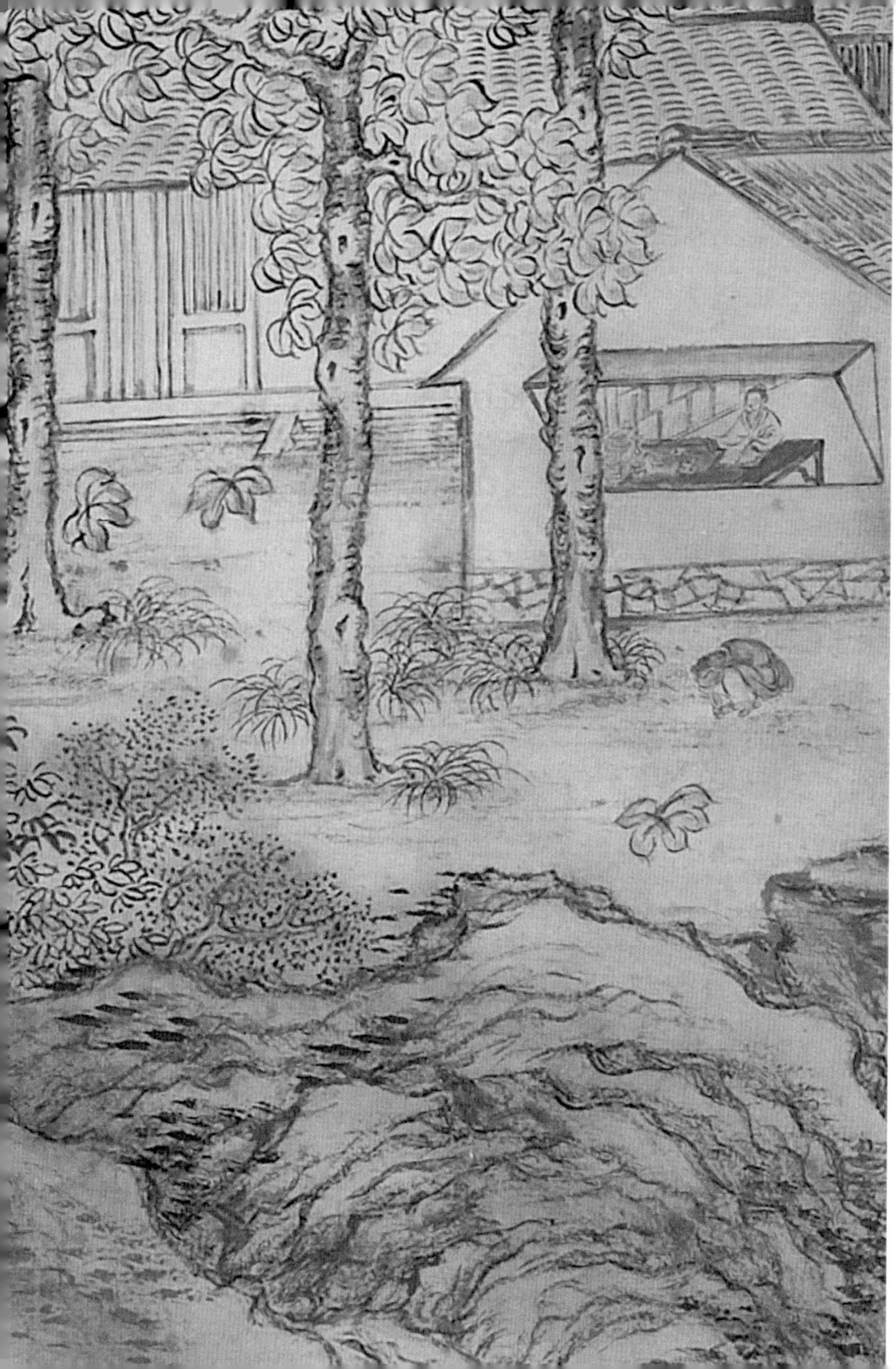

三候

# 寒蟬鳴

## Cicadas Sing of Coming Autumn

The third harbinger is "Cicadas Sing of Coming Autumn" (寒蟬鳴 Hán Chán Míng), their sounds becoming a poignant reminder of summer's end and the encroaching coolness.

# 暑

Shǔ

Chǔ

處

# END OF HEAT

August 23 (± 1)

# 處暑

## End of Heat

Chŭshŭ: Marking the waning of summer heat, with "chŭ" indicating a cessation or decline. It signifies the retreat of the intense summer heat and the onset of milder autumnal weather. Agricultural activities adapt to this change, with the sowing of autumn crops and the second planting of rice. While temperatures remain high, the decrease in heat is notable, leading to the emergence of a more temperate climate.

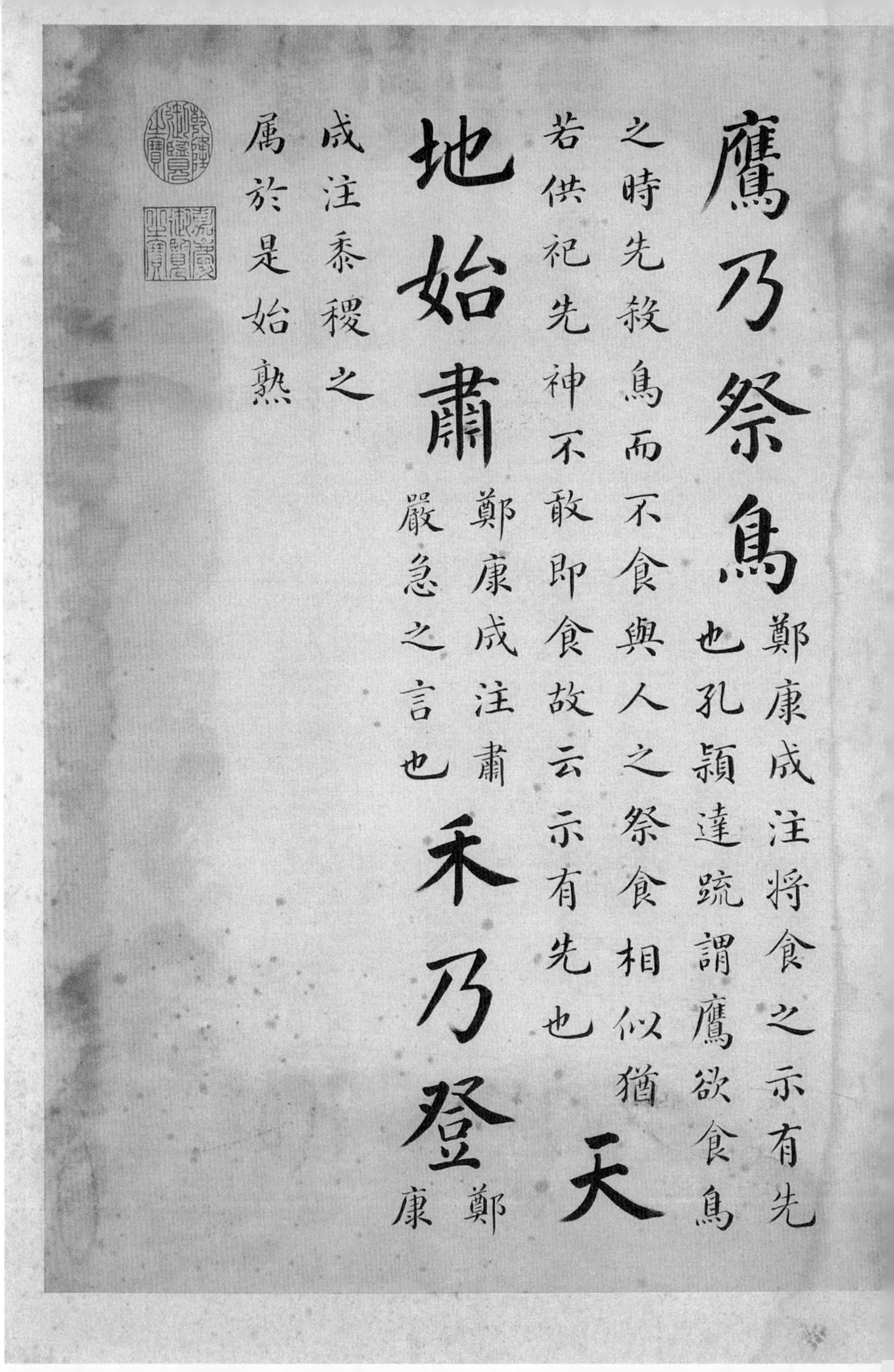

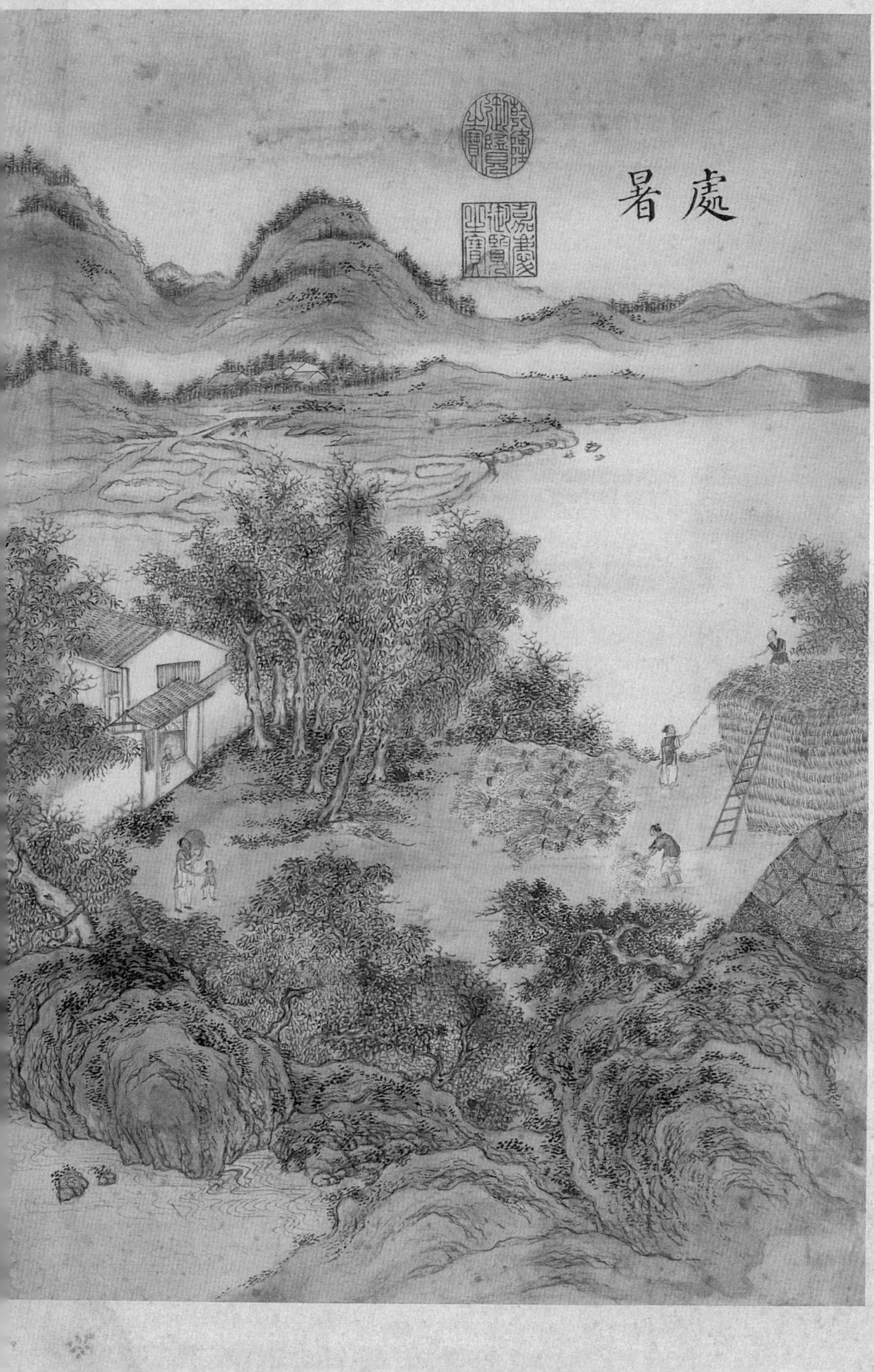


初候

## 鷹乃祭鳥
## Eagles Present Their Prey

二候

## 天地始肅
## Nature Turns Solemn

三候

## 禾乃登
## All Grains Ripen

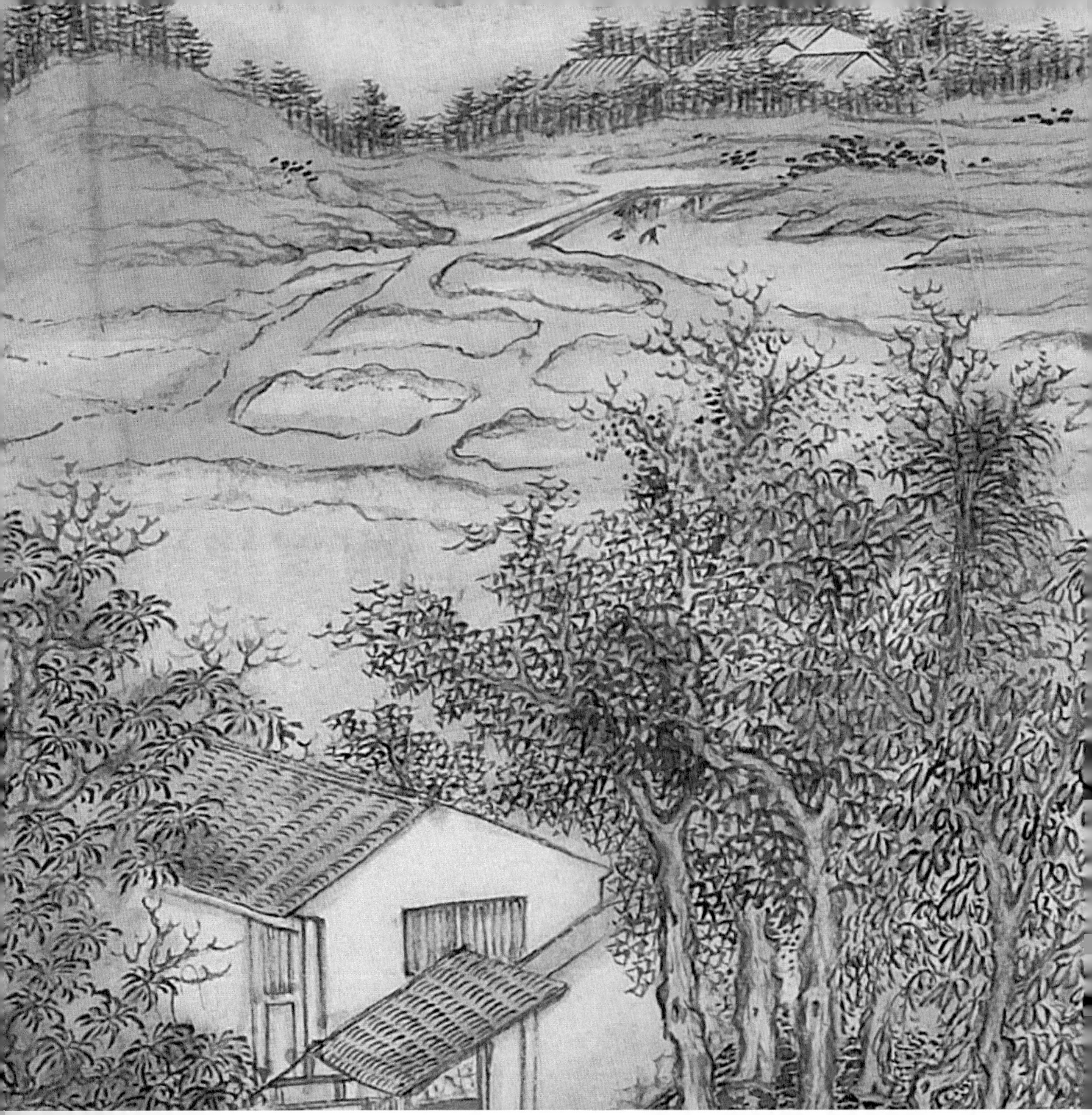

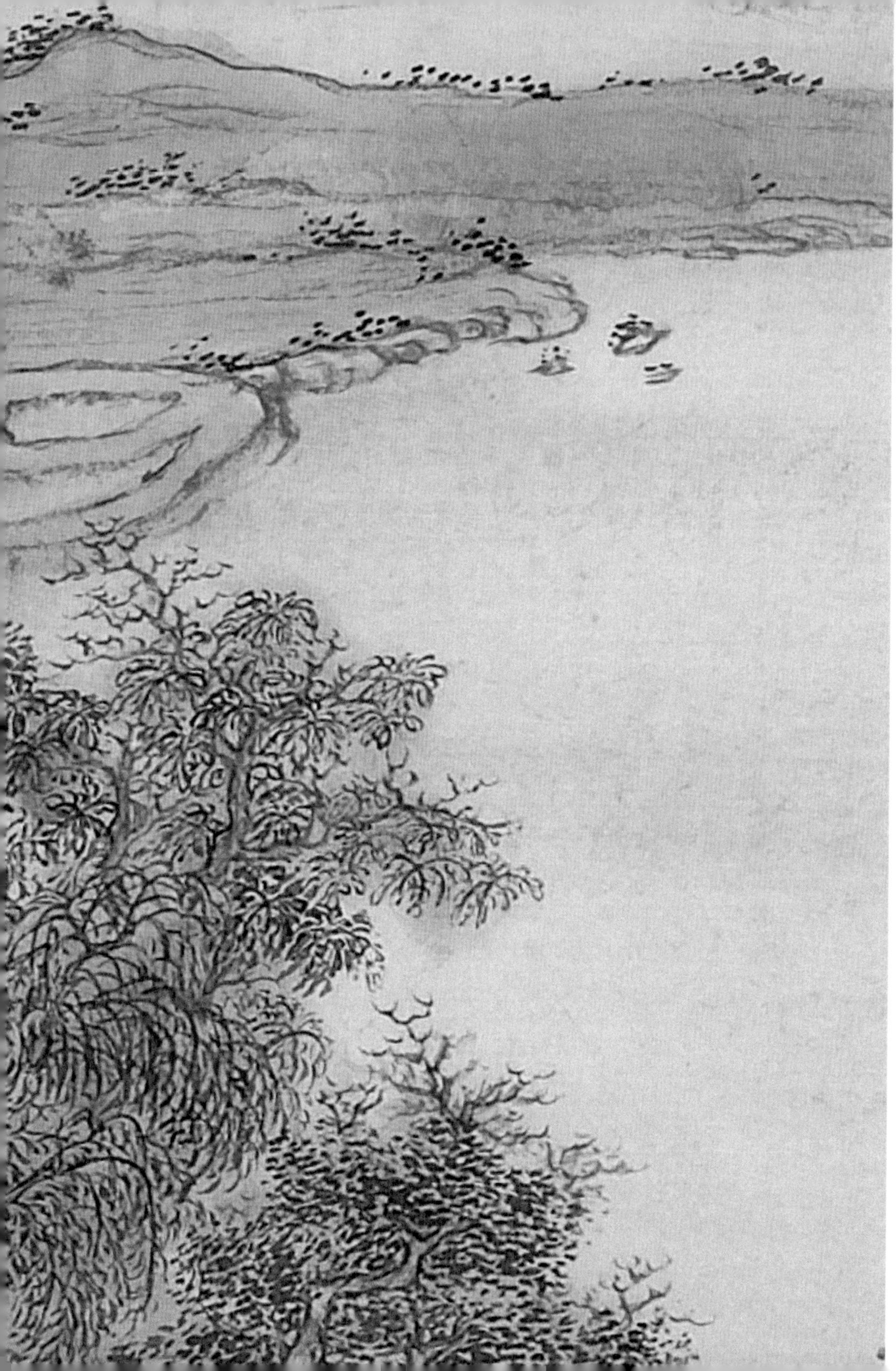

初候

# 鷹乃祭鳥

## Eagles Present Their Prey

The first omen of Chùshǔ is "Eagles Present Their Prey" (鷹乃祭鳥 Yīng Nǎi Jì Niǎo), a display of might and the cycle of life in the natural world. This sign describes the behavior of eagles displaying their catch prominently before consuming it, a ritual-like act that may symbolize respect or a form of offering for their prey. This behavior also indicate eagles during the Xiǎoshǔ period have now grown into mature eagles , have grown to become adept at hunting at Chùshǔ.

二候

# 天地始肅
## Nature Turns Solemn

The second cue is "Nature Turns Solemn" (天地始肅 Tiān Dì Shǐ Sù): As the heat of summer fades, a more solemn and austere energy begins to permeate heaven and earth. It serves as a reminder of the seriousness of the season, urging people to be cautious, introspective, and to align with the sobering qualities of the approaching autumn.

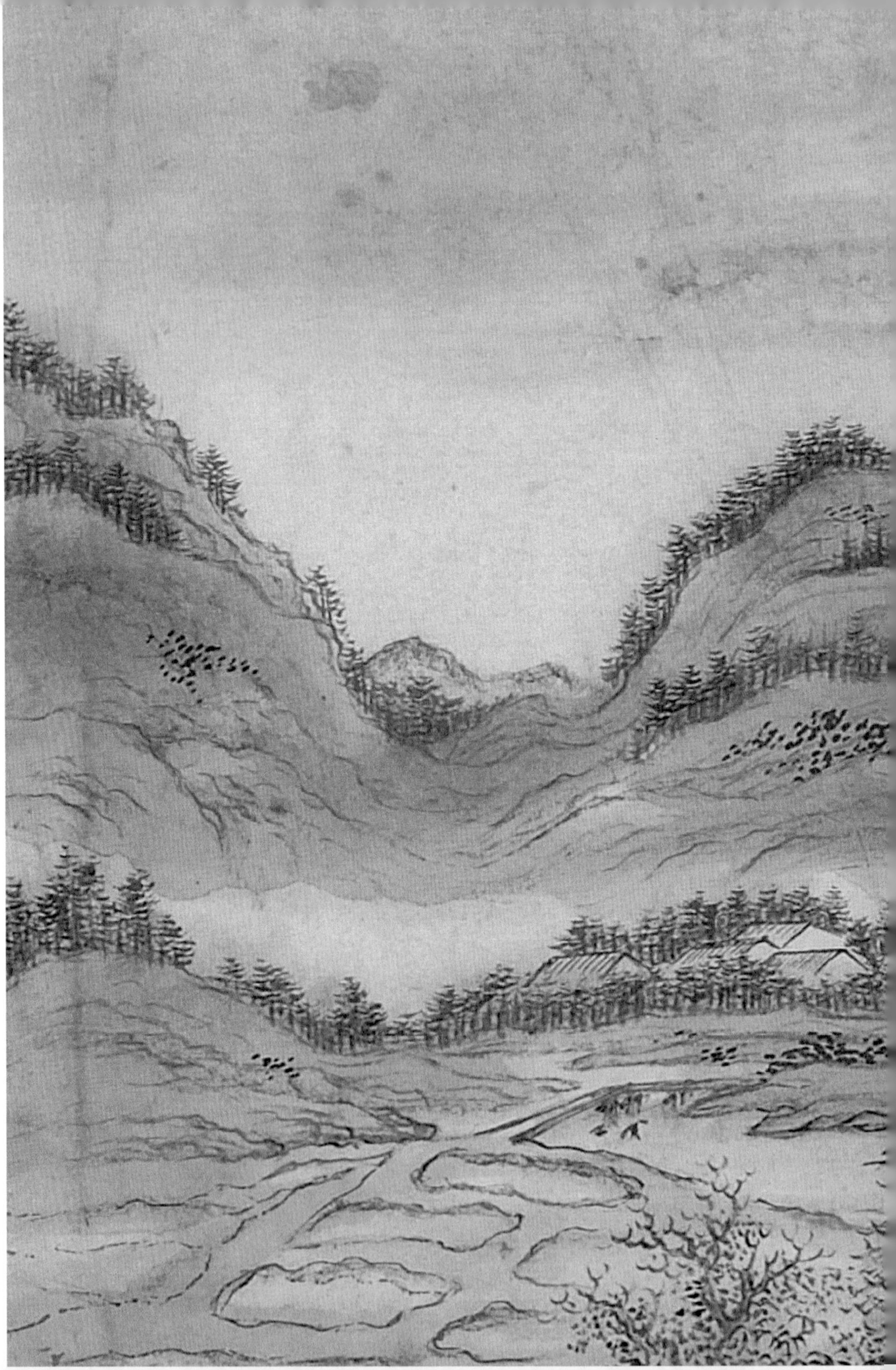

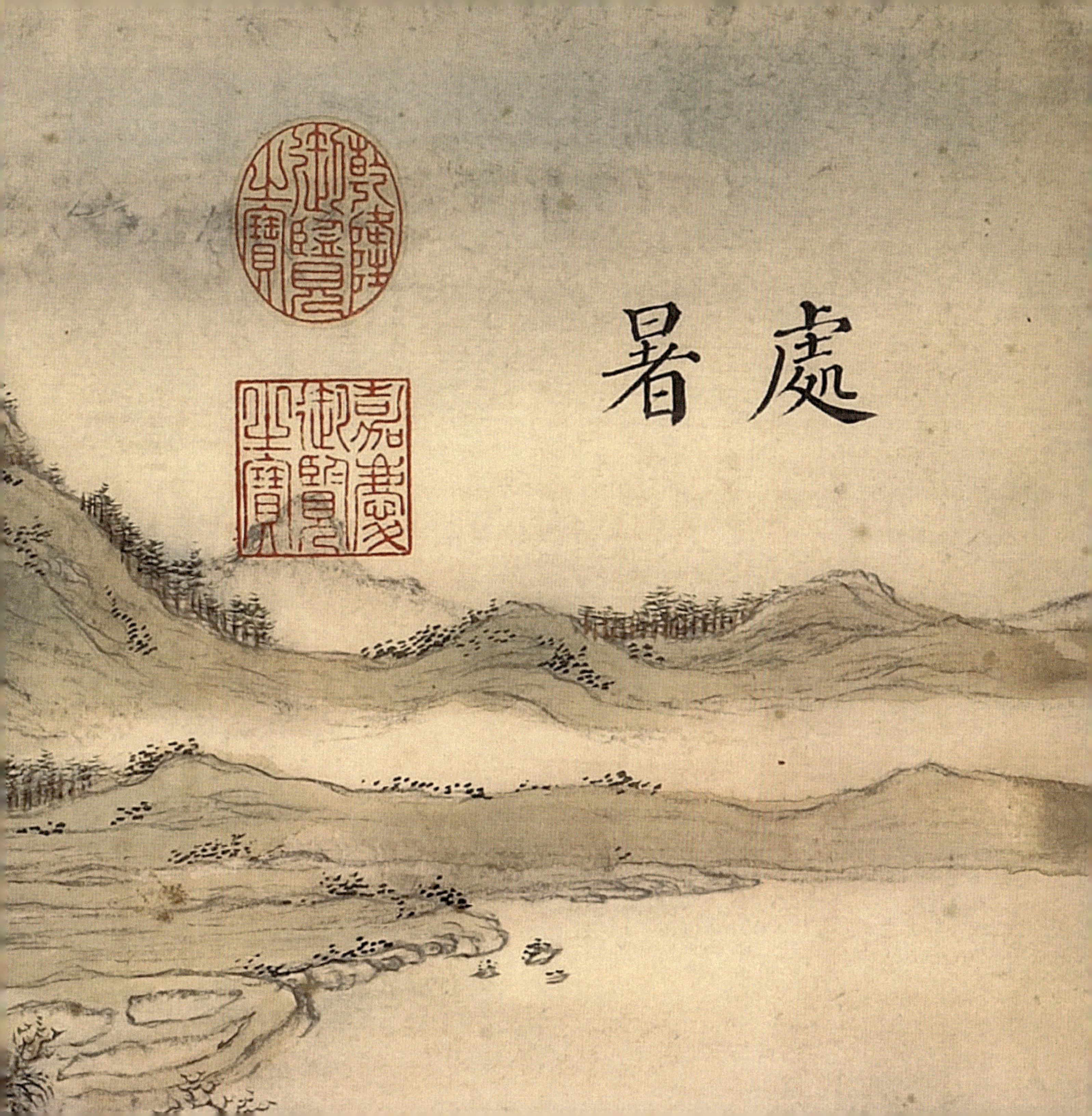
處暑

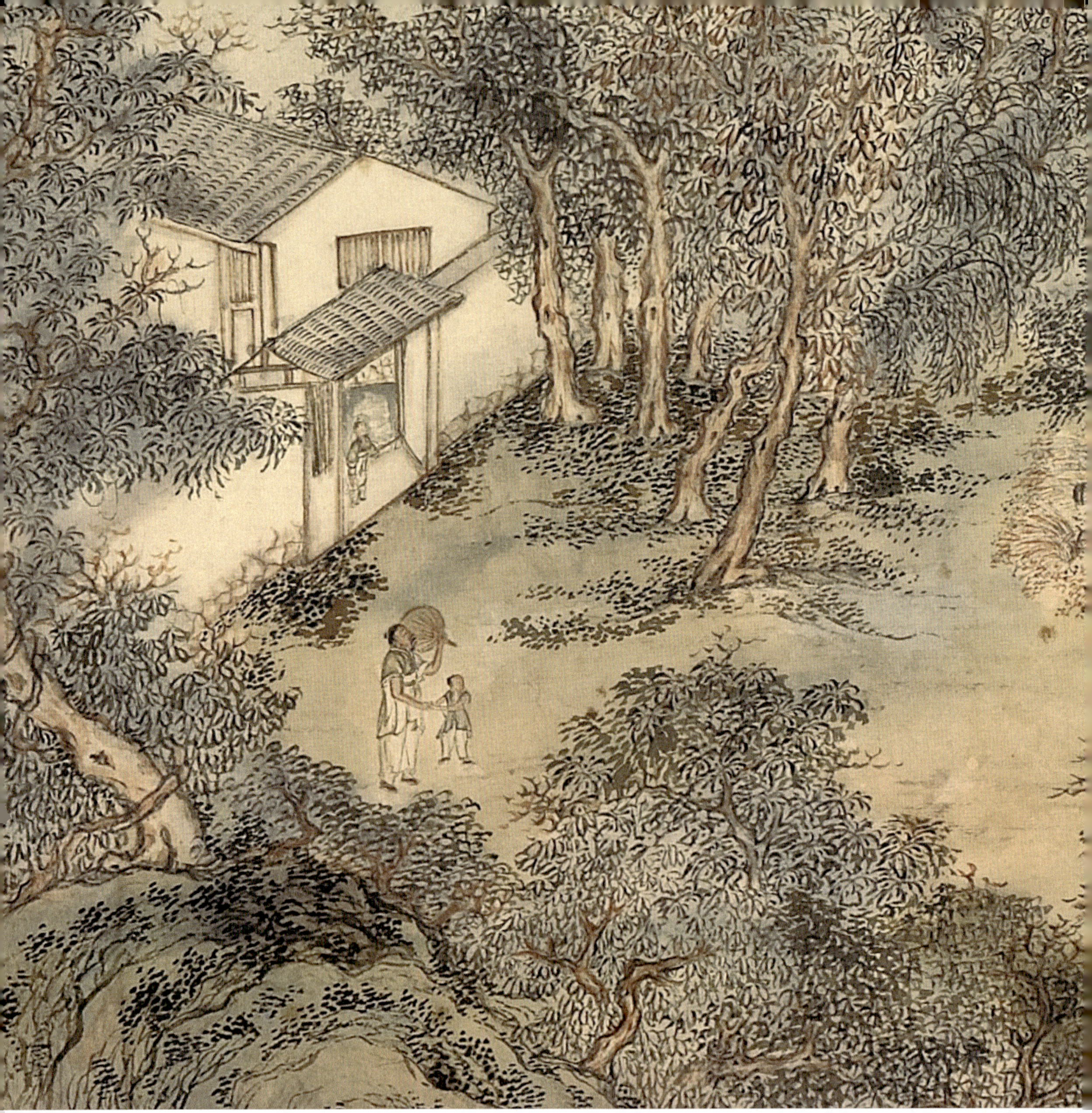

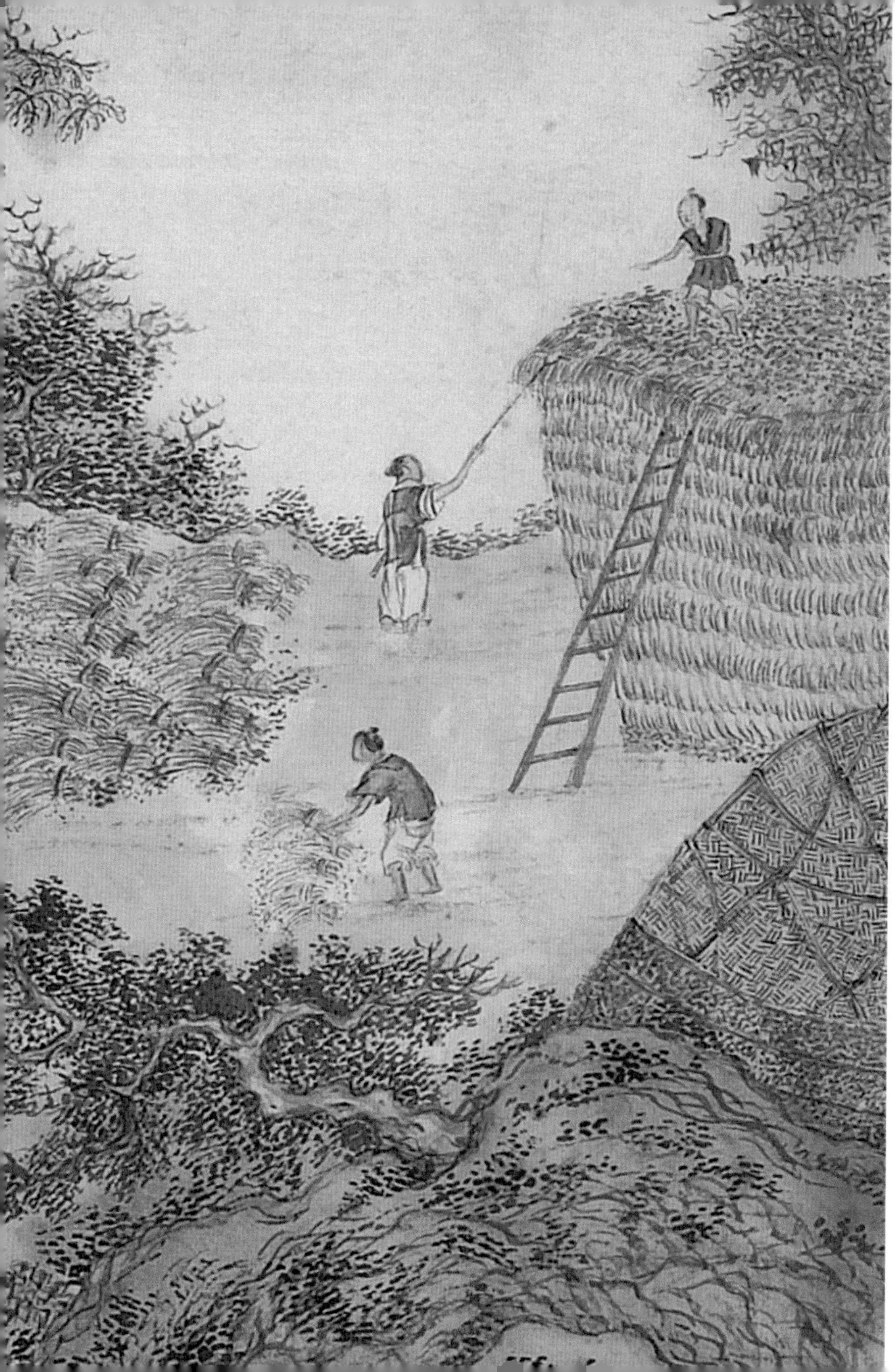

三候

# 禾乃登

## All Grains Ripen

The third inkling is "All Grains Ripen" (禾乃登 Hé Nǎi Dēng):
This sign marks the ripening and readiness for harvest of various grains such as millet, sorghum, rice, and other cereals. It' s a time of abundance and the culmination of agricultural efforts, where the mature grains are sometimes presented as offerings in ancient rituals. "Grains Ripen" signifies the success of the season' s labors and the beginning of gathering and storing for the future, embodying the cycle of sowing, growing, and harvesting that defines agrarian life.

# 露

Lù

Bái

白

# WHITE DEW

September 8 (± 1)

# 白露

## White Dew

Báilù : It is characterized by the significant temperature differences between day and night, leading to the condensation of water vapor into dew. As the weather turns cooler, dew forms on objects, particularly in the early mornings. This term symbolizes the progression of autumn and the impending arrival of colder days. “Bái” refers to the white color of the dew, reflecting the autumn element of metal in the Wu Xing, and the increasing chill in the air.

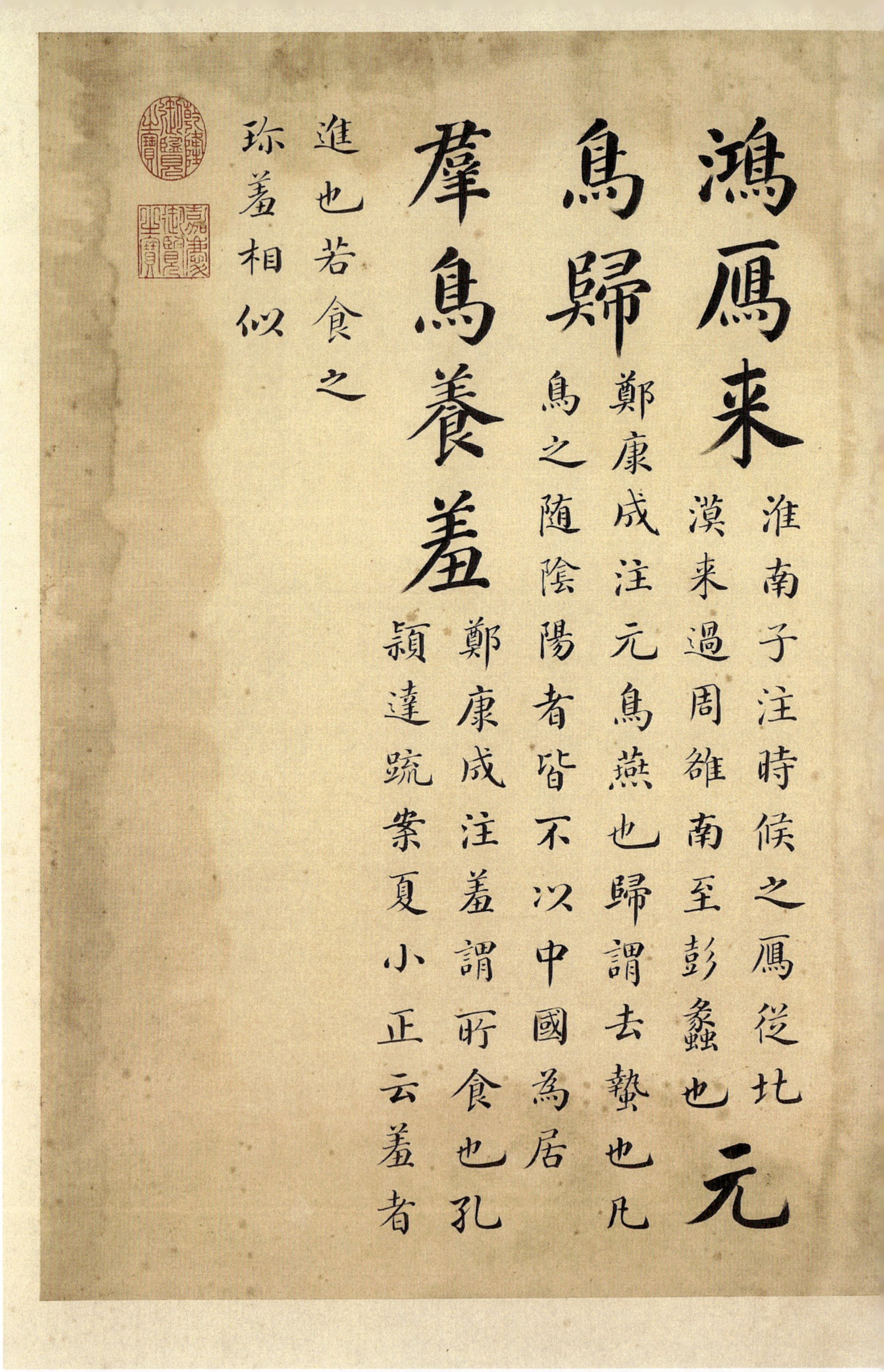

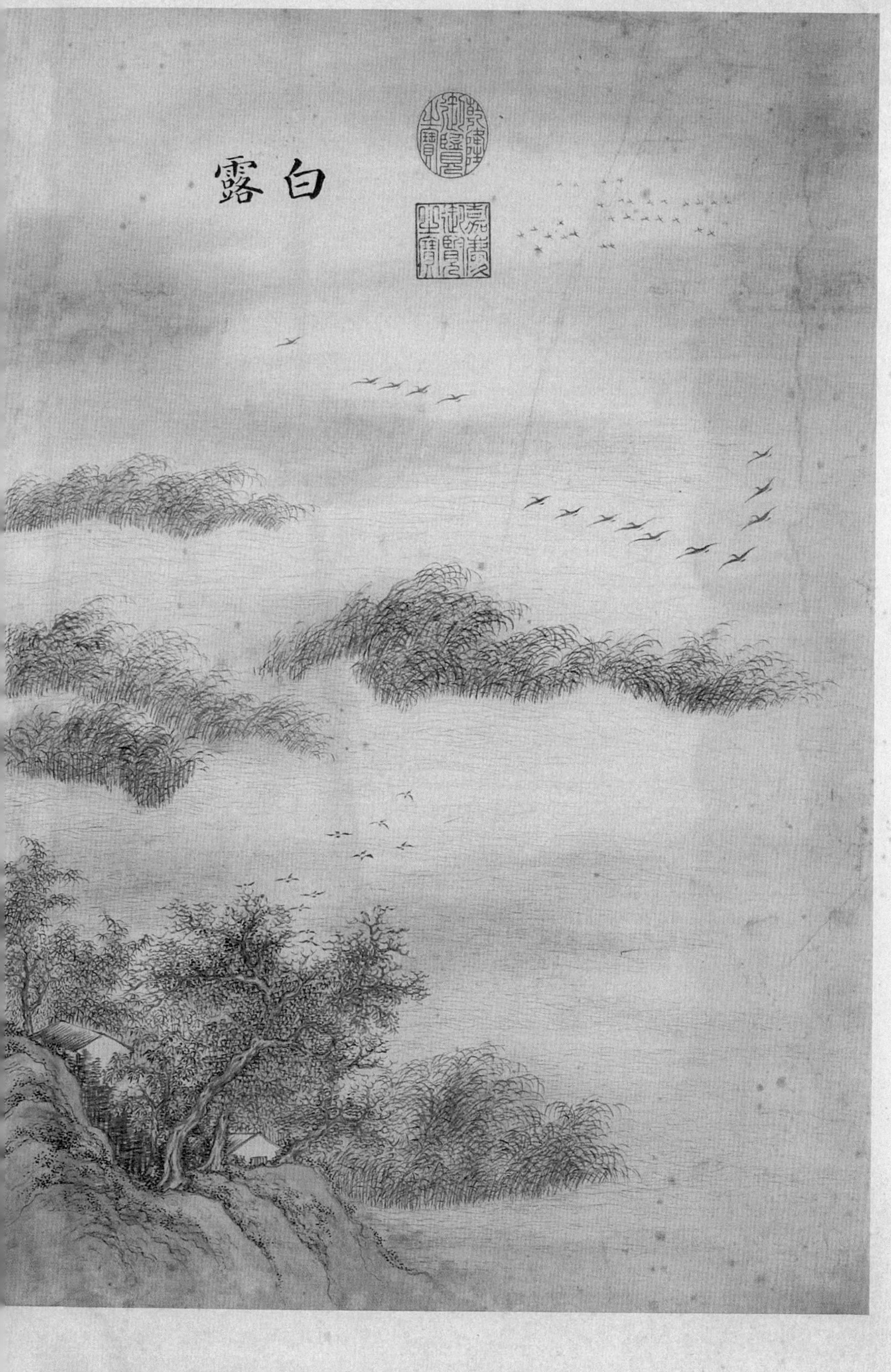


初候

## 鴻雁來
## Wild Geese Arrive

二候

## 元鳥歸
## Swallows Return

三候

## 群鳥養羞
## Birds Prepare for Winter

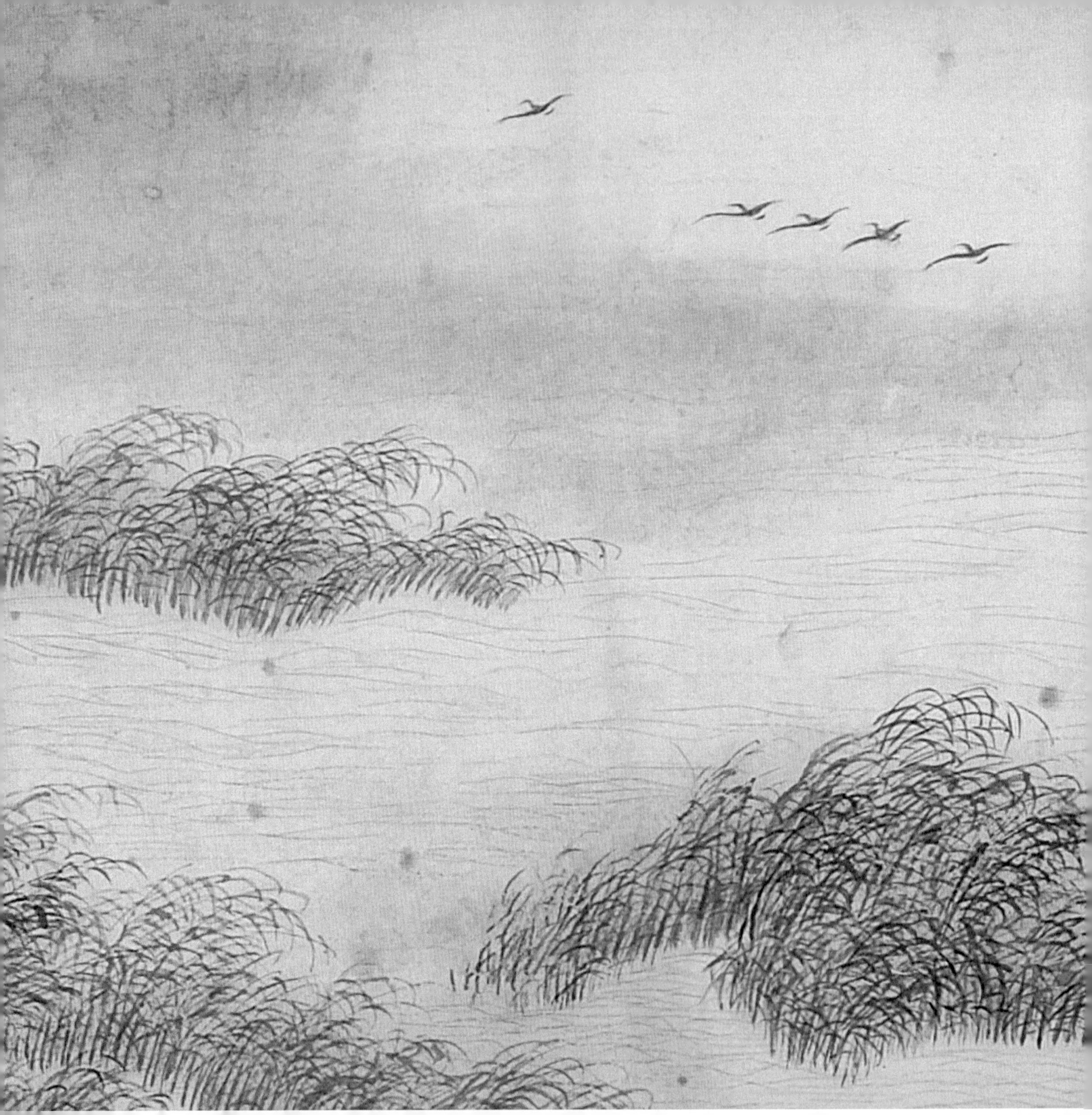

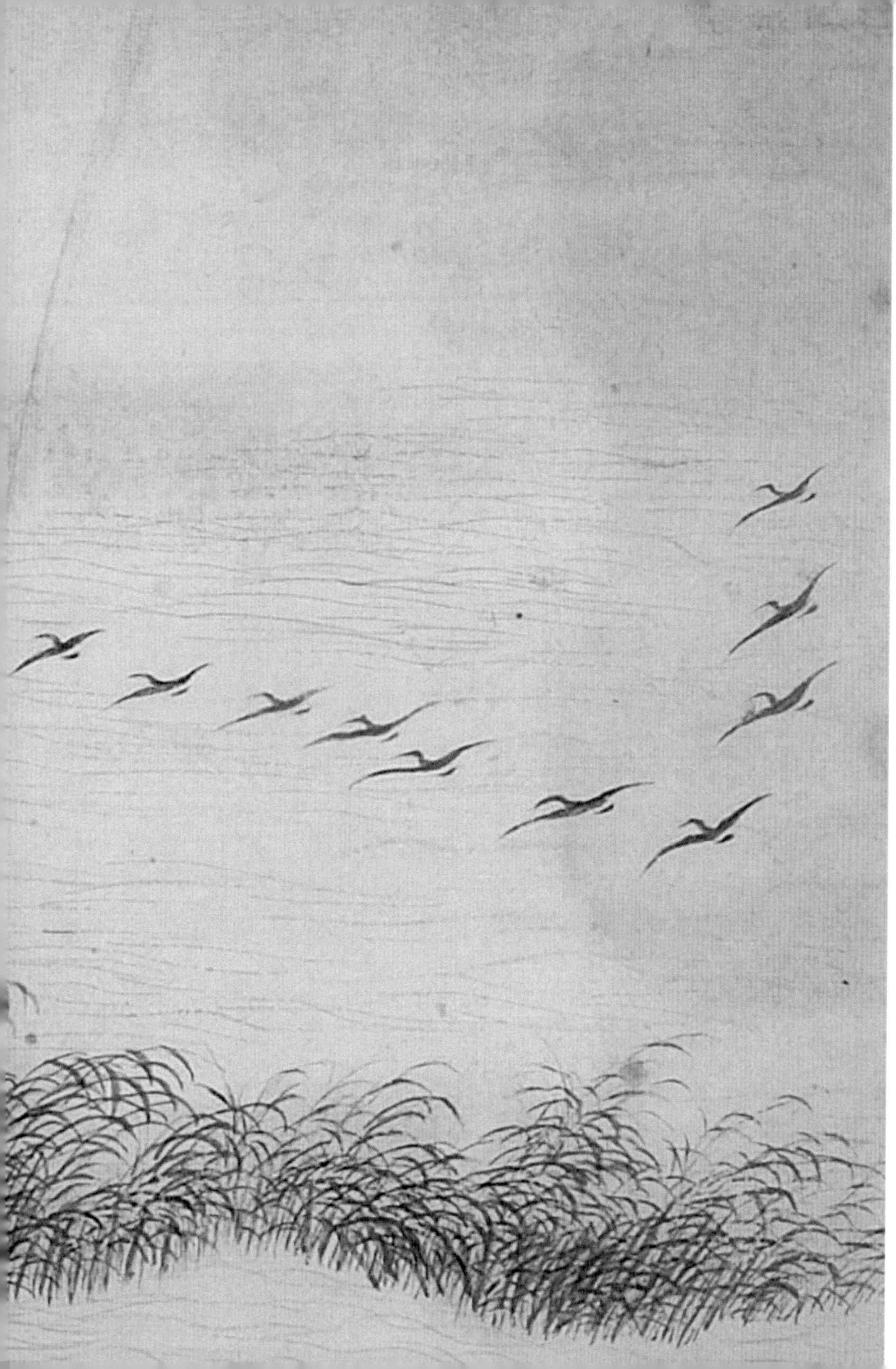

初候

# 鴻雁來

## Wild Geese Arrive

The first signal of Báilù is "Wild Geese Arrive" (鴻雁來 Hóng Yàn Lái), their migration towards the south to avoid cold is a clear indication of the seasonal changes.

二候

# 元鳥歸

## Swallows Return

The second sign, “Swallows Return” (元鳥歸 Yuán Niǎo Guī), signals the migration of swallows back towards the south as autumn approaches. This sign is a counterpart to the “Swallows Arrive” observed at the Spring Equinox. It reflects the cyclical and migratory patterns of swallows, mirroring the shift in Yin and Yang energies.

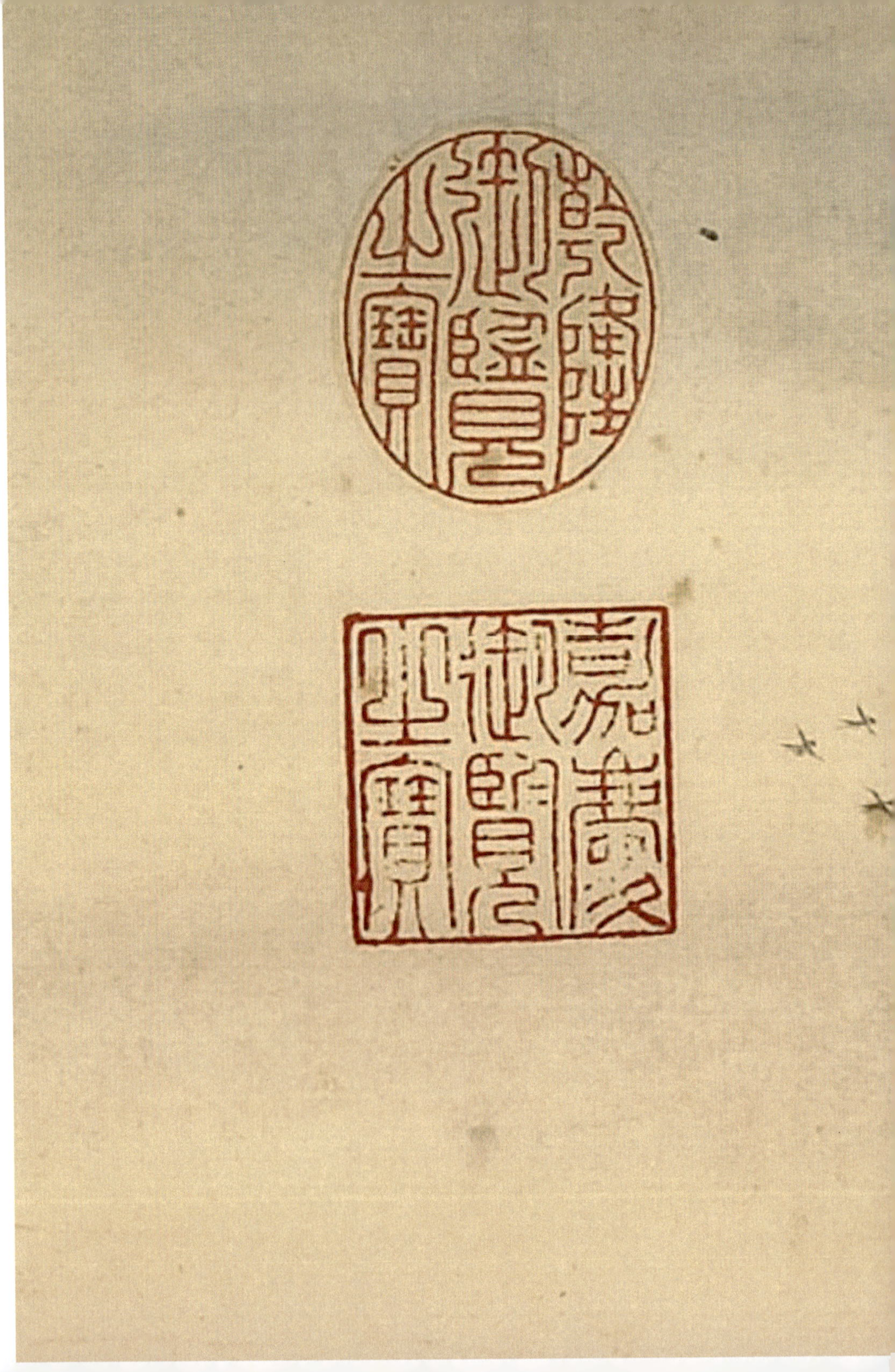

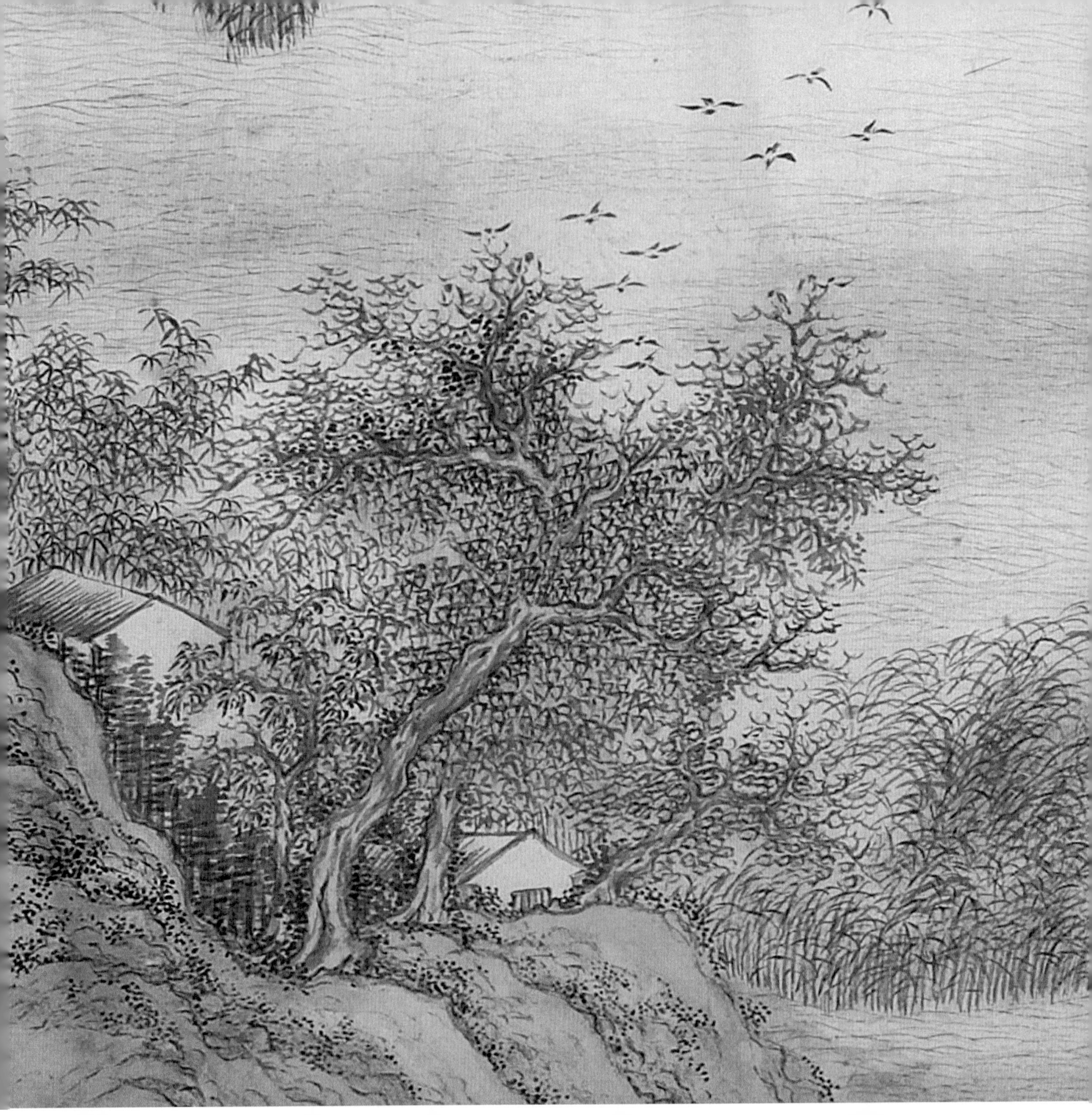

三候

# 群鳥養羞

## Birds Prepare for Winter

The third vision is “Birds Prepare for Winter” (群鳥養羞 Qún Niǎo Yǎng Xiū), gathering sustenance and growing thicker feathers in anticipation of the cold.

# 分

# Fēn

Qiū

秋

# THE AUTUMN EQUINOX

September 23 (± 1)

# 秋分

## The Autumn Equinox

Qiūfēn : The Autumn Equinox, is a time of balance between day and night, mirroring the Spring Equinox. From this point, days begin to shorten, and nights lengthen, signaling a gradual shift towards colder weather. It's a critical period for agriculture, as crops like rice reach maturity. Qiūfēn represents a time of equilibrium and transition, reflecting the constant change inherent in nature.

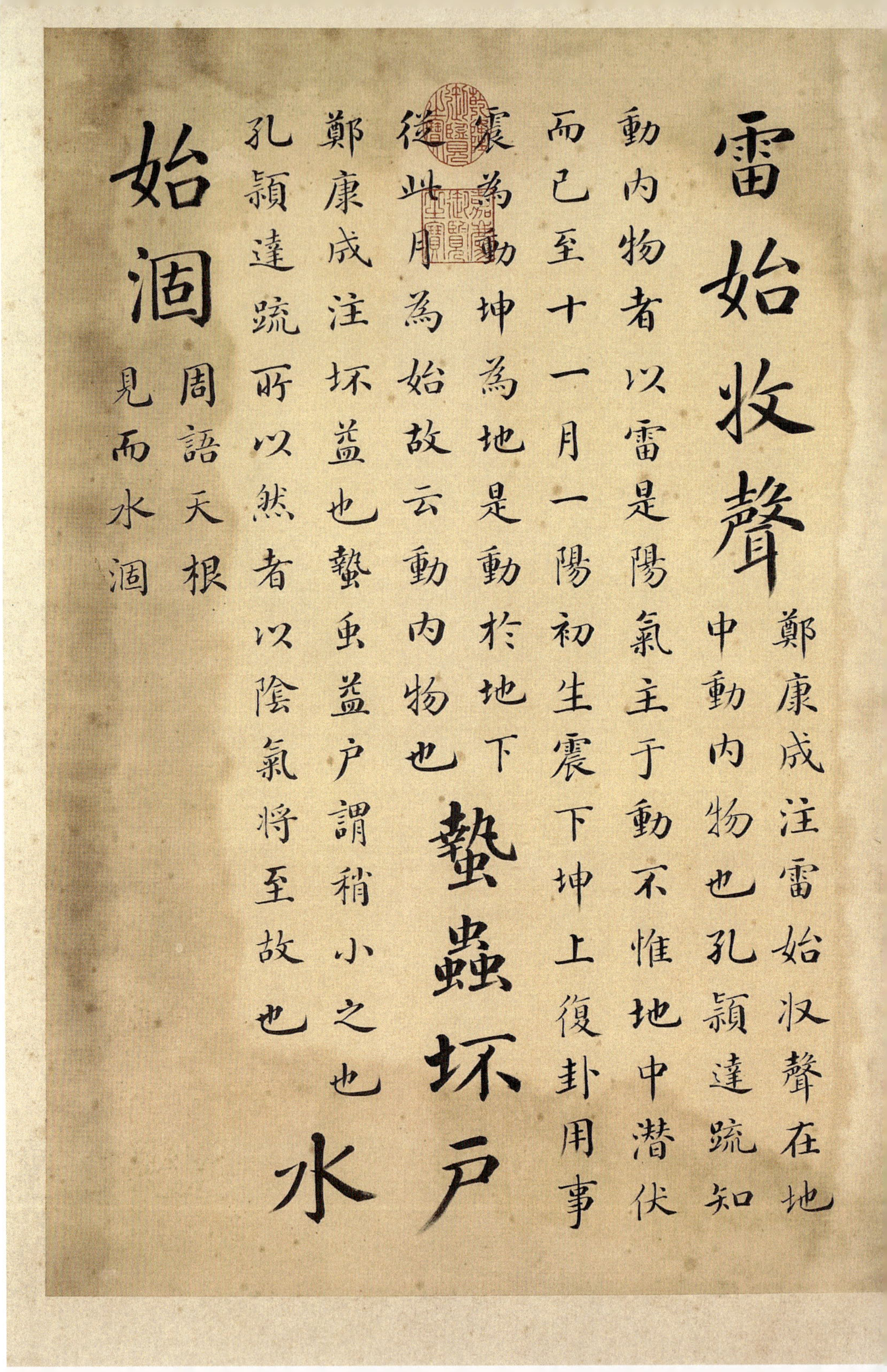
雷始收聲

鄭康成注雷始收聲在地中動內物也孔穎達疏知動內物者以雷是陽氣主于動不惟地中潛伏而已至十一月一陽初生震下坤上復卦用事震為動坤為地是動於地下從此月為始故云動內物也

蟄蟲坏戶

鄭康成注坏益也蟄虫益戶謂稍小之也孔穎達疏所以然者以陰氣將至故也

水始涸

周語天根見而水涸

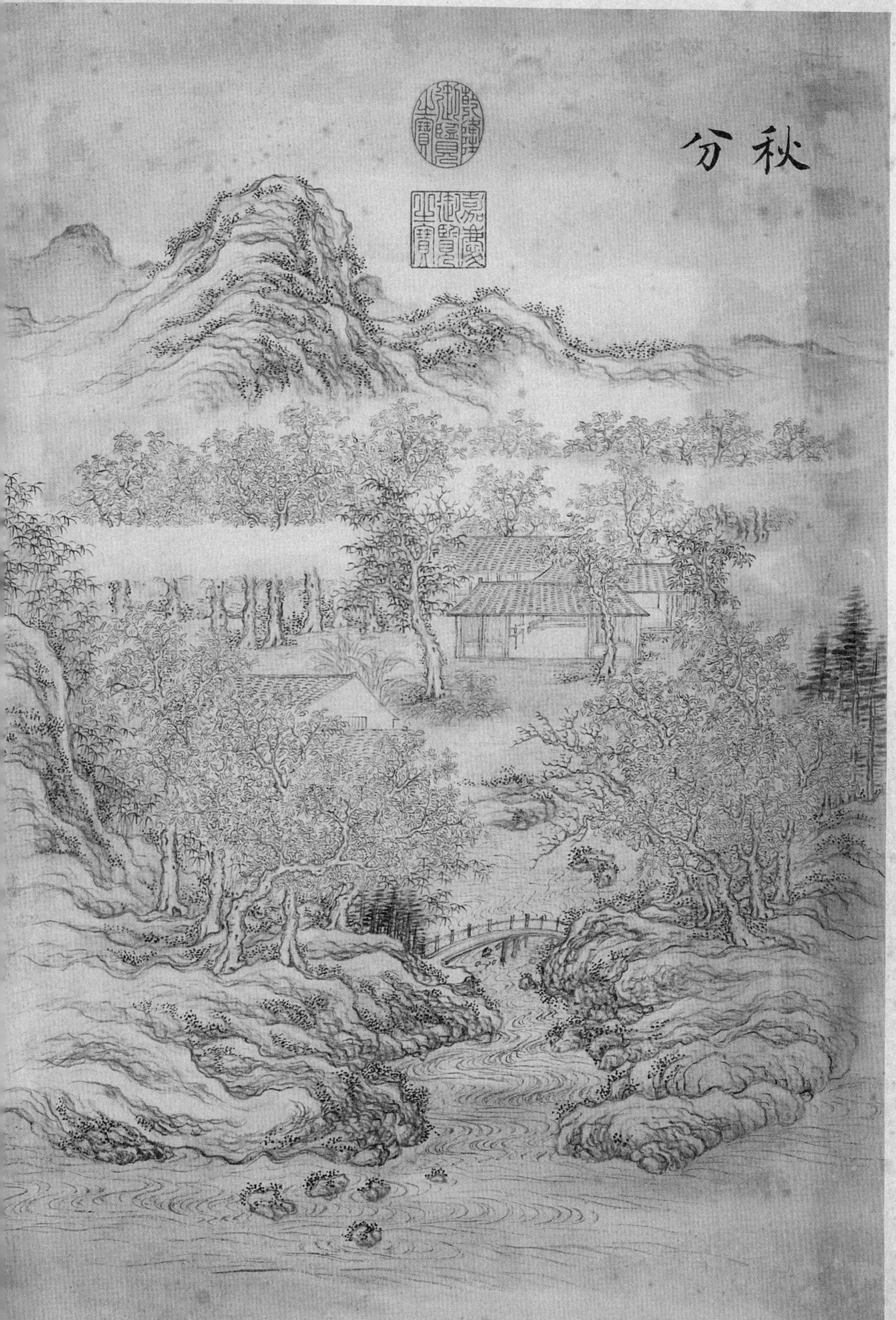


初候

## 雷始收聲
## Thunder Retires

二候

## 蟄蟲坯戶
## Insects Seal Their Burrows

三候

## 水始涸
## Waters Recede

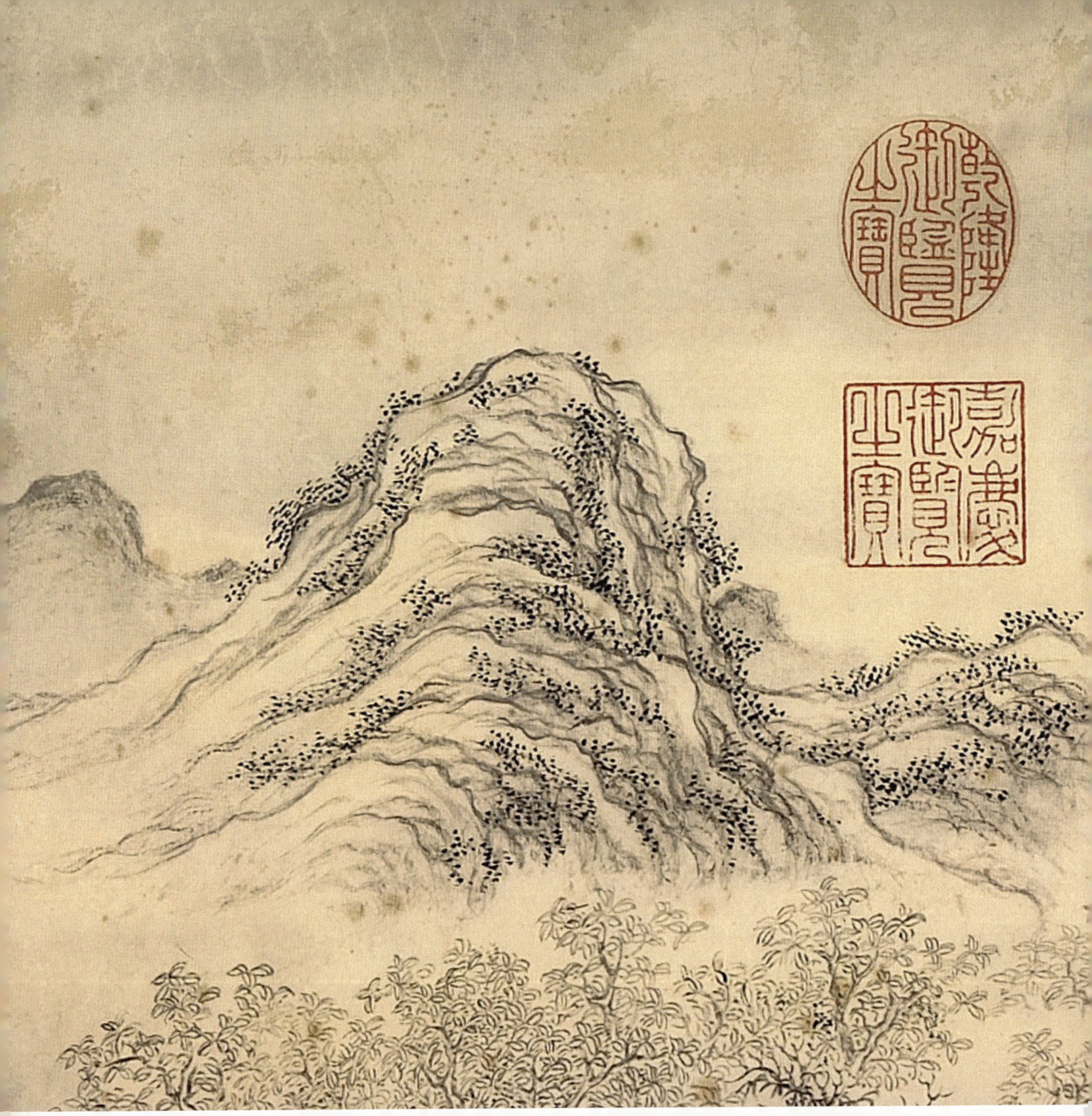
乾隆御覽之寶
嘉慶御覽之寶

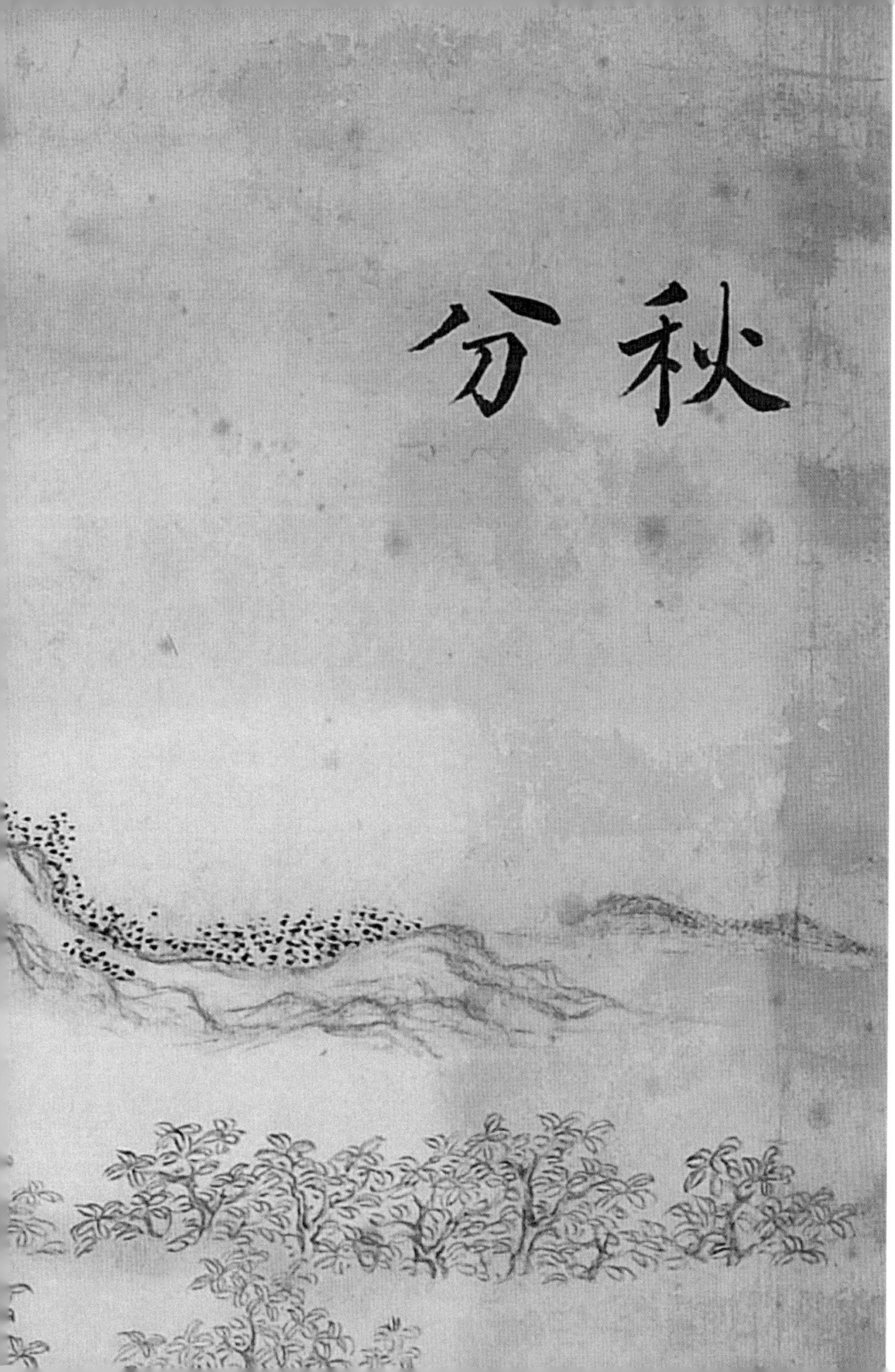


初候

# 雷始收聲
## Thunder Retires

The first whisper of Qiūfēn is “Thunder Retires” (**雷始收聲** Léi Shǐ Shōu Shēng), marking a quieter, more reflective period in the natural world.

二候

# 蟄蟲坯戶

## Insects Seal Their Burrows

The second hint is "Insects Seal Their Burrows" (蟄蟲坯戶 Zhé Chóng Pī Hù), "坯 (Pī)" means fine earth or clay. During this period, many insects have already burrowed into the ground to take refuge from the cold. They seal their burrows with fine soil to insulate themselves against the cold and maintain a stable environment as they enter a state of dormancy.

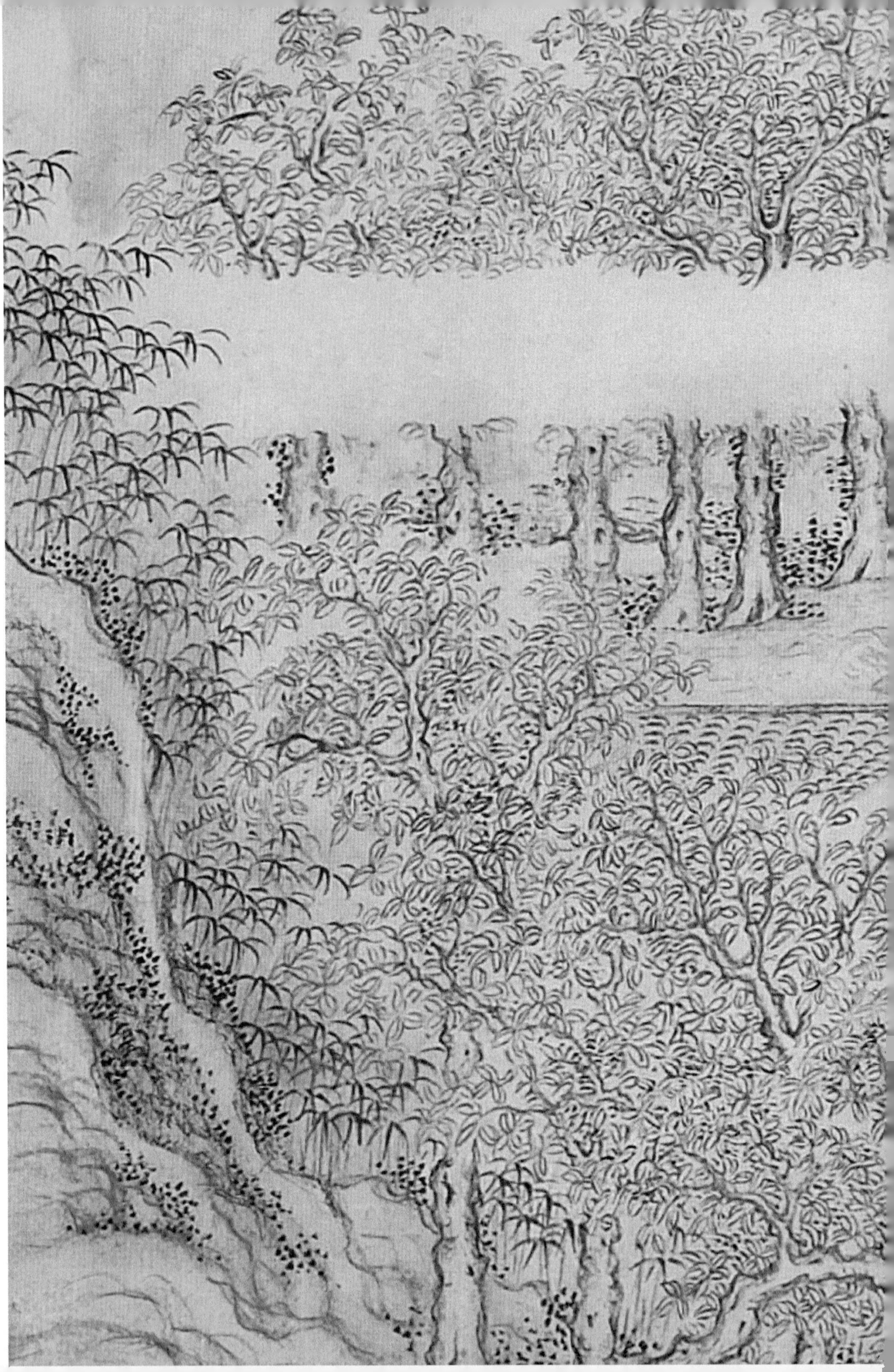

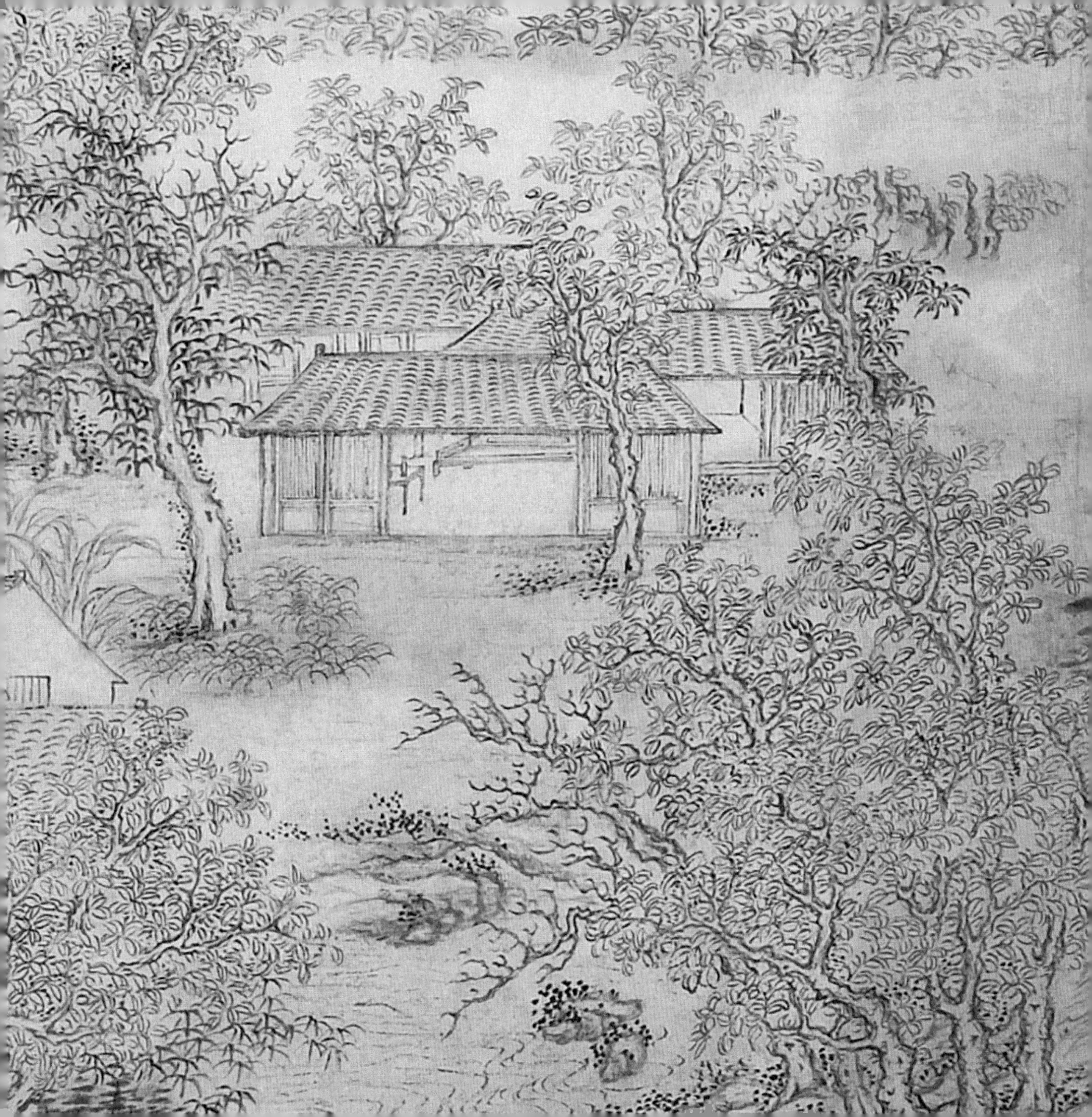

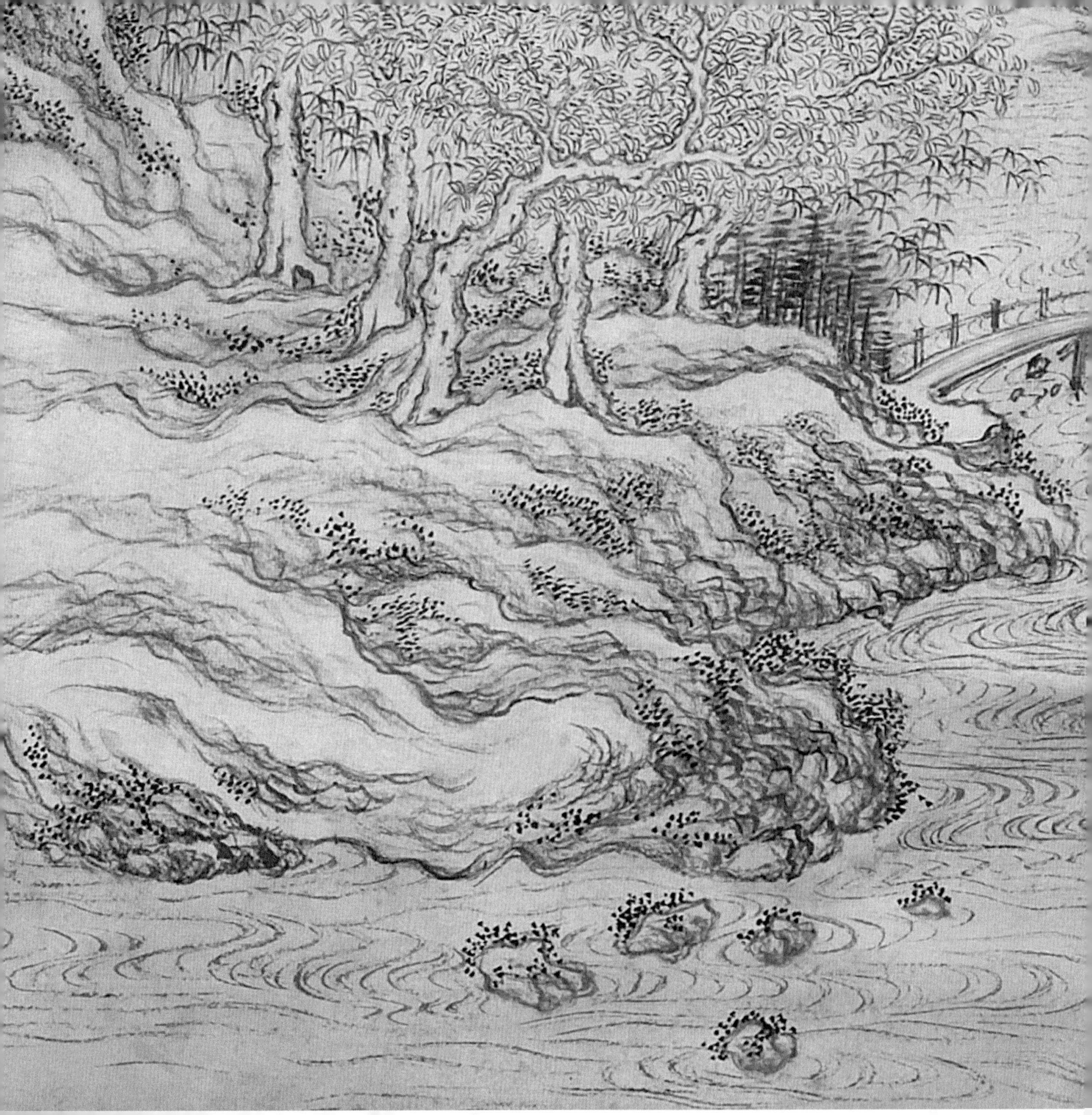

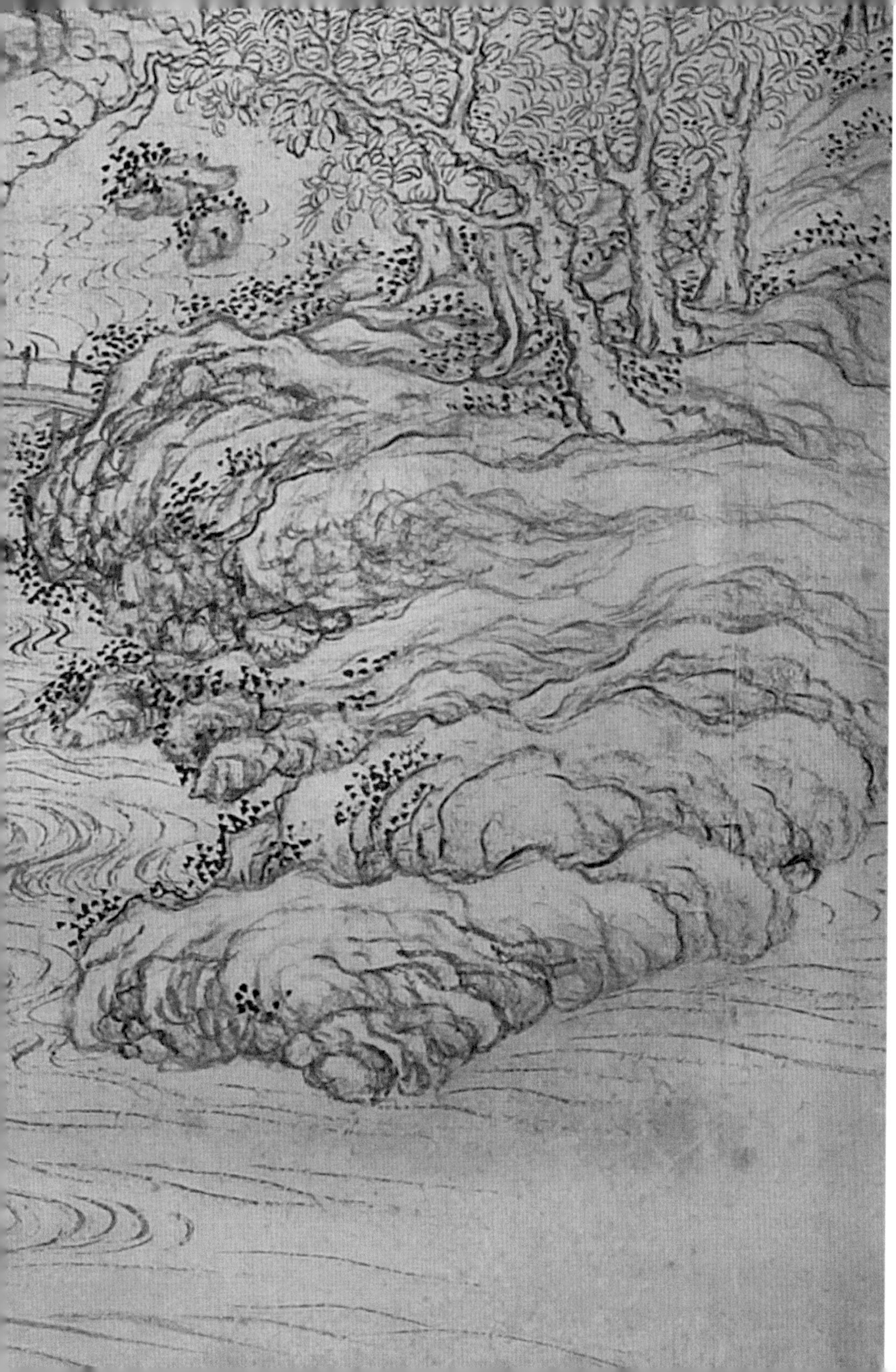

三候

# 水始涸

## Waters Recede

The third indication is "Waters Recede" (水始涸 Shuǐ Shǐ Hé). As autumn deepens during Qiūfēn, this sign indicates the beginning of drier conditions, particularly noticeable in North China where spring and summer are typically marked by ample rainfall. Streams and rivers start to diminish, and the moisture in the air lessens, leading to clearer, cloudless nights. This period is also culturally significant for the Mid-Autumn Festival, where moon gazing and offerings are traditional practices, celebrated more distinctly under the clear, dry autumn sky.

露

Lù

Hán

寒

# COLD DEW

October 8 (± 1)

# 寒露

## Cold Dew

Hánlù : Meaning "Cold Dew," signifies a further decrease in temperature and the appearance of colder, more pronounced dew. It' s a period where the chill begins to set in more firmly, leading to the formation of frost as the next phase. Hánlù is an indicator of the deepening autumn and the gradual move toward early winter, marked by the ripening of certain fruits and grains.

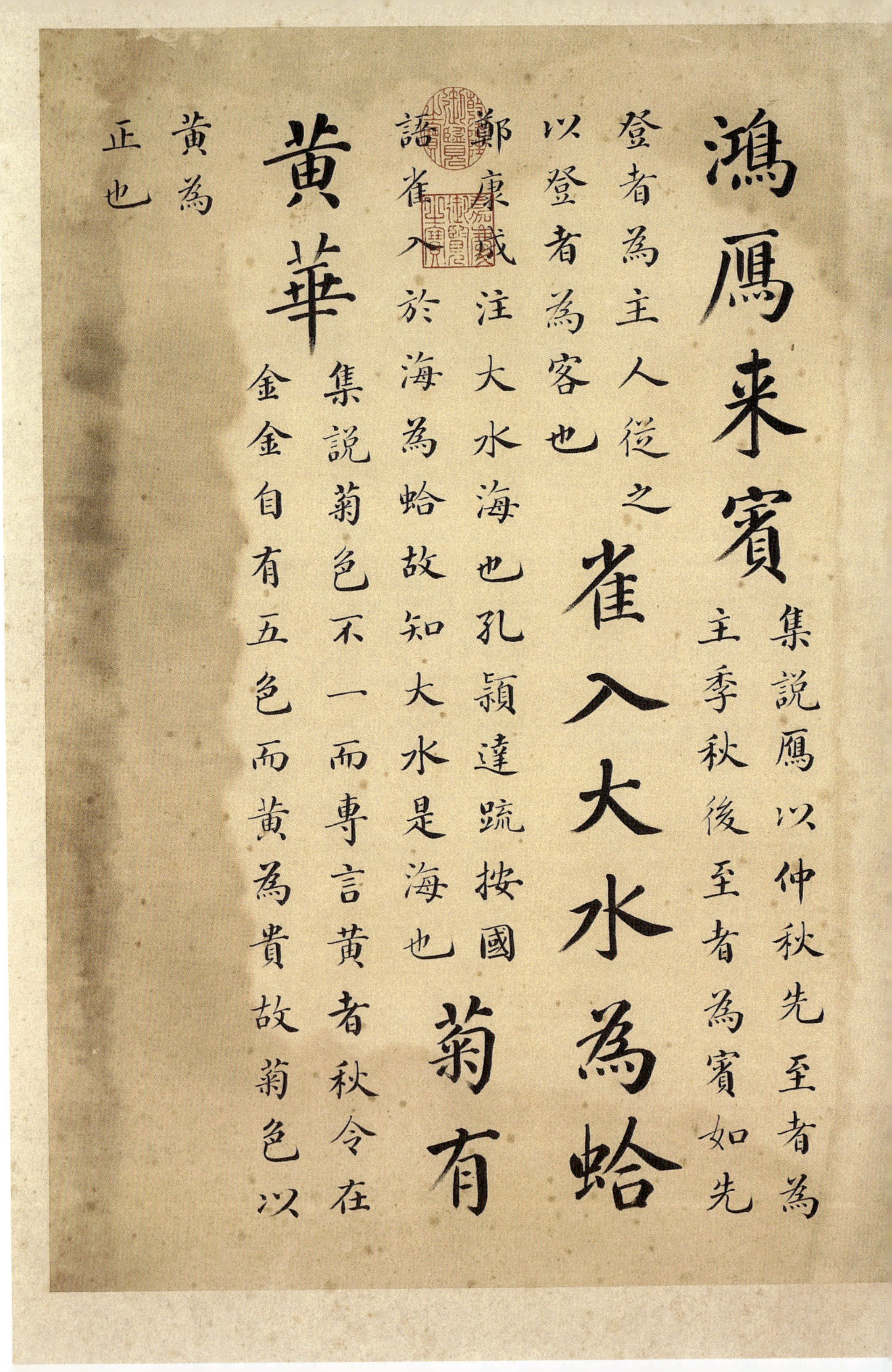

鴻鴈来賓

集說鴈以仲秋先至者為主季秋後至者為賓如先登者為主人從之以登者為客也

雀入大水為蛤

鄭康成注大水海也孔穎達疏按國語雀入於海為蛤故知大水是海也

菊有黄華

集說菊色不一而專言黄者秋令在金金自有五色而黄為貴故菊色以黄為正也

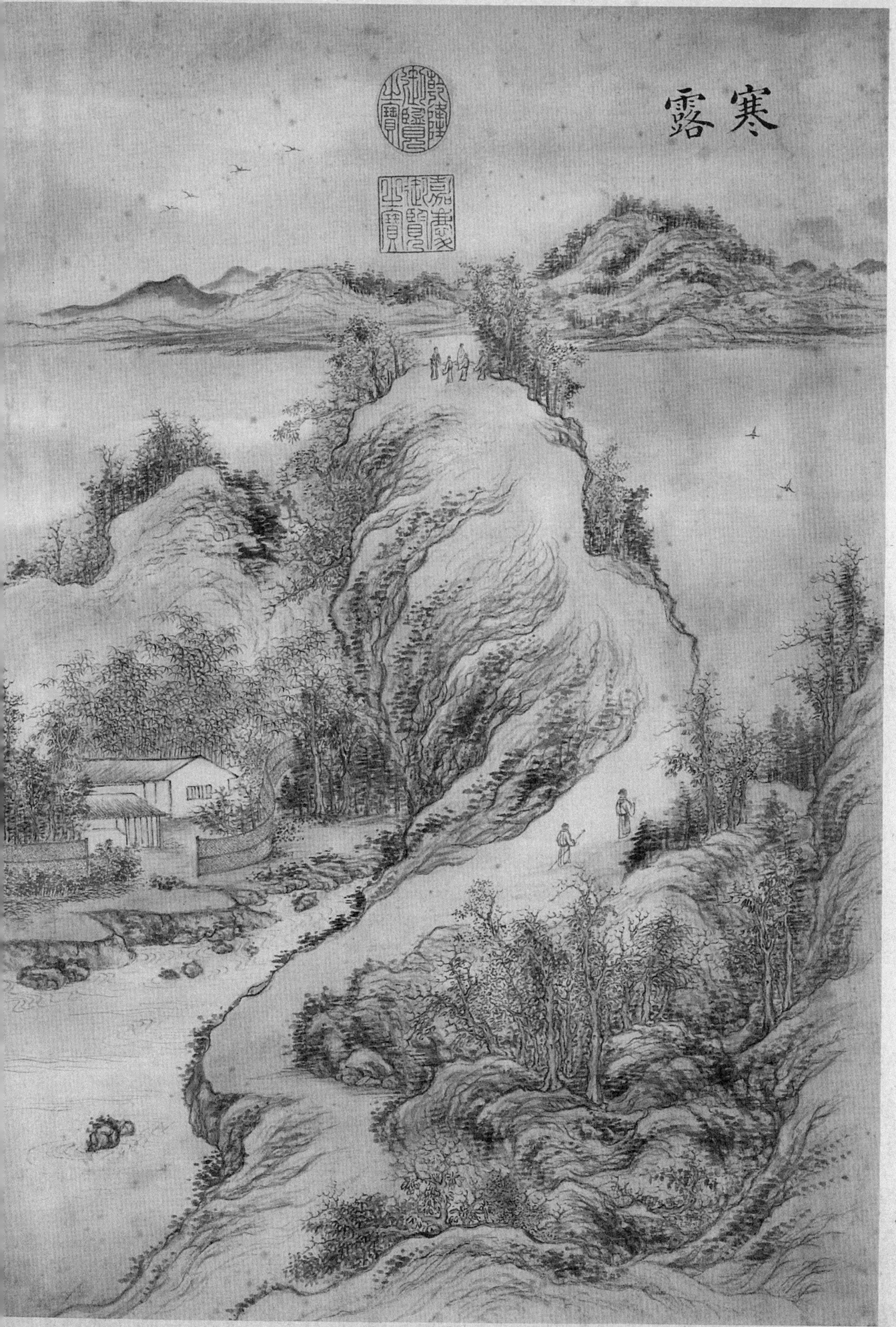


初候

## 鴻雁來賓
## Arrival of the Wild Geese

二候

## 雀入大水為蛤
## Sparrows Transform into Clams

三候

## 菊有黃華
## Chrysanthe-mums Bloom

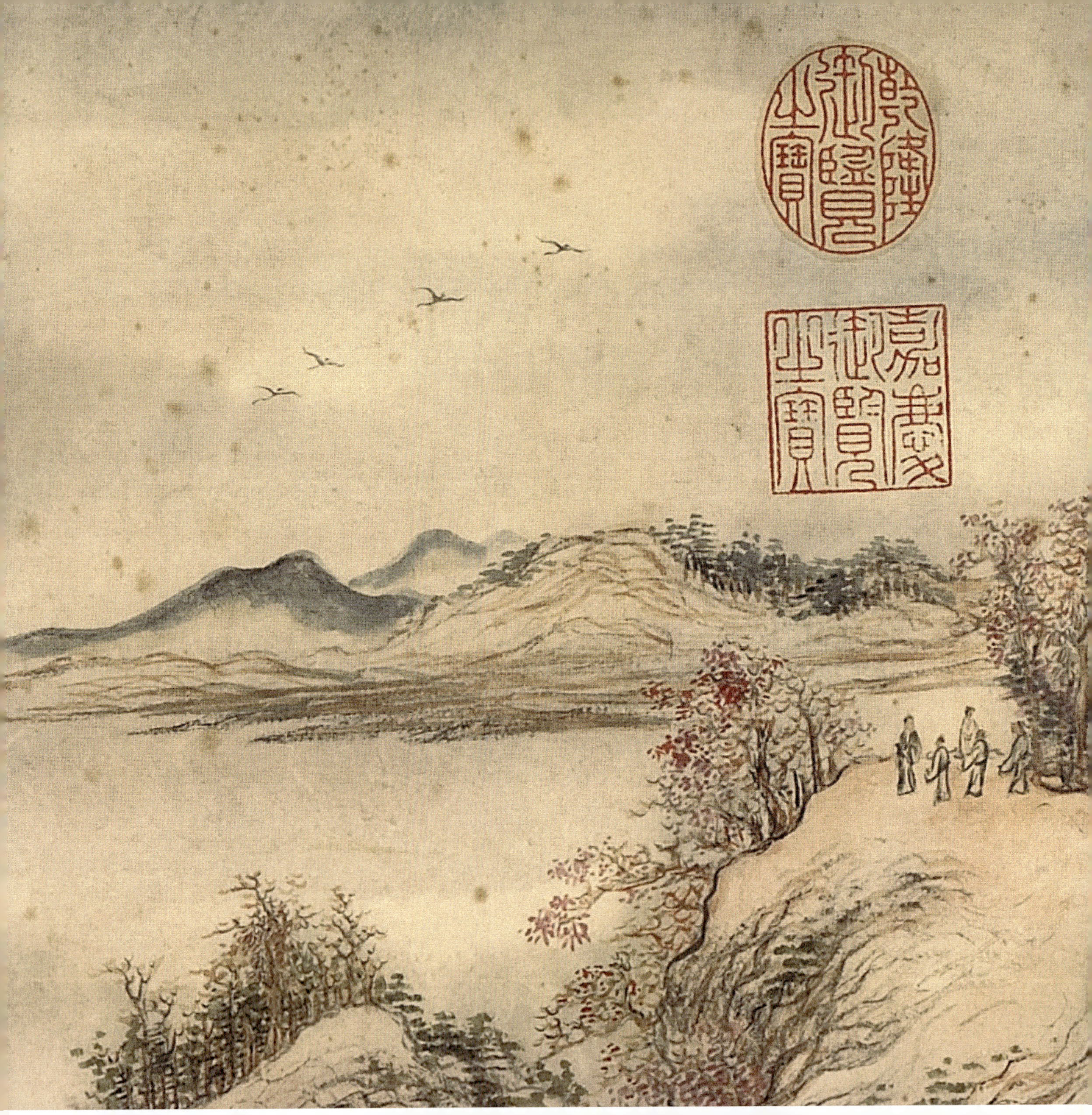
乾隆御覽之寶
嘉慶御覽之寶

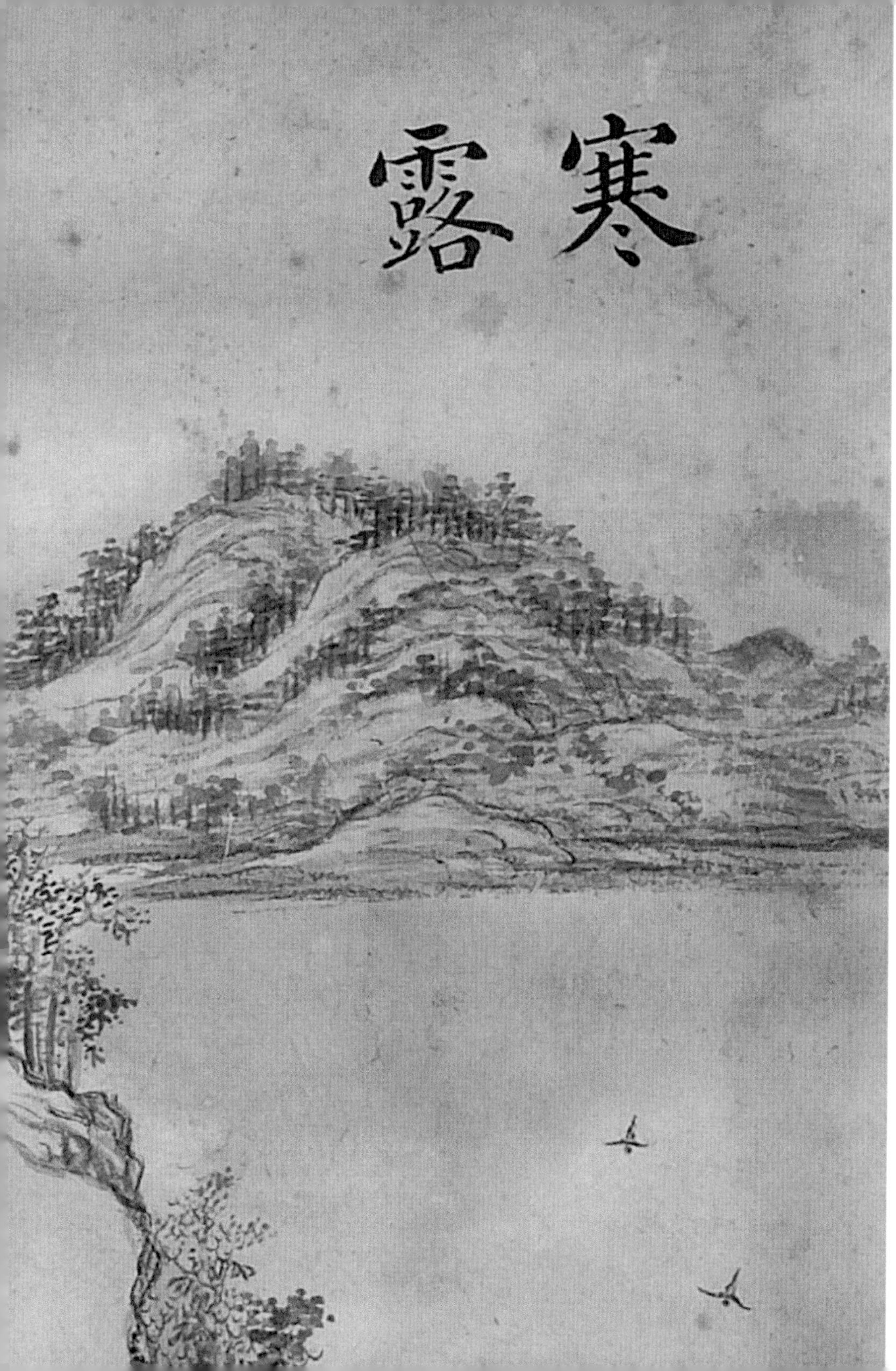


初候

# 鴻雁來賓
## Arrival of the Wild Geese

Theinitial marker of Hánlù is "Arrival of the Wild Geese" (鴻雁來賓 Hóng Yàn Lái Bīn).
Hánlù marks a time when the air turns crisper and cooler, signaling wild geese to continue their migration southward. The term "Wild Geese Become Guests" reflects the phenomenon of these birds arriving in new habitats as temporary visitors, embodying migratory nature. Their presence is a clear sign of autumn' s progression, as they move to warmer areas in a display of nature' s cyclical and adaptive behaviors. This period is a reminder of the ongoing changes in the natural world, with each species responding according to the rhythm of the seasons.

二候

# 雀入大水為蛤

## Sparrows Transform into Clams

The second sign of Hánlù is "Sparrows Transform into Clams" (雀入大水為蛤 Què Rù Dà Shuǐ Wèi Gé). This sign is rooted in an ancient metaphor where it was believed that sparrows would dive into the sea and transform into clams as autumn deepened. The stripes and coloration of clams were thought to resemble the birds, leading to this poetic association. It reflects a time of transformation and adaptation, symbolizing the preparations that all creatures make for the colder months ahead, even if in mythic and poetic terms.

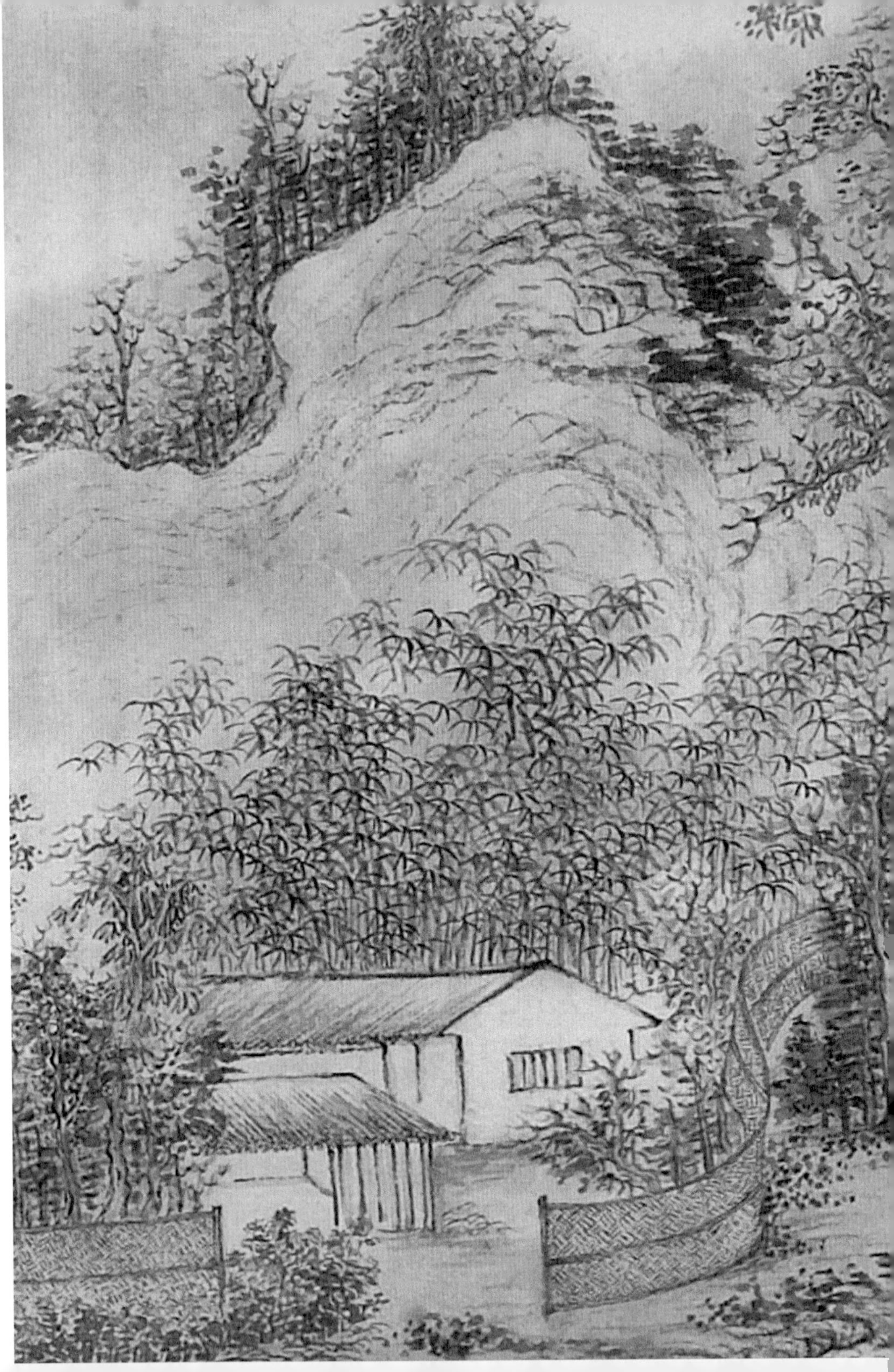

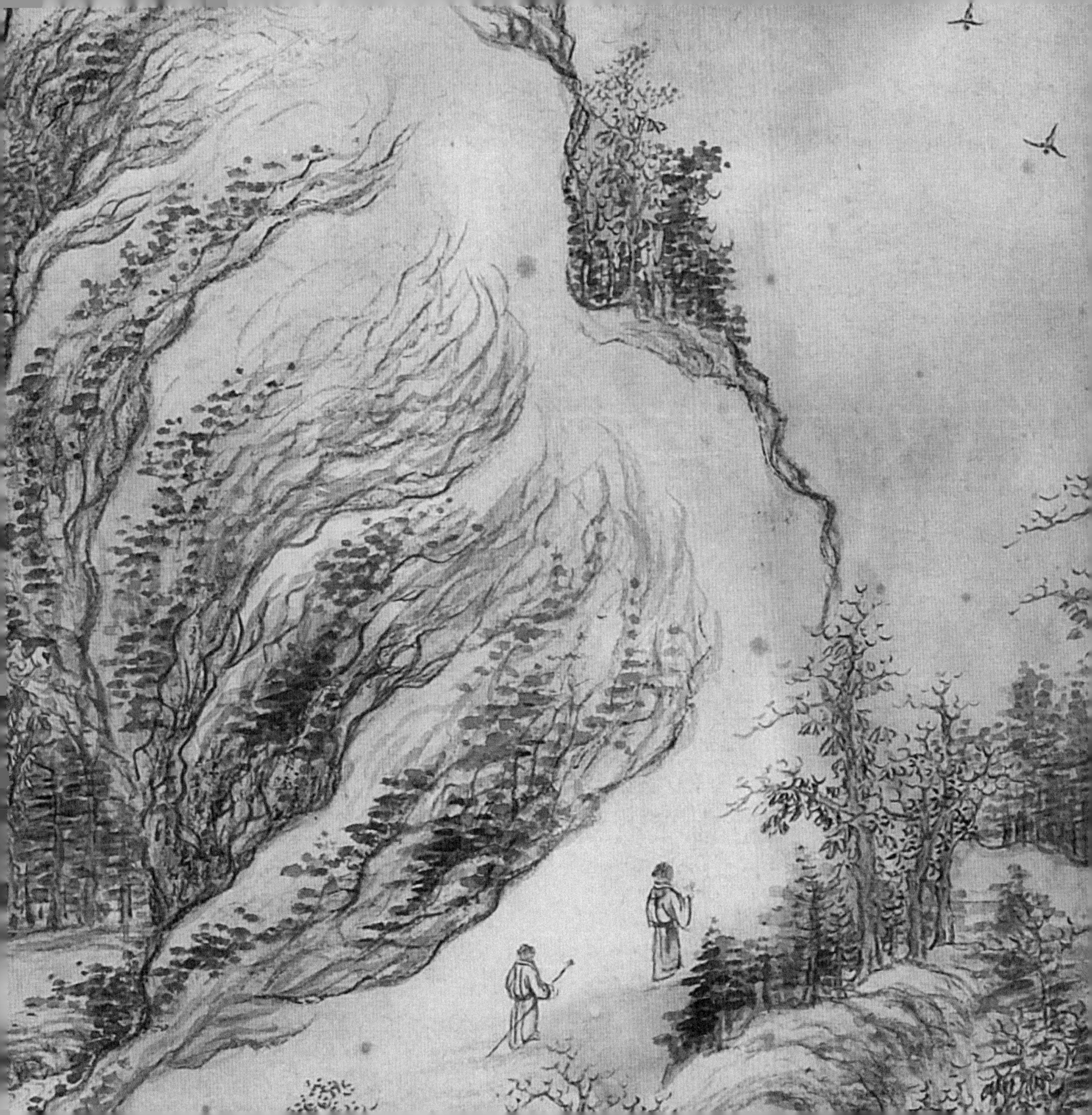

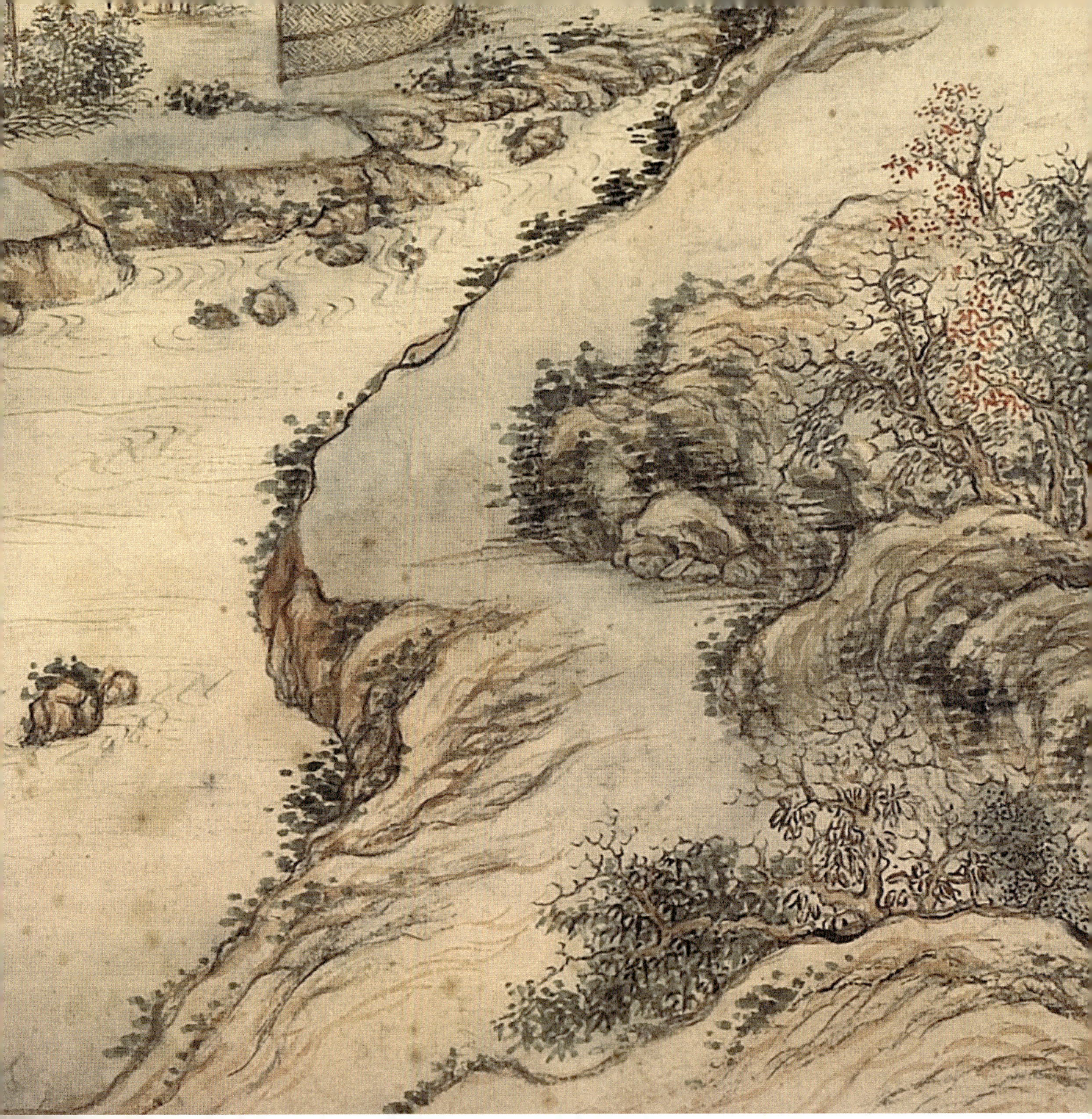

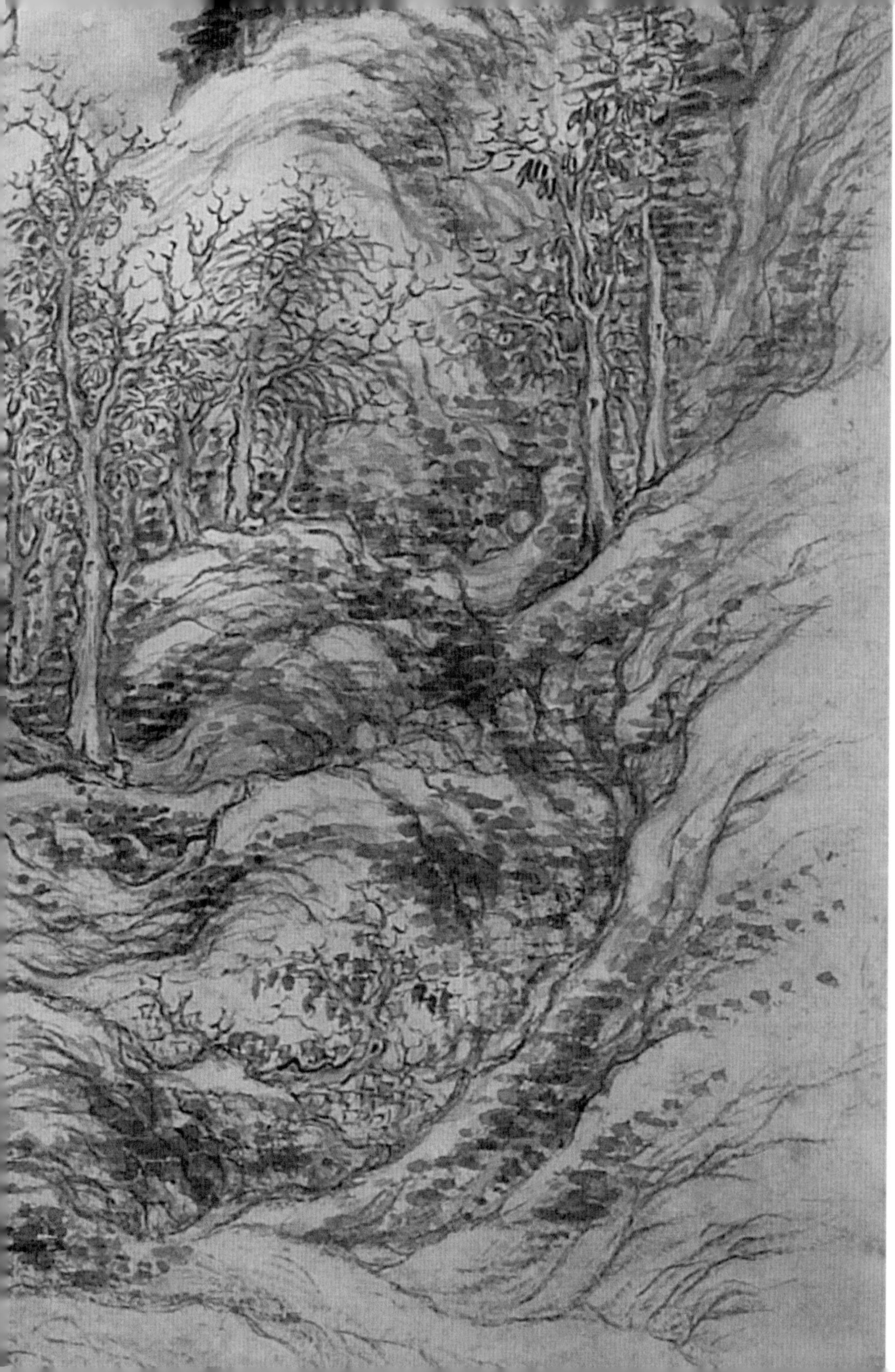

三候

# 菊有黃華

## Chrysanthemums Bloom

The third sign of Hánlù is "Chrysanthemums Bloom" (**菊有黃華** Jú Yǒu Huáng Huá):

As autumn progresses, yellow chrysanthemums begin to bloom, bringing vibrant color to the increasingly sparse landscape. This sign signifies the peak of autumnal beauty, with chrysanthemums symbolizing resilience and the ability to thrive as cooler days arrive. The blooming of these flowers is celebrated in Chinese culture with traditional activities like crab feasting and chrysanthemum appreciation, marking a time of enjoyment before the onset of winter.

# 降

Jiàng

Shuāng

# 霜

## FROST' S DESCENT

October 23 (± 1)

# 霜降

## Frost’s Descent

Shuāngjiàng: Signaling the time when the chill in the air turns dew into frost, indicating a significant deepening into the autumn season. It marks an increase in Yin energy, leading to colder mornings and the first appearances of frost. Shuāngjiàng is a sign of nature’s preparation for the cold winter ahead, as the environment adapts to the dropping temperatures and the landscape takes on a frost-covered appearance.

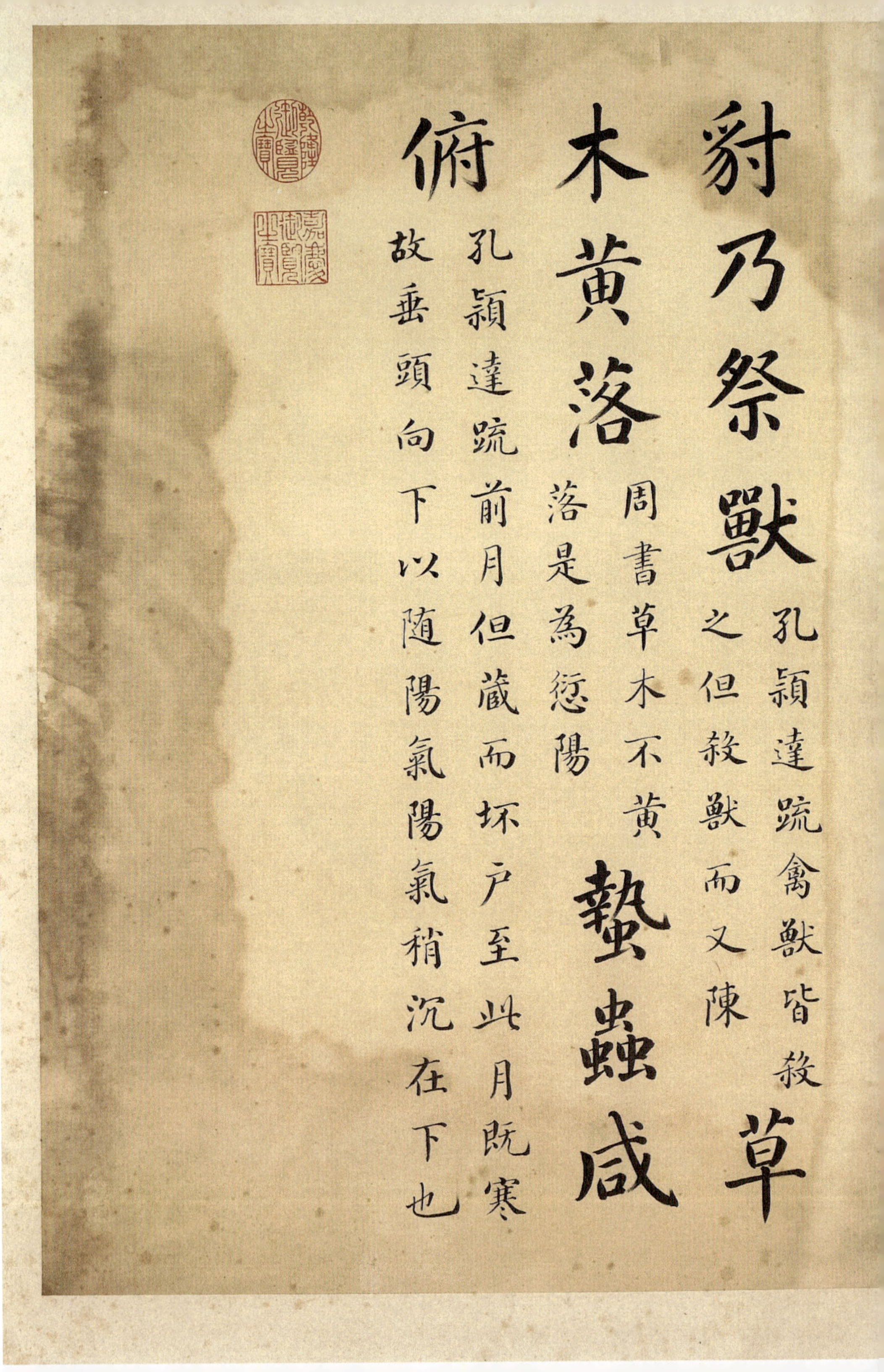

豺乃祭獸
孔穎達疏禽獸皆殺之但殺獸而又陳
草木黄落
周書草木不黄落是為愆陽
蟄蟲咸俯
孔穎達疏前月但藏而坏户至此月既寒故垂頭向下以隨陽氣陽氣稍沉在下也

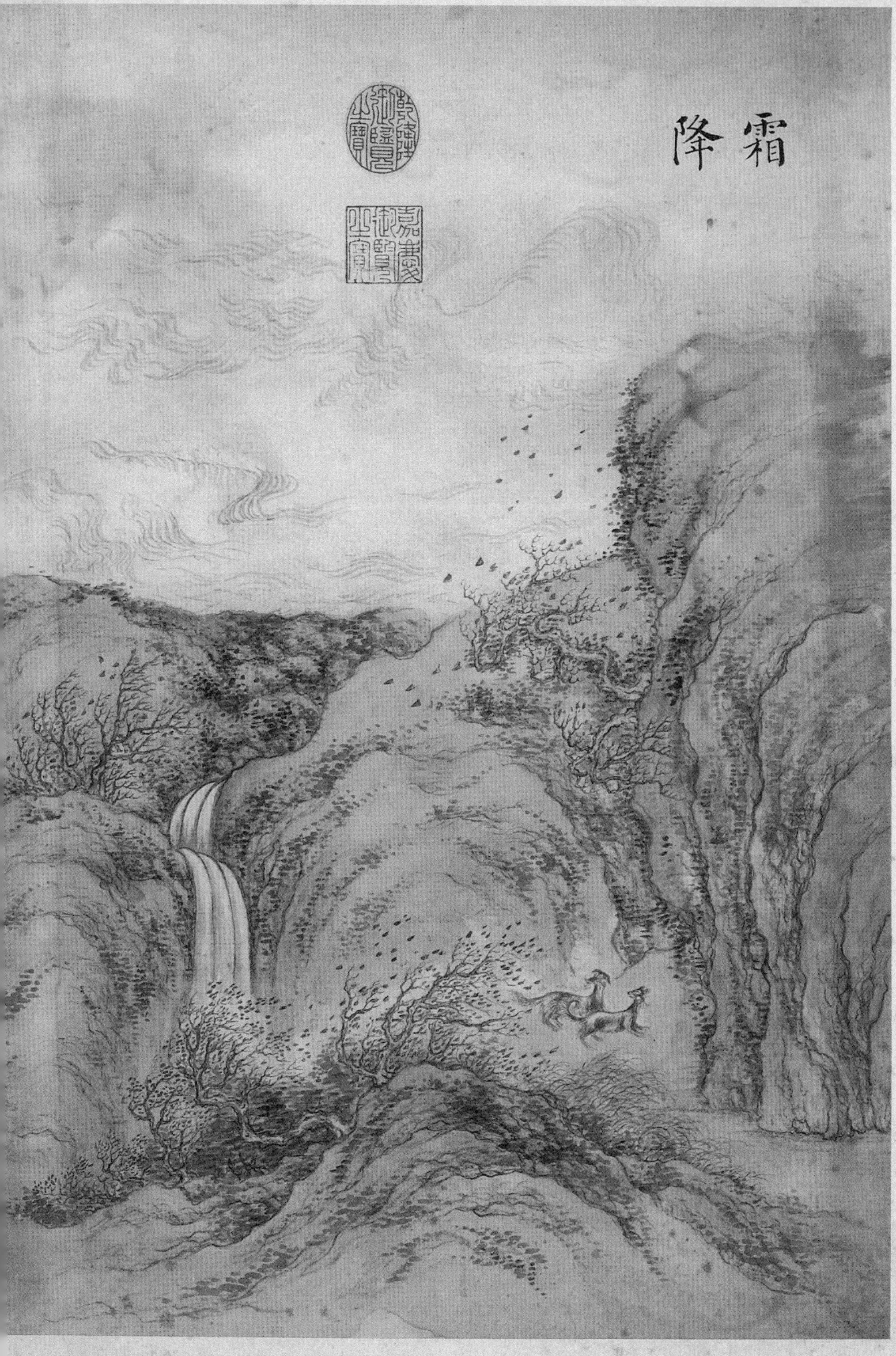


初候

## 豺乃祭獸

## Jackals Honor Their Hunt

二候

## 草木黃落

## Vegetation Turns Yellow and Falls

三候

## 蟄蟲咸俯

## Insects Enter Hibernation

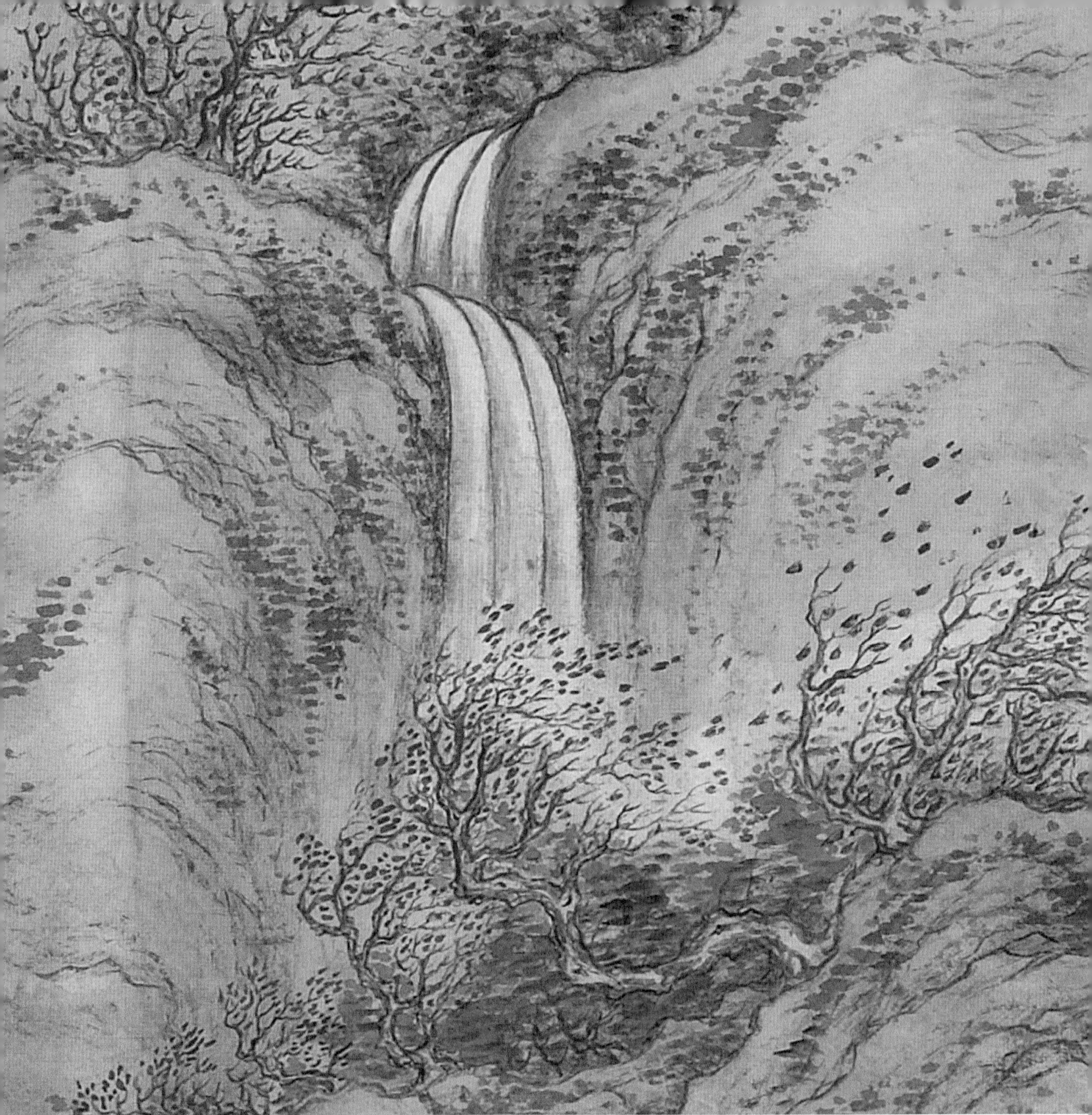

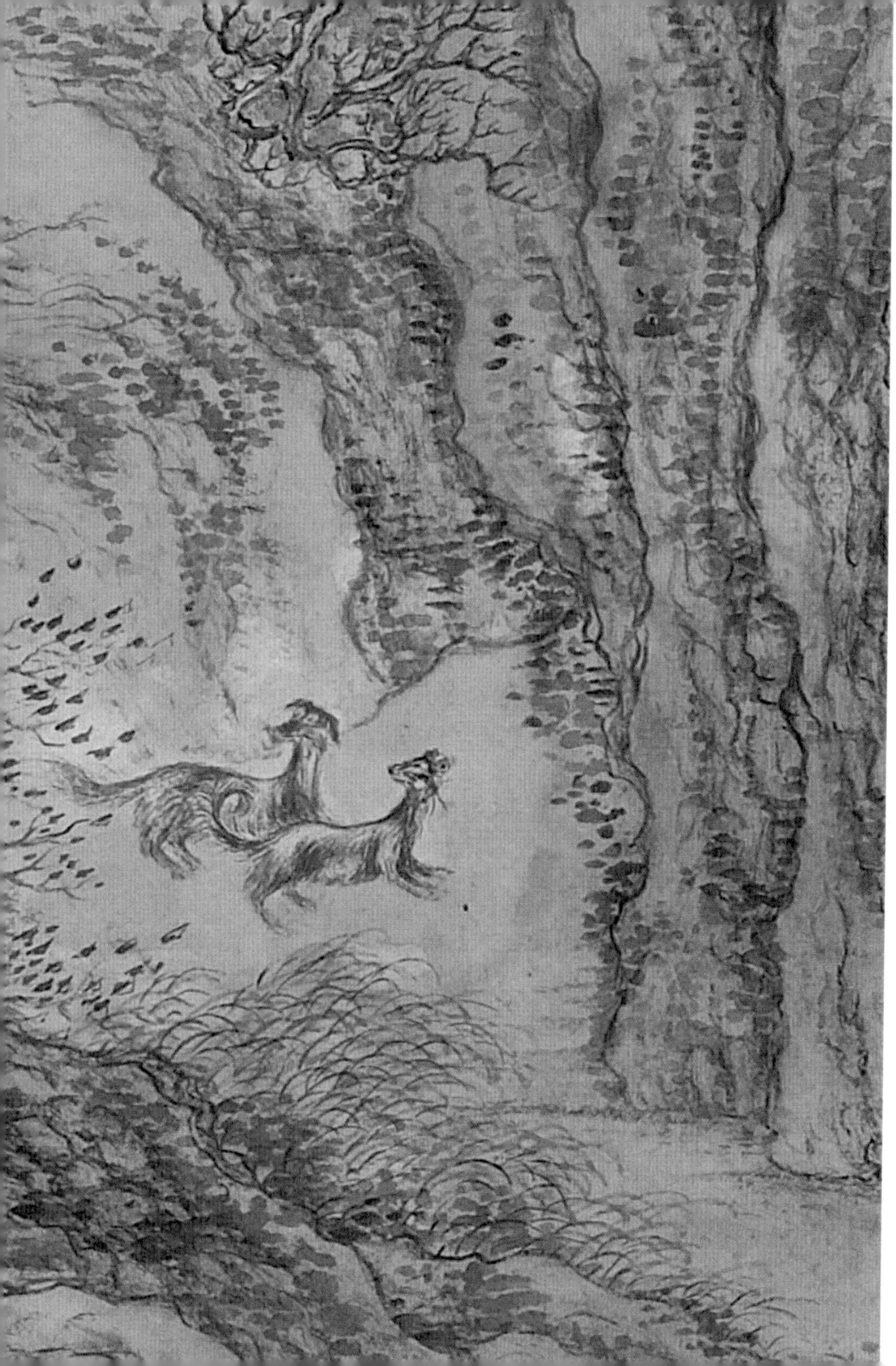

初候

# 豺乃祭獸

## Jackals Honor Their Hunt

The First Sign of Shuāngjiàng is "Jackals Honor Their Hunt" (豺乃祭獸 Chái Nǎi Jì Shòu).

In this first sign of Shuāngjiàng, jackals meticulously display their captured prey, engaging in a ritualistic practice before consumption. This sign is part of a significant triad within the 72 pentads or periods, including "Otters Perform Fish Ritual" (獺祭魚) representing creatures of the water in spring, "Eagles Present Their Prey" (鷹乃祭鳥) representing creatures of the sky in summer, and "Jackals Honor Their Hunt" (豺乃祭獸 ) representing creatures of the land in autumn. These three 'offerings' symbolically cover the domains of water, air, and land, reflecting the comprehensive and interconnected nature of life across different environments. This thematic grouping across seasons emphasizes the continuous and universal respect for life and death seen in nature' s rituals, celebrating the diversity and adaptability of living beings within their respective realms throughout the year.

二候

# 草木黄落

## Vegetation Turns Yellow and Falls

The second sign is "Vegetation Turns Yellow and Falls" (**草木黄落** Cǎo Mù Huáng Luò).
As autumn deepens, this sign signifies the transition of foliage and grasses from their verdant green to a faded yellow, leading to their eventual falling. This natural process marks the preparation of the plant world for the coming winter, signaling a time of decline in growth and vitality. Trees shed their leaves in a final display of autumnal beauty, conserving energy and resources as they enter a state of dormancy. It's a widespread adaptation that underscores the cycle of life, death, and rebirth inherent in nature.

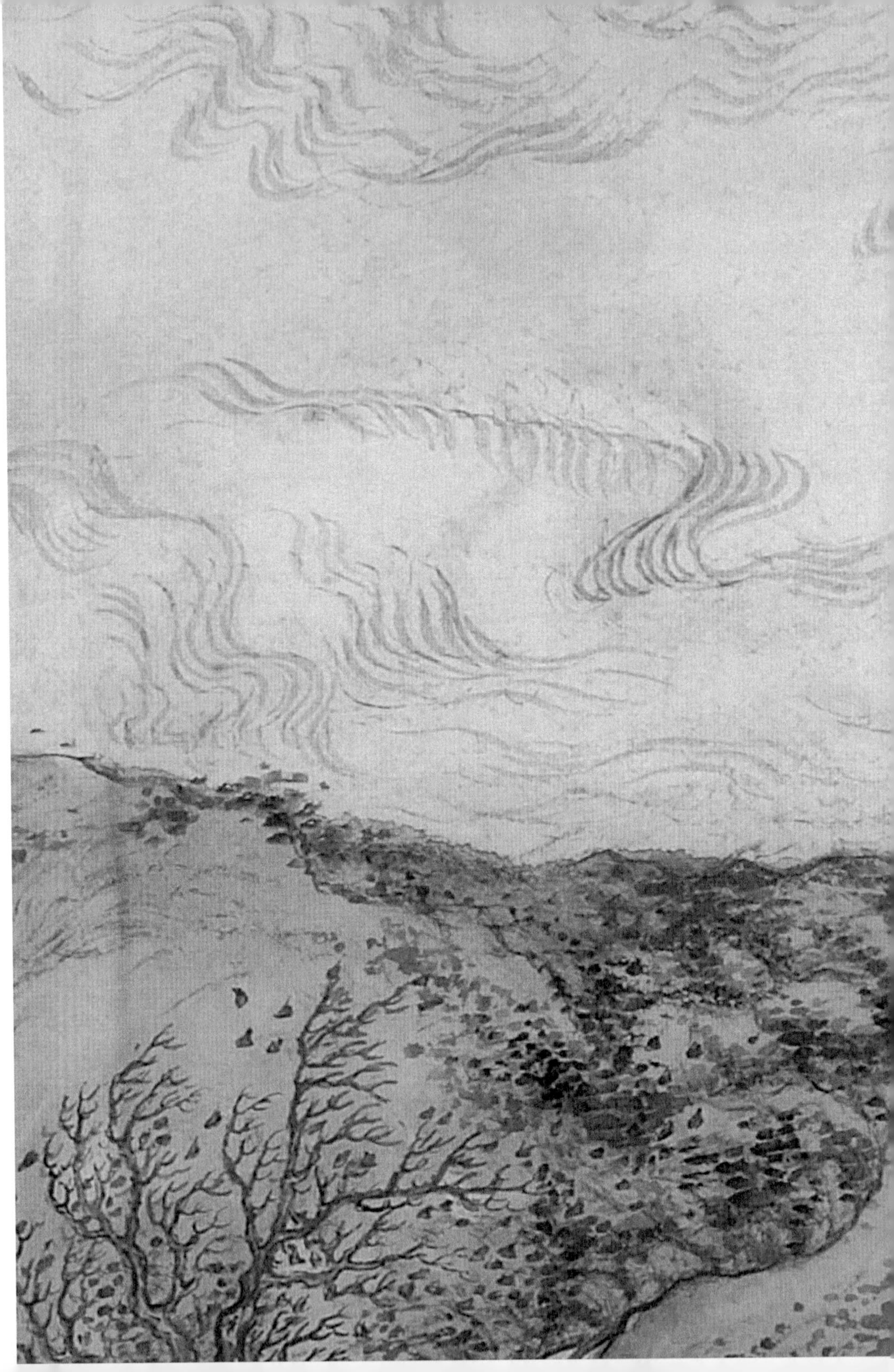

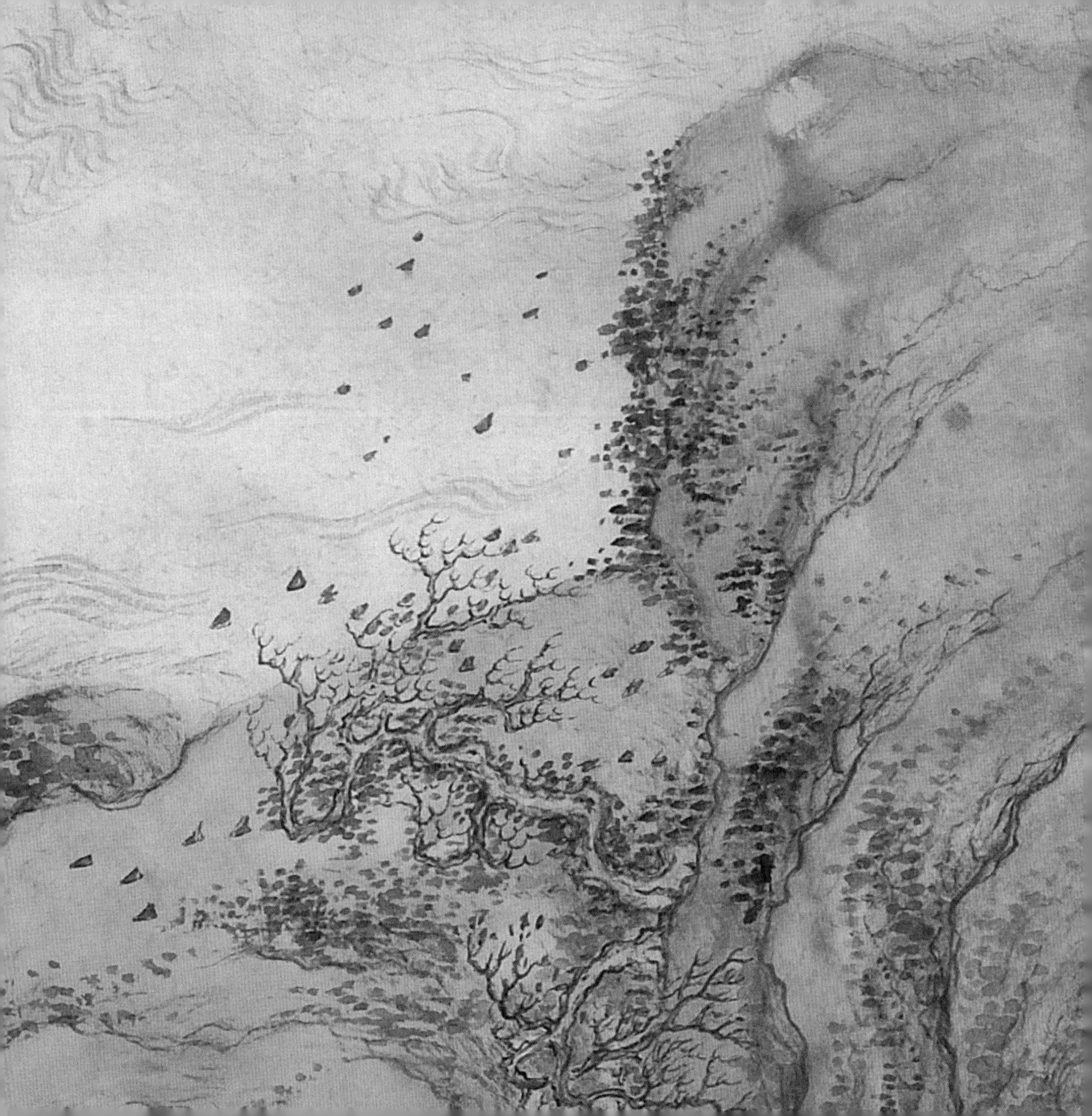

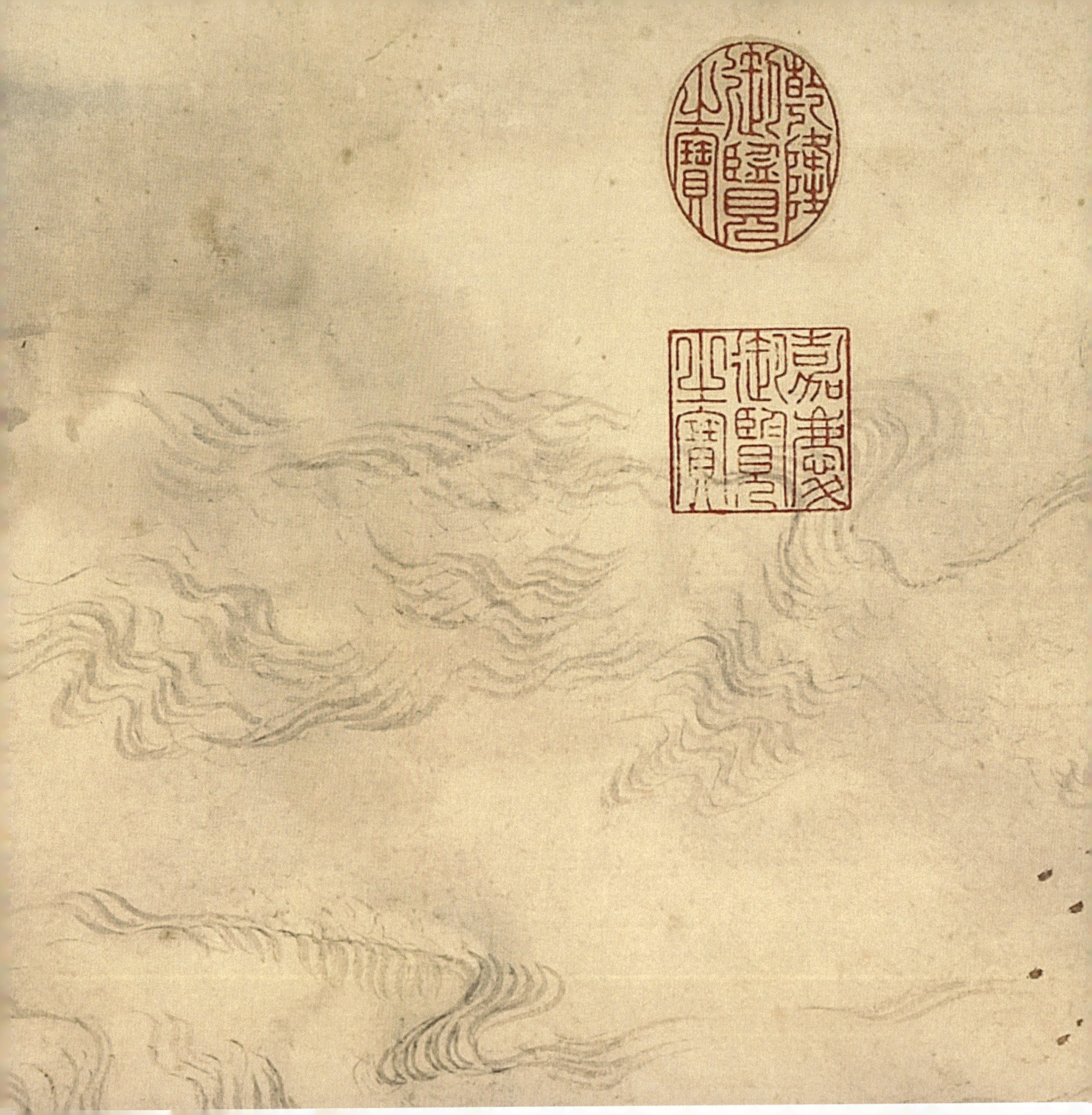
乾隆御覽之寶
嘉慶御覽之寶

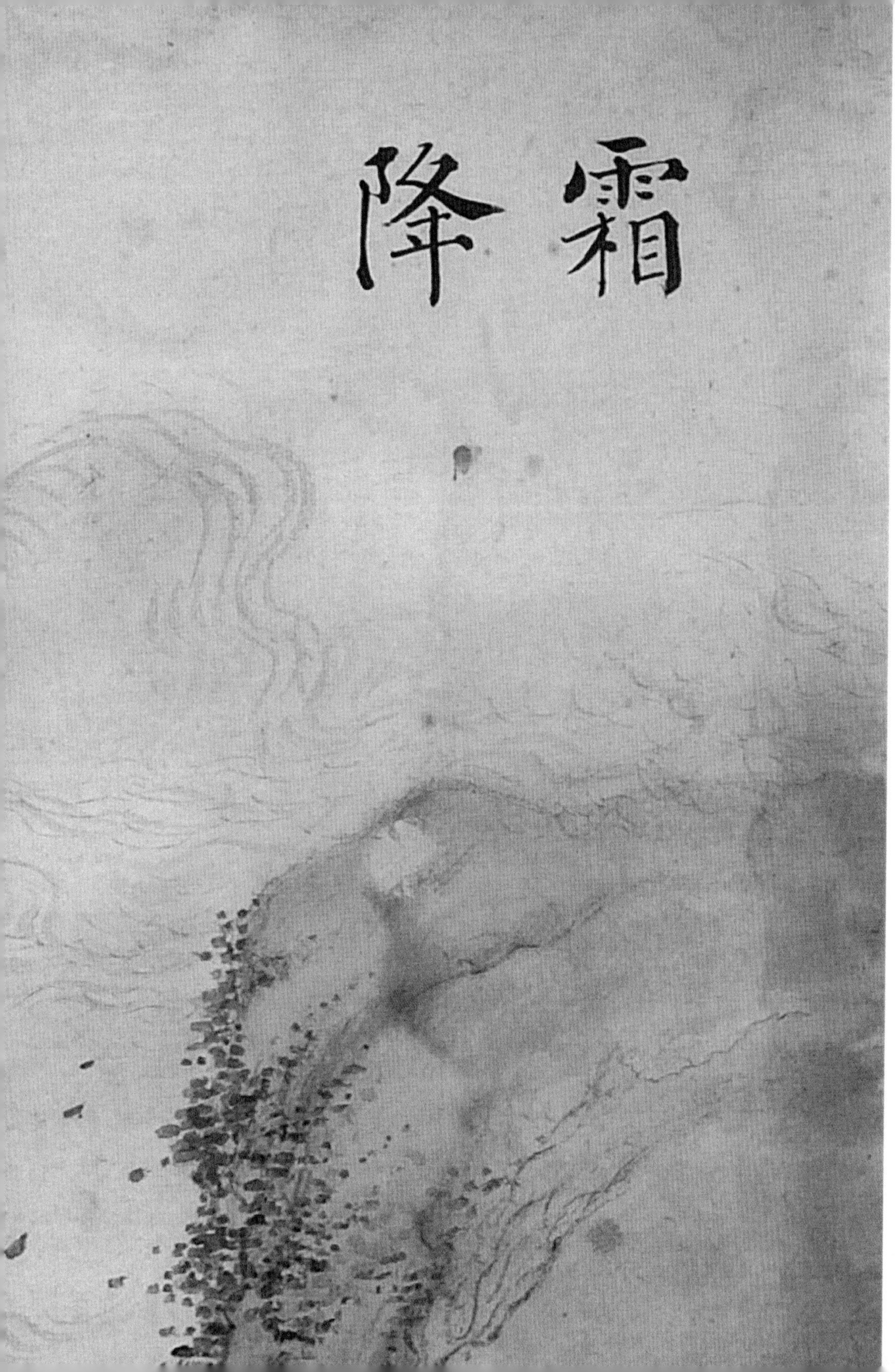


三候

# 蟄蟲咸俯

## Insects Enter Hibernation

The third sign is “Insects Enter Hibernation” (蟄蟲咸俯 Zhé Chóng Xián Fǔ). During this period, overwintering insects adopt a state of stillness, withdrawing into their nests or burrows. “咸俯” refers to their motionless, bowed posture as they enter hibernation. This sign symbolizes a deep pause in the natural world, with insects ceasing their activity to conserve energy for the cold months ahead. It’ s a time of quiet and stillness across the ecosystem, as all living beings brace for the winter’ s chill, reflecting the cycle of activity and dormancy that governs the natural world.

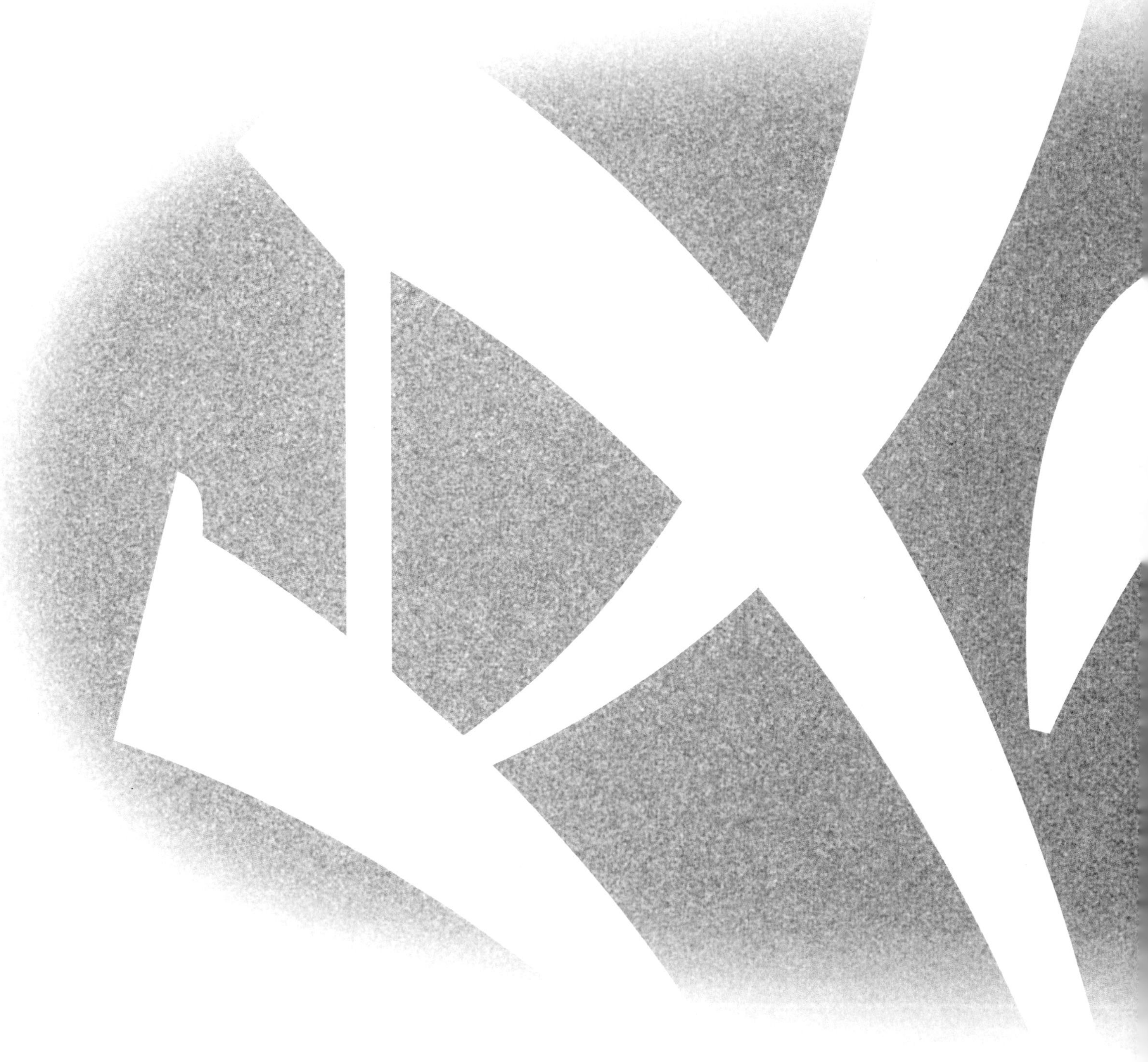

# WINTER

# 冬

# Dōng

Lì

立

# BEGINNING OF WINTER

November 7 (± 1)

# 立冬

## Beginning of Winter

Lìdōng : Marking the onset of winter and symbolizes conclusion and introspection, as “冬 (dōng)” is associated with the end“終 (zhōng)”. It signifies the year’ s end and the retreat of autumn, welcoming the initial presence of winter. After enduring the extremes of the year, this period is crucial for rejuvenation and nourishment, following the tradition of fortifying the body to restore vitality. It’ s also a time when the harvest concludes, and preparations for winter crops or green manure begin, reflecting the cycle of agricultural life.

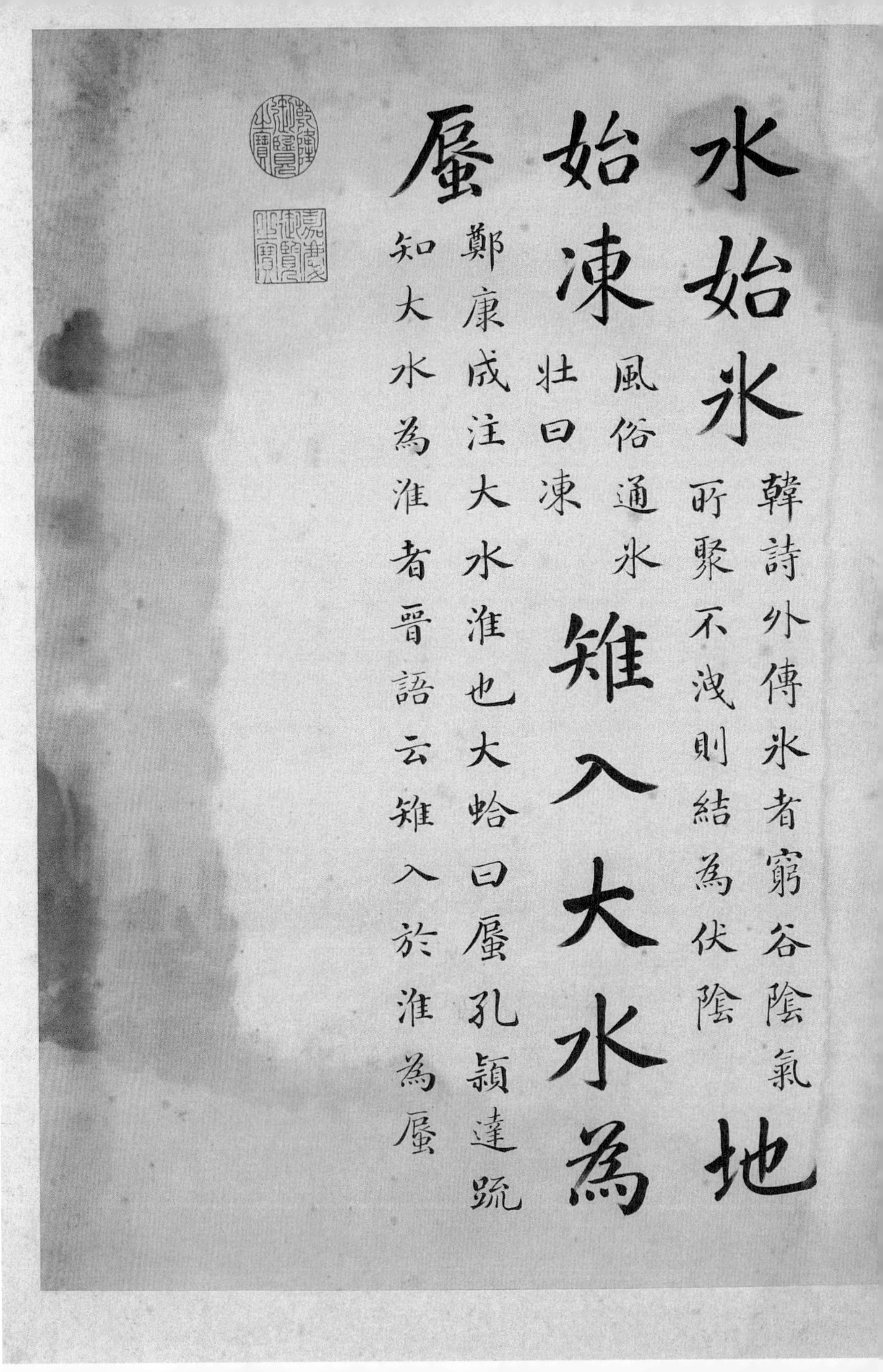

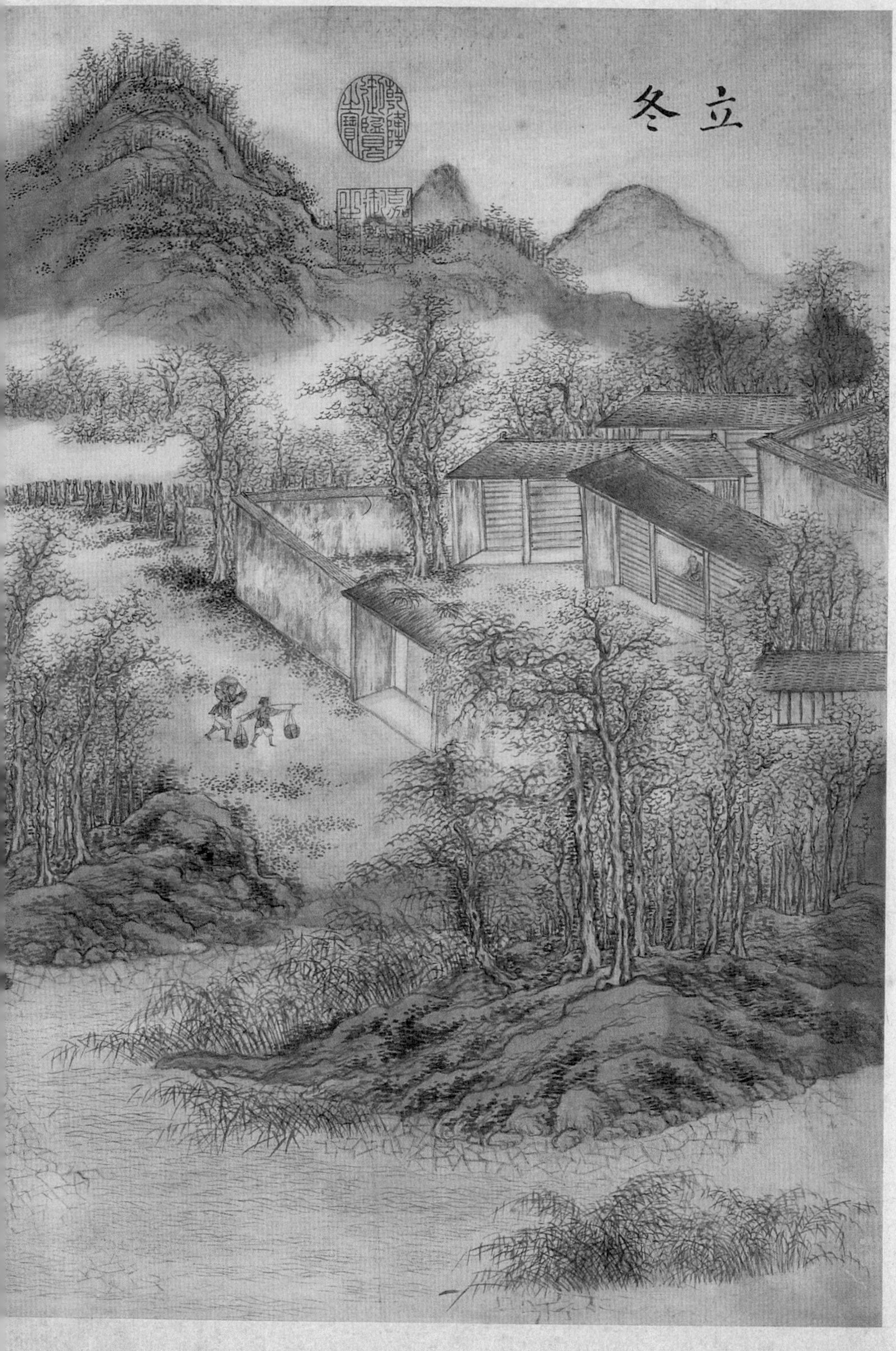


初候

## 水始冰
## Water Begins to Ice

二候

## 地始凍
## Ground Begins to Freeze

三候

## 雉入大水為蜃
## Pheasants Turn to Clams

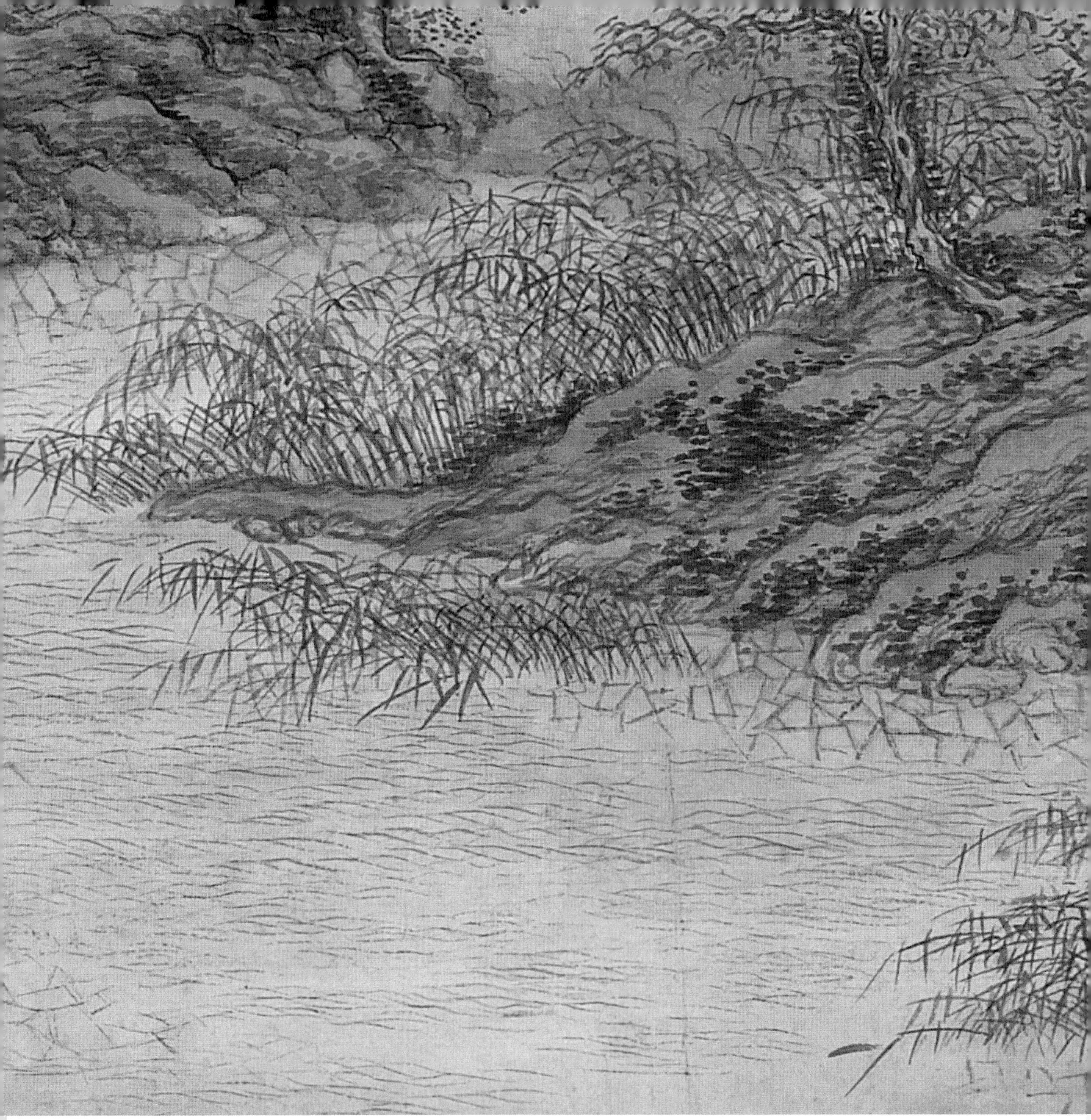

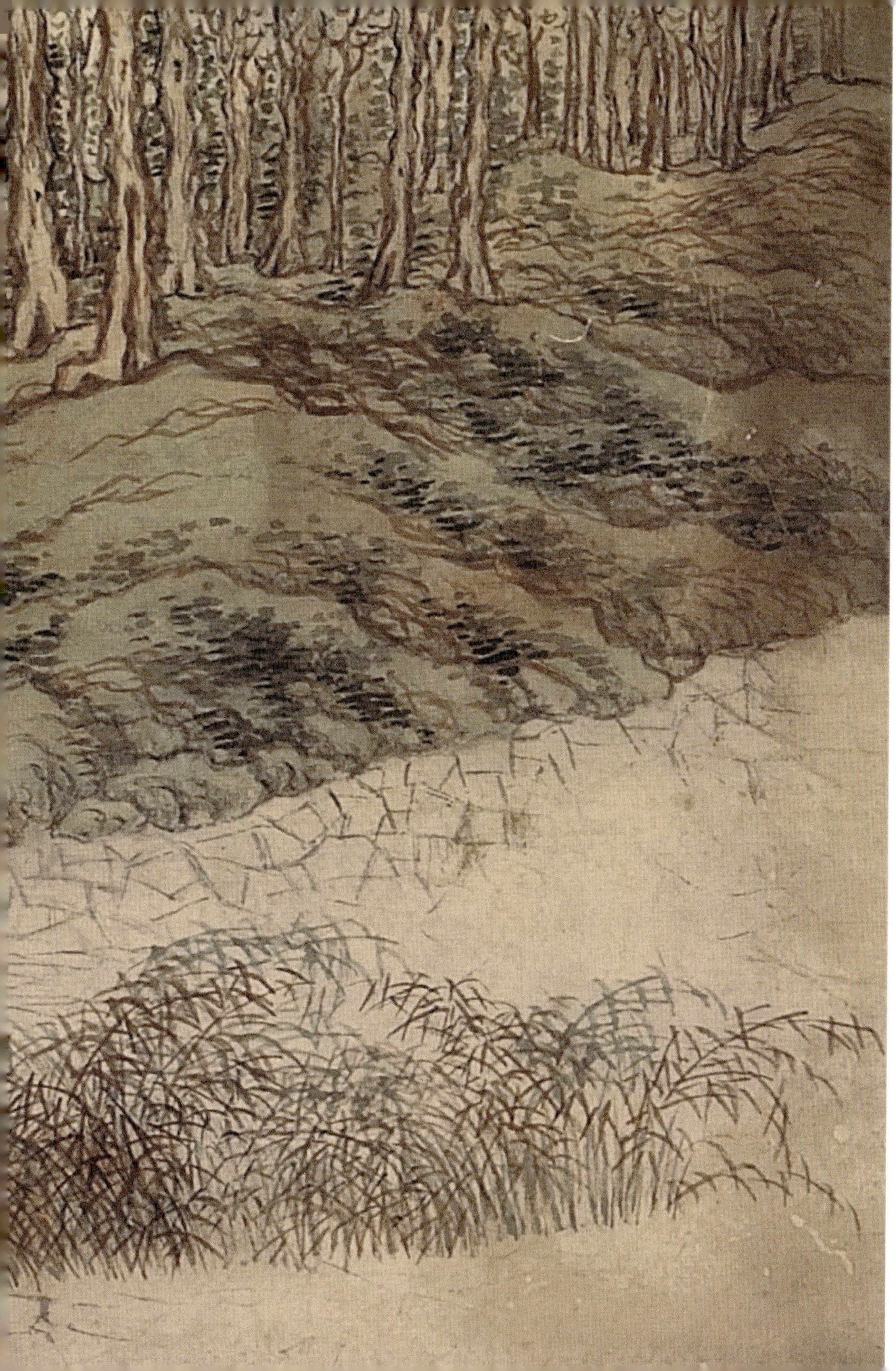

初候

# 水始冰

## Water Begins to Ice

The first sign, "Water Begins to Ice" (水始冰 Shuǐ Shǐ Bīng), marks the beginning of ice formation on bodies of water, heralding the deeper cold of winter. It signifies the initial solidification of water surfaces as temperatures drop, foreshadowing the coming of the coldest season.

二候

# 地始凍

## Ground Begins to Freeze

The second becon is "Ground Begins to Freeze" (地始凍 Dì Shǐ Dòng). As winter approaches, this sign marks the beginning of soil freezing. The moisture within the earth starts to solidify due to dropping temperatures, leading to the hardening of the ground.

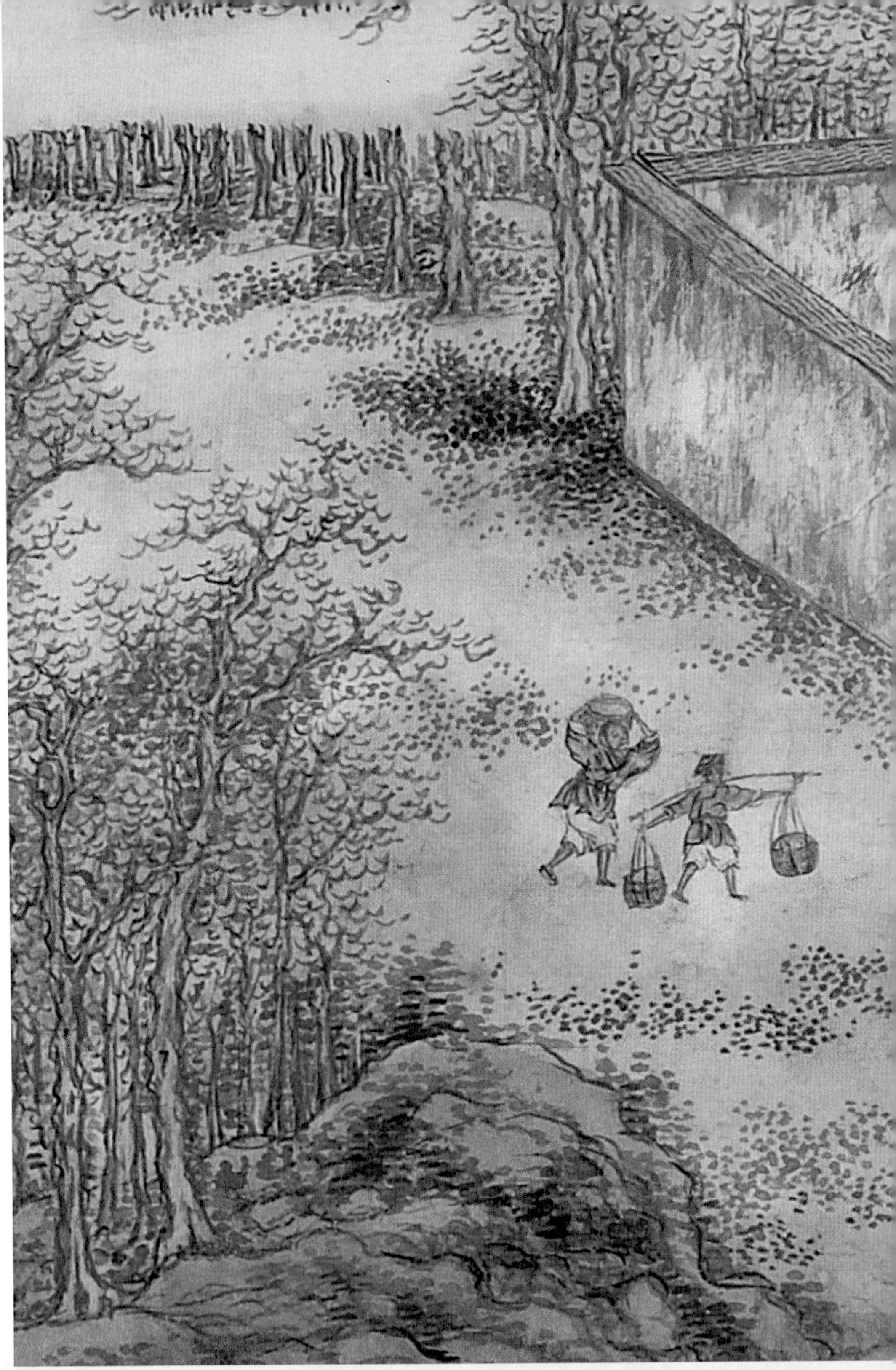

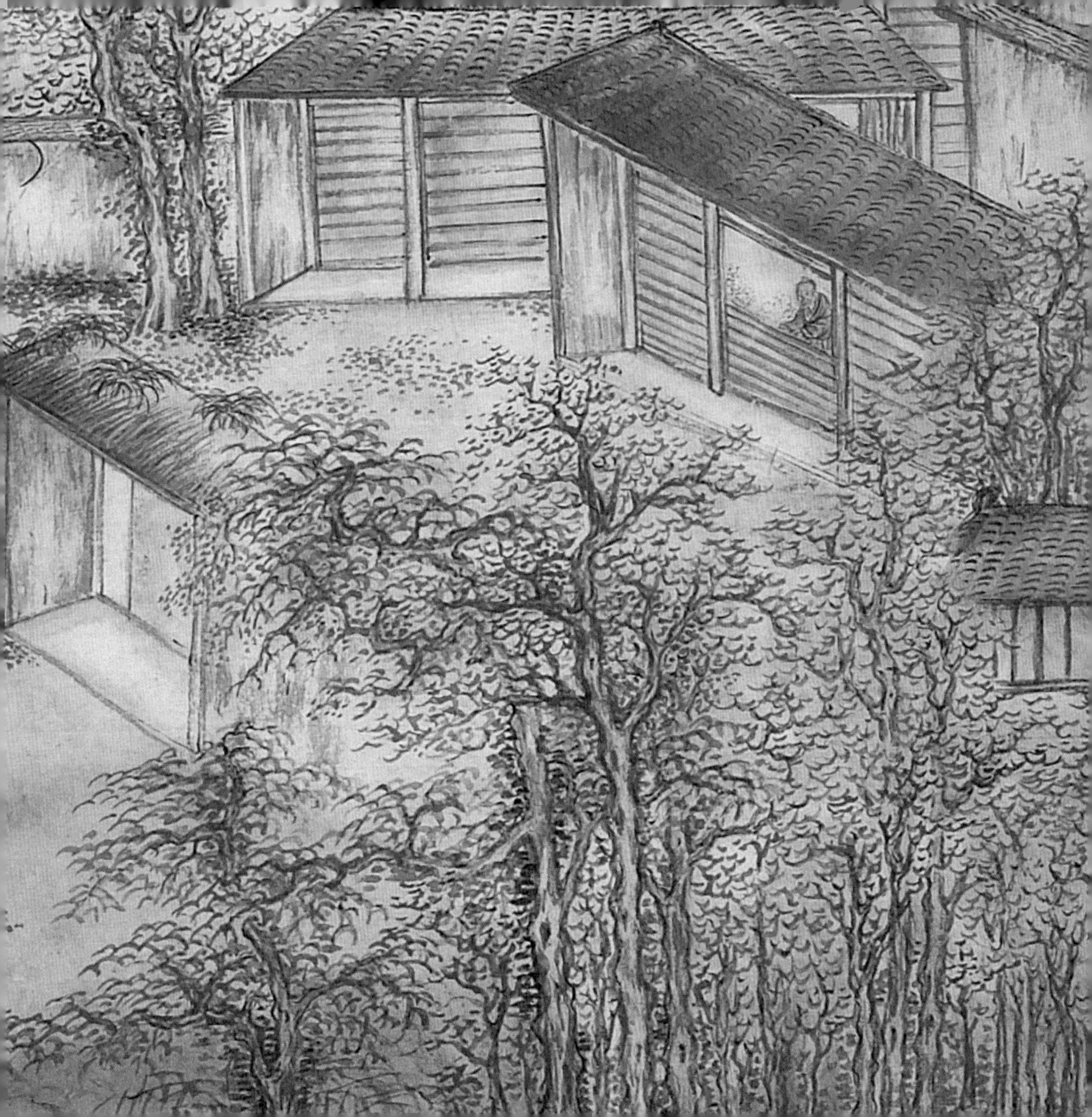

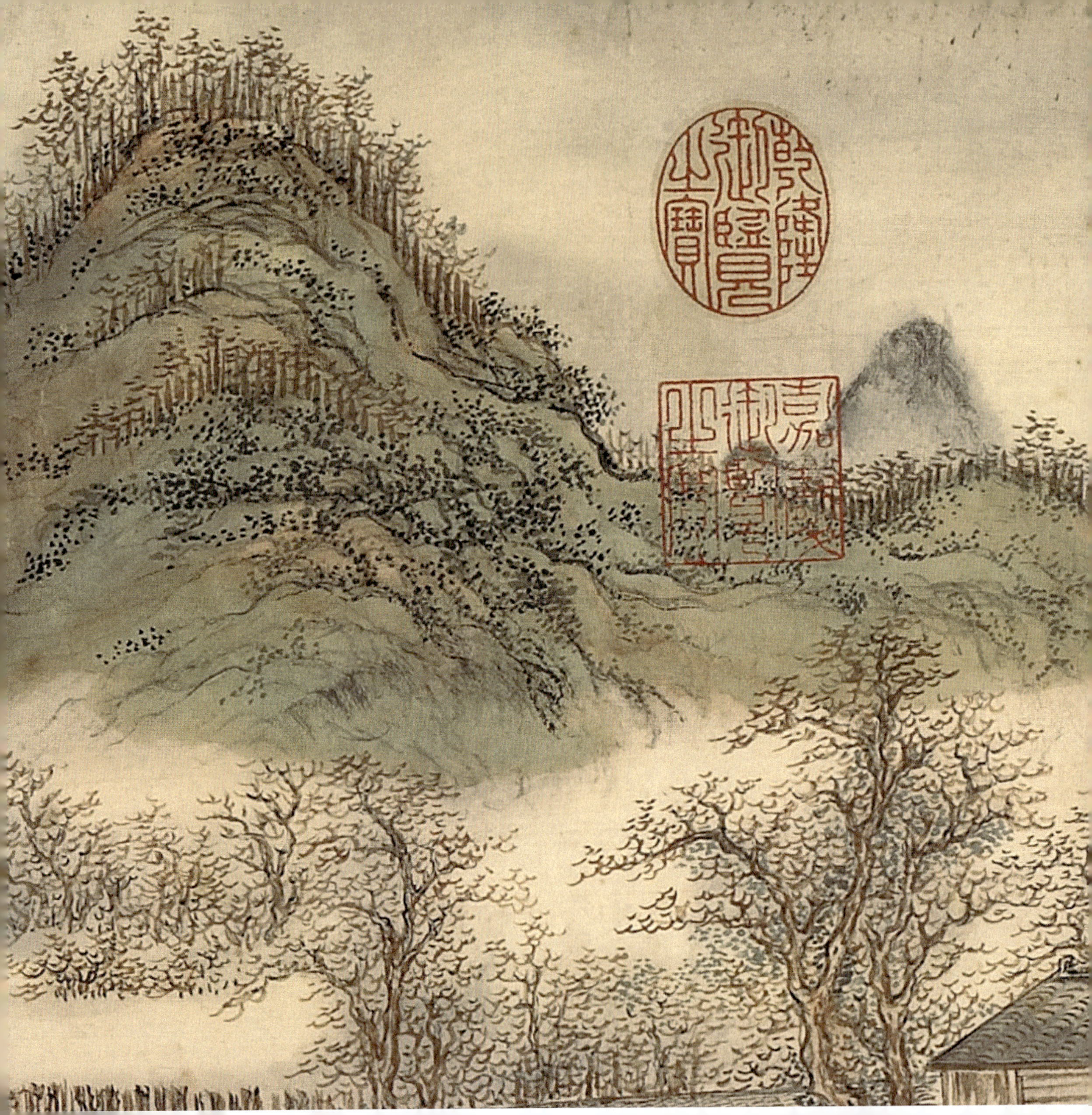
乾隆御覽之寶
嘉慶御覽之寶

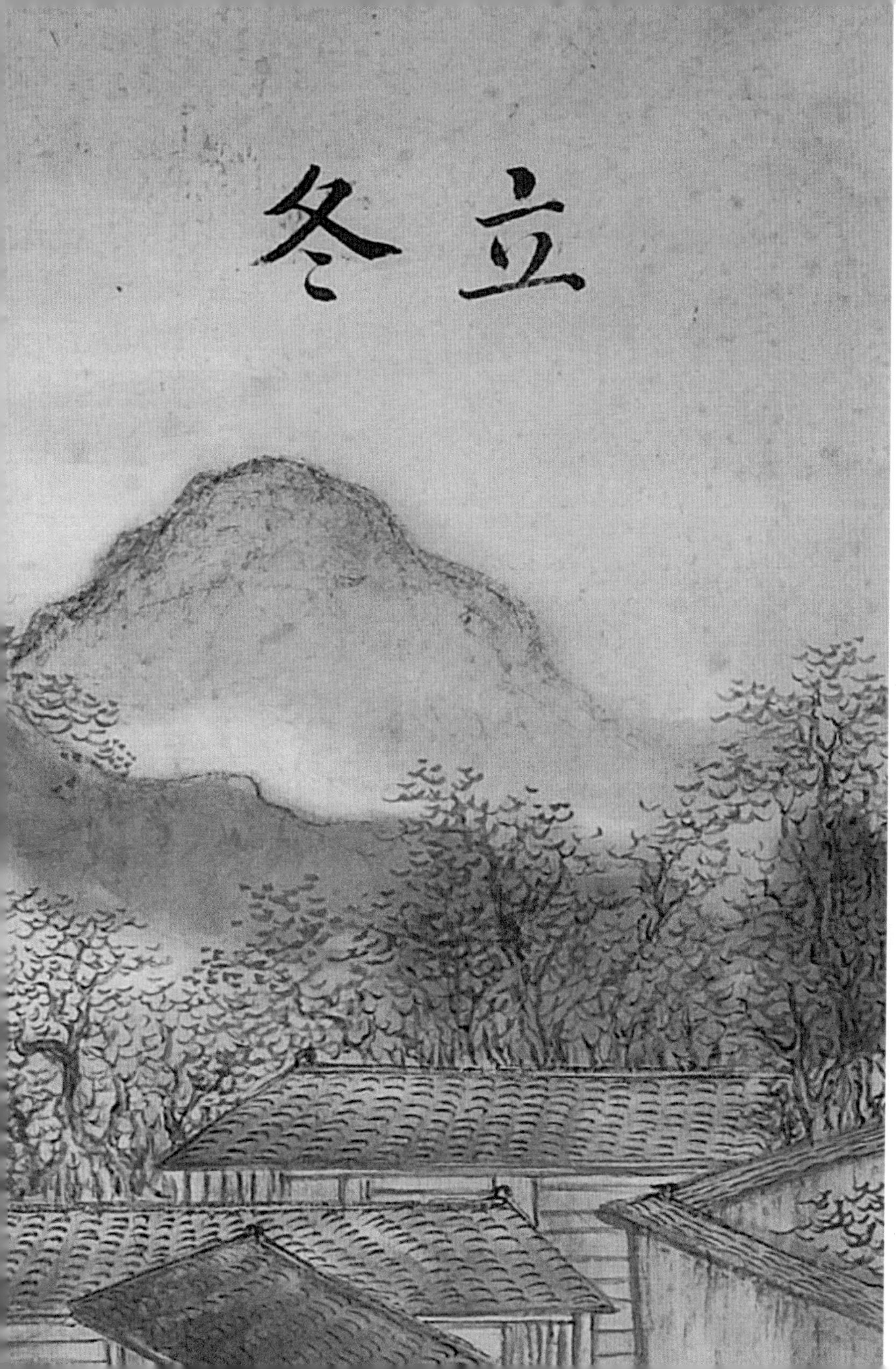


三候

# 雉入大水為蜃

## Pheasants Turn to Clams

The Third Sign of Lidong “Pheasants Turn to Clams” (雉入大水為蜃 Zhì Rù Dà Shuǐ Wèi Shèn):
This sign is a continuation of the metaphorical transformation seen in the animal kingdom, similar to the “Sparrows Enter the Water as Clams” from the second sign of Hánlù. In this lore, it is believed that pheasants, larger than sparrows, transform into large clams known as “蜃” as winter sets in. While a mythological expression, it reflects the ancient understanding of the profound changes in nature during the cold season, symbolizing deep transformation and adaptation.

Xuě

Xiǎo

小

# LESSER SNOW

November 22 (± 1)

# 小雪

## Lesser Snow

Xiǎoxuě : Representing the beginning of light snow, signaling the transition deeper into winter. The term reflects the arrival of colder weather, characterized by the first gentle snowflakes that set the stage for the coming colder months. It's a time when the landscape begins to be dusted with snow, marking a shift in the season's atmosphere and the gradual progression into a quieter, more contemplative time of year.

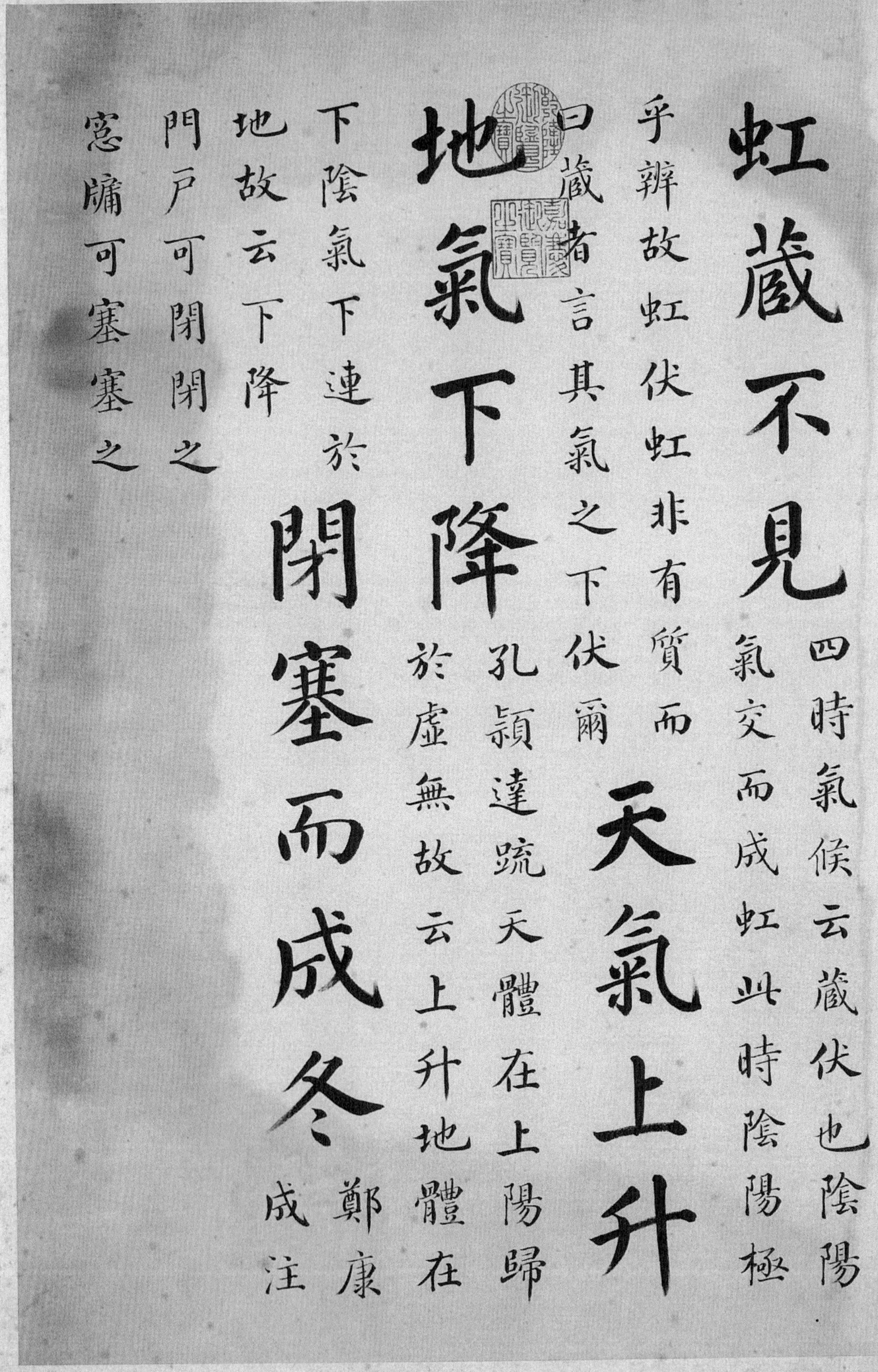

虹藏不見
四時氣候云藏伏也陰陽氣交而成虹此時陰陽極乎辨故虹伏虹非有質而曰藏者言其氣之下伏爾
天氣上升
孔穎達疏天體在上陽歸於虛無故云上升地體在下陰氣下連於地故云下降
地氣下降
閉塞而成冬
門戶可閉閉之窓牖可塞塞之
鄭康成注

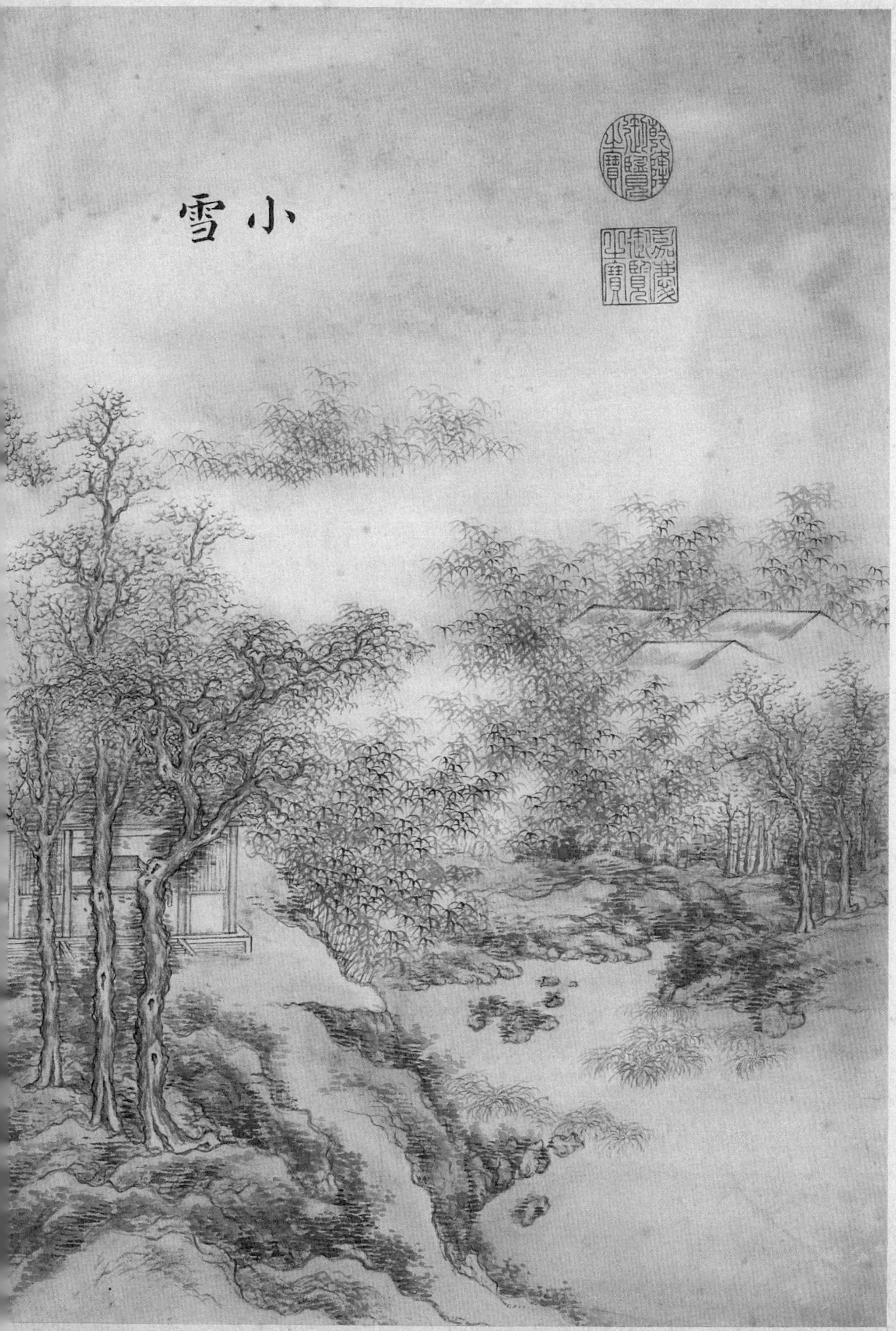


初候

## 虹藏不見

## Rainbows Disappear

二候

## 天氣上升 地氣下降

## Separation of Heavenly and Earthly Qi

三候

## 閉塞而成冬

## All Becomes Closed and Winter is Established

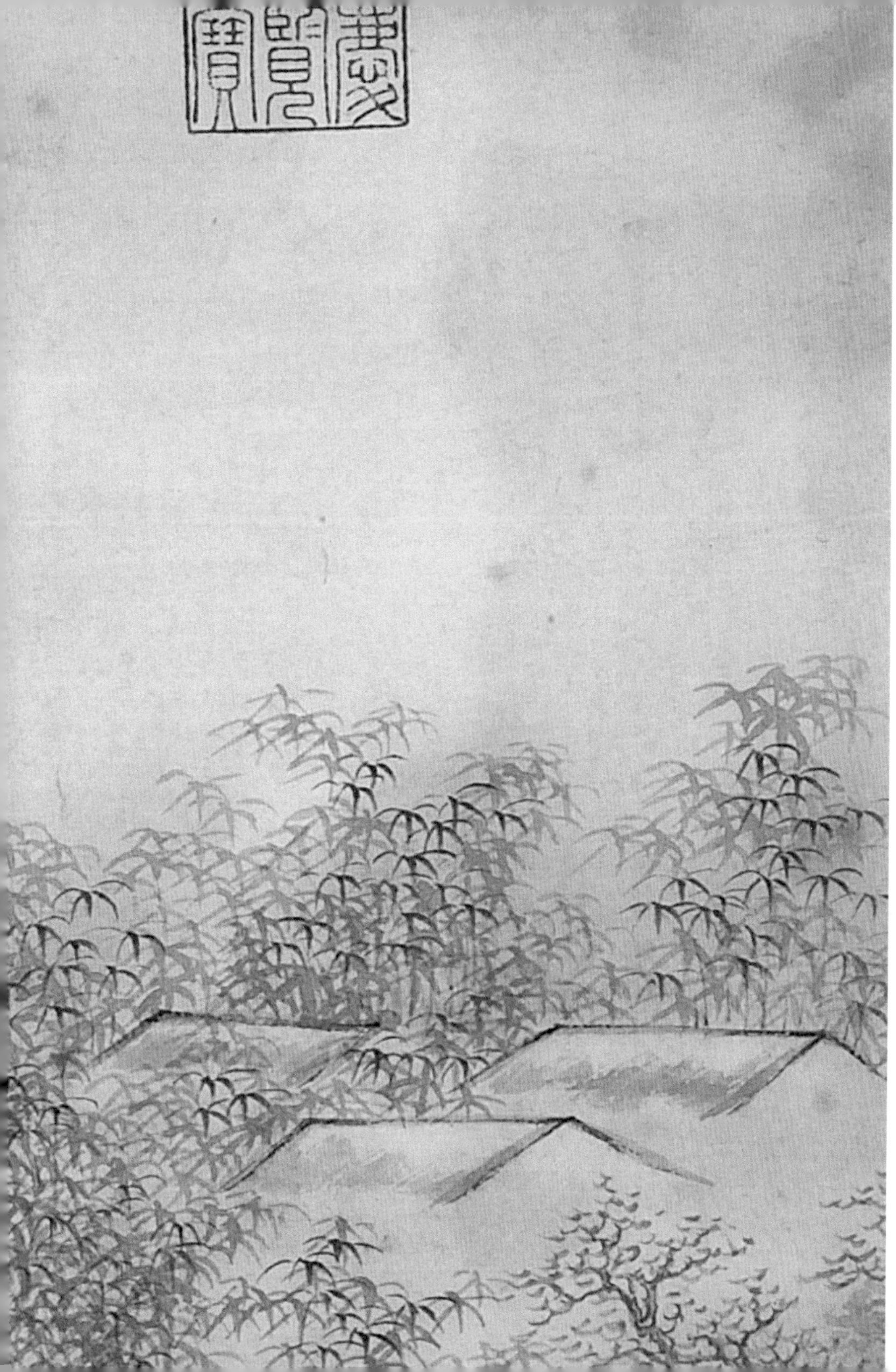

初候

# 虹藏不見

## Rainbows Disappear

The first hint of Xiǎoxuě is "Rainbows Disappear" (**虹藏不見** Hóng Cáng Bù Jiàn).
With the arrival of early winter, the vibrant rainbows that once adorned the sky become a rare sight. This sign symbolizes the dominance of yin energy as the yang retreats, leading to a less colorful and more subdued natural environment. The hidden rainbows mark a time of lessened activity and vibrancy in the world, as nature conserves its energy during the colder and darker months.

二候

# 天氣上升
# 地氣下降

## Separation of Heavenly and Earthly Qi

The second cue is Xiǎoxuě "Separation of Heavenly and Earthly Qi" (天氣上升地氣下降 Tiān Qì Shàng Shēng Dì Qì Xià Jiàng).

As winter deepens, there is a perceptible shift in the qi or vital energies of heaven and earth. The yang, or active heavenly qi, ascends, while the yin, or passive earthly qi, descends. This sign reflects the increasing separation and specialization of energies, leading to a more pronounced stillness and quiet in the natural world. It's a time when the vibrant activity of life gives way to a period of rest and conservation, preparing all beings for the regeneration that comes with the eventual return of spring.

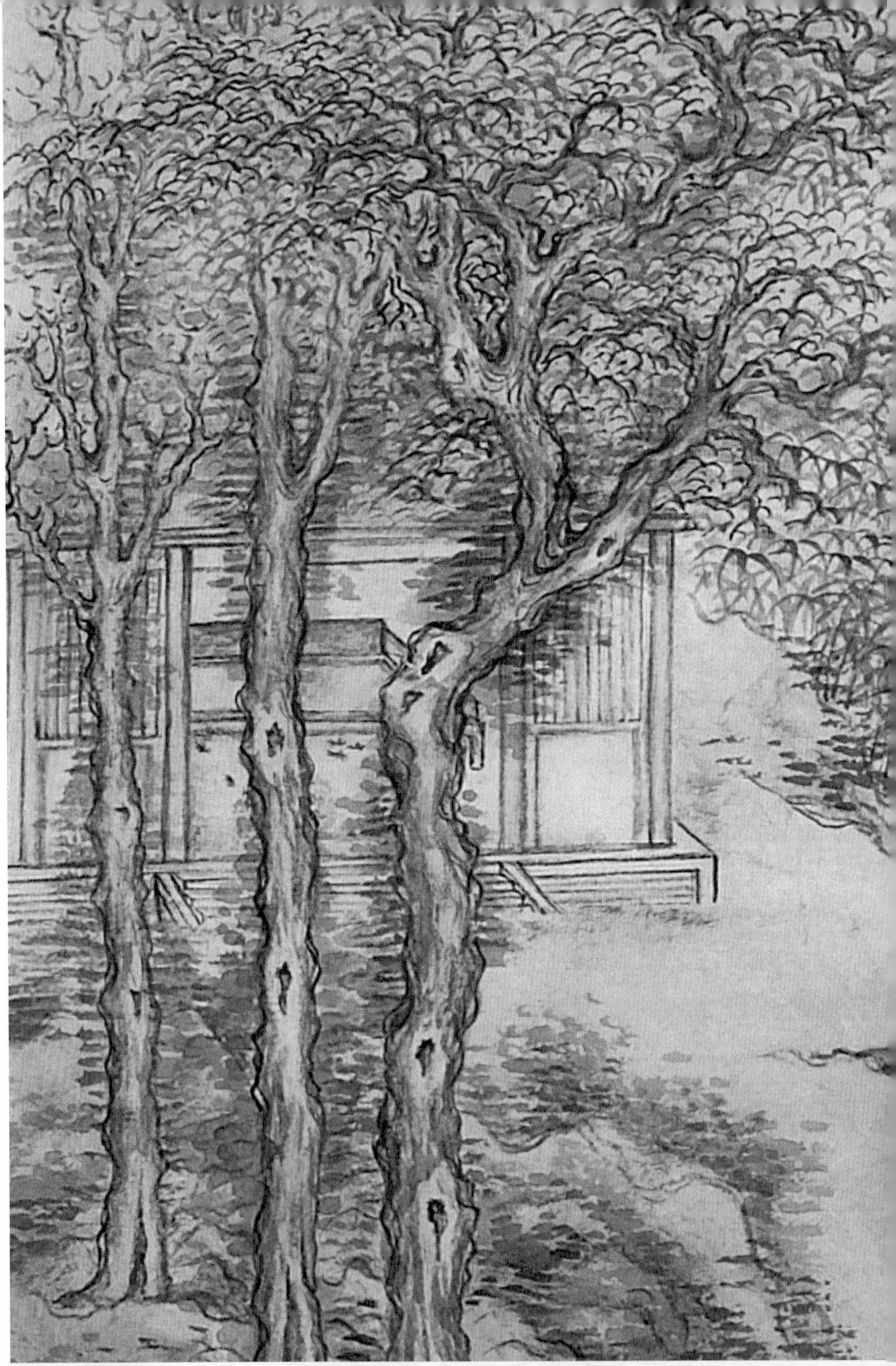

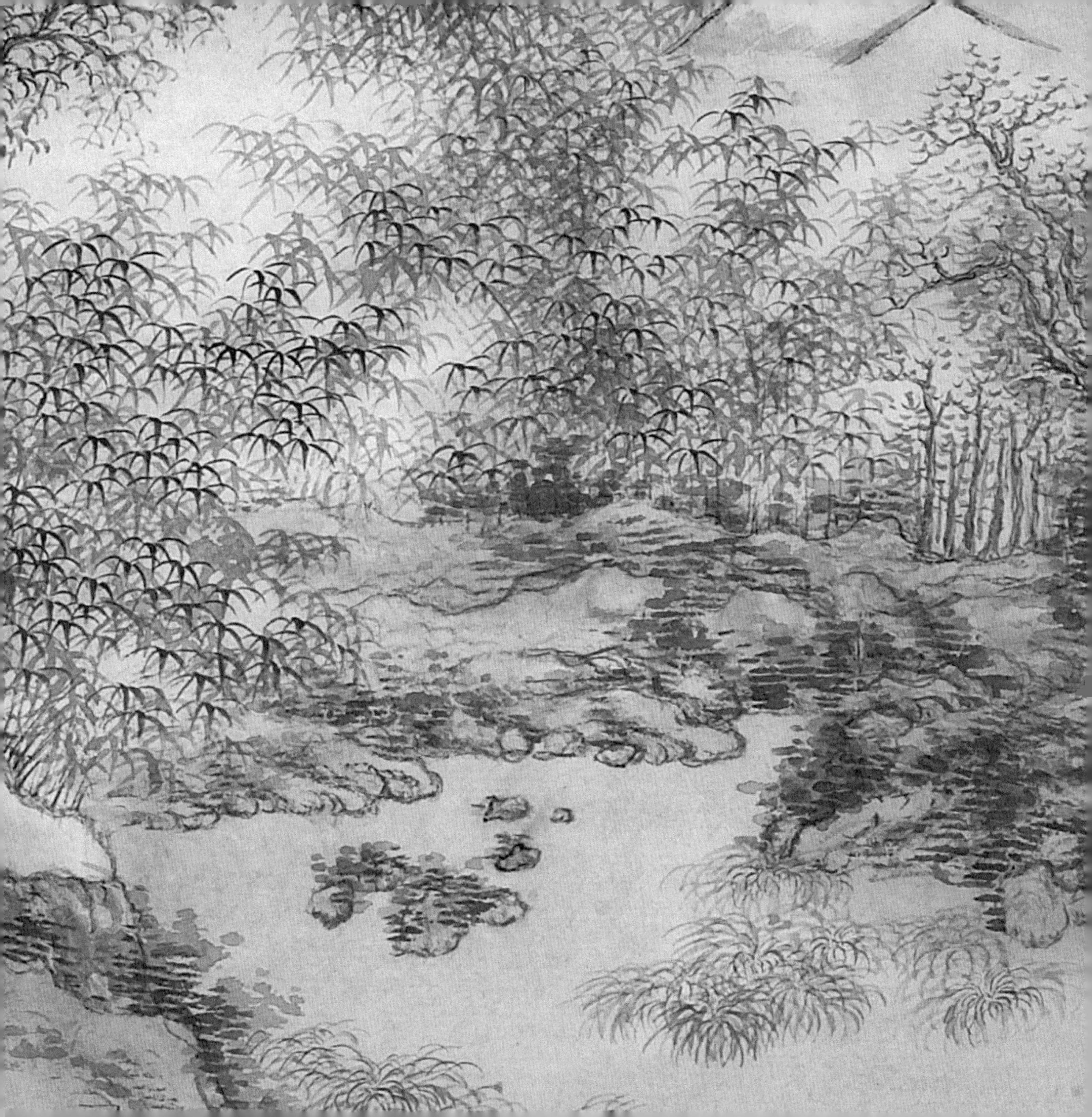

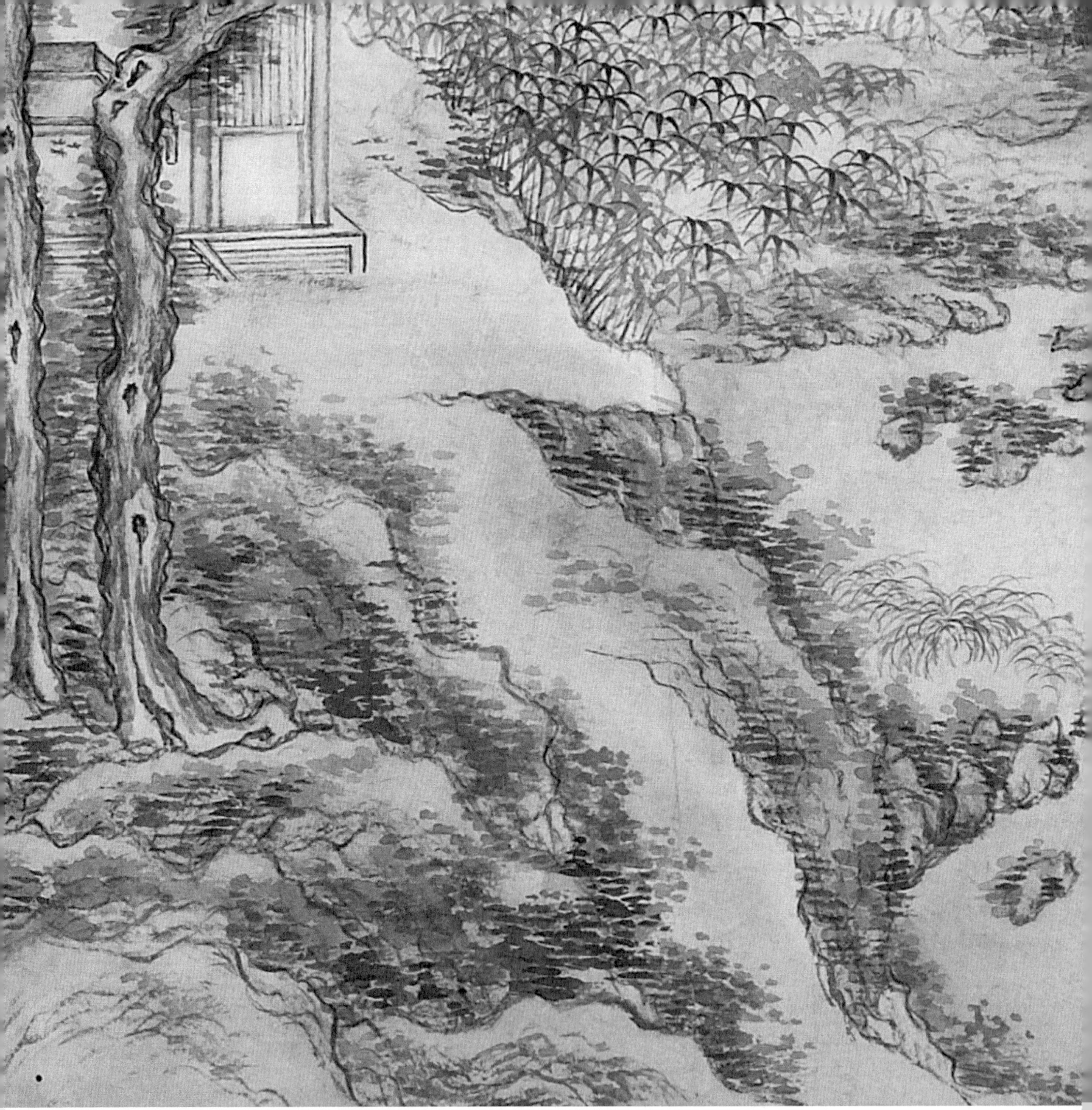

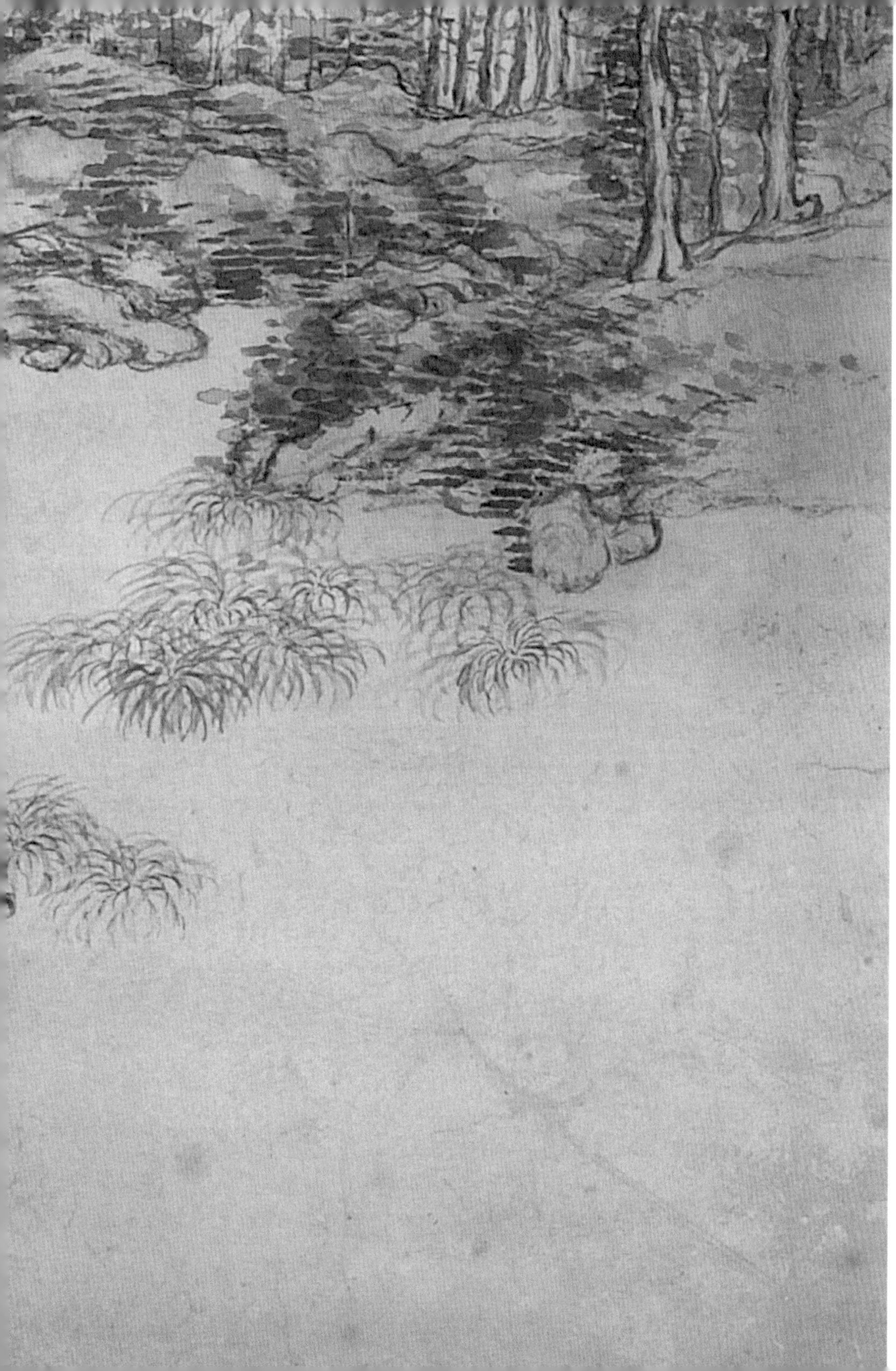

三候

# 閉塞而成冬

## All Becomes Closed and Winter is Established

The third indication is “All Becomes Closed and Winter is Established” (閉塞而成冬 Bì Sè Ér Chéng Dōng), marking the full arrival of winter’s stillness and dormancy.

This sign signifies the moment when the natural world fully succumbs to winter’s embrace. The closure of the earth and sky’s qi leads to a landscape devoid of vitality. It is a time when life retreats into the deepest state of hibernation and conservation, marking the full establishment of winter with its characteristic stillness and dormancy. This period reflects the completion of the transition into the cold and quiet season, where all activities are minimized, and the world awaits the eventual return of warmth.

Xuě

Dà

大

# GREATER SNOW

December 7 (± 1)

# 大雪

## Greater Snow

Dàxuě : Denoting a period of significant snowfall, embodying the full embrace of winter. During this time, continental high pressures prevail, and cold winds bring substantial snow, transforming the landscape into a white expanse. It' s a time characterized by the beauty and challenges of deeper winter, with activities and life rhythms adjusting to the colder and snowier conditions.

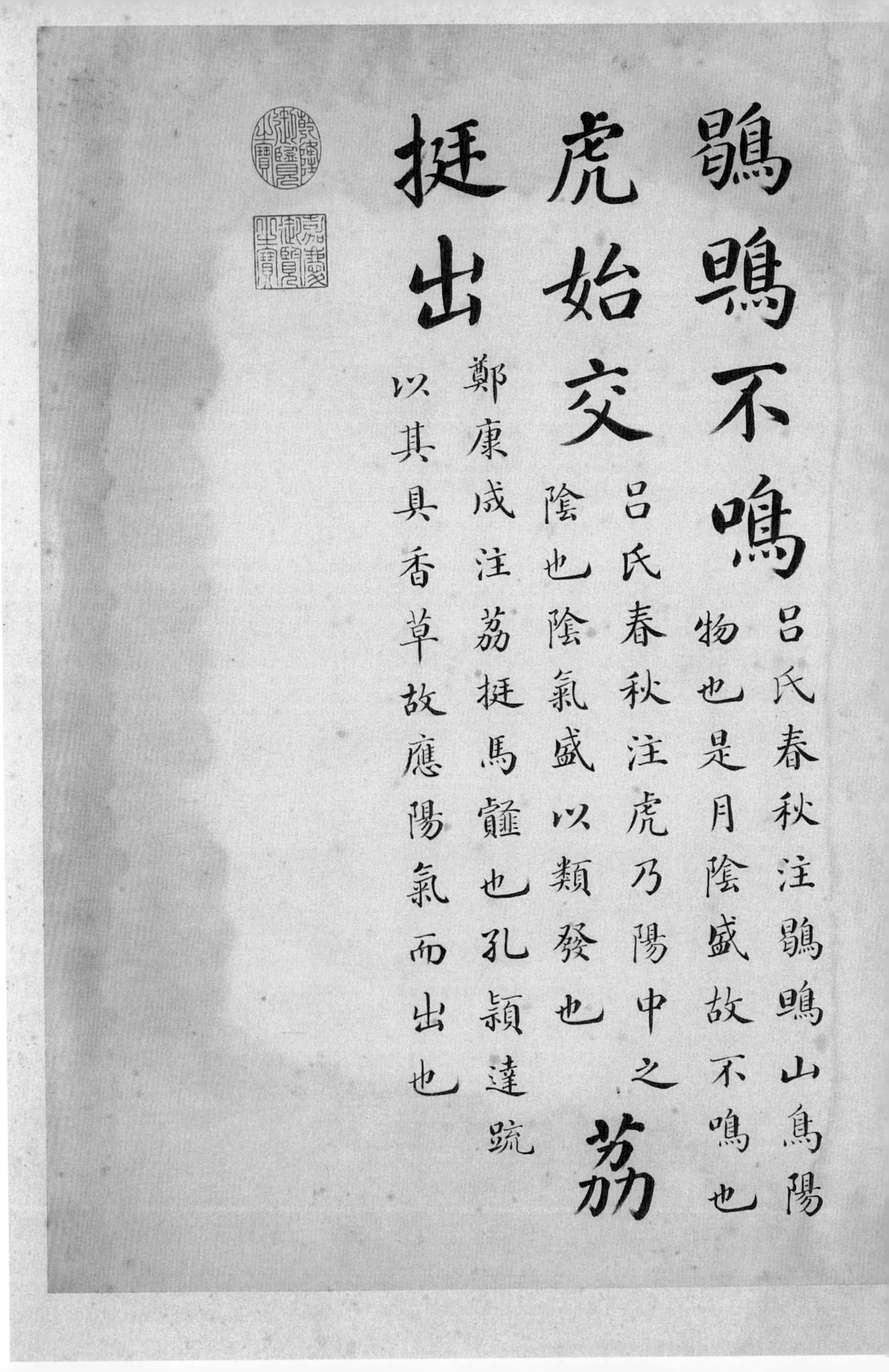

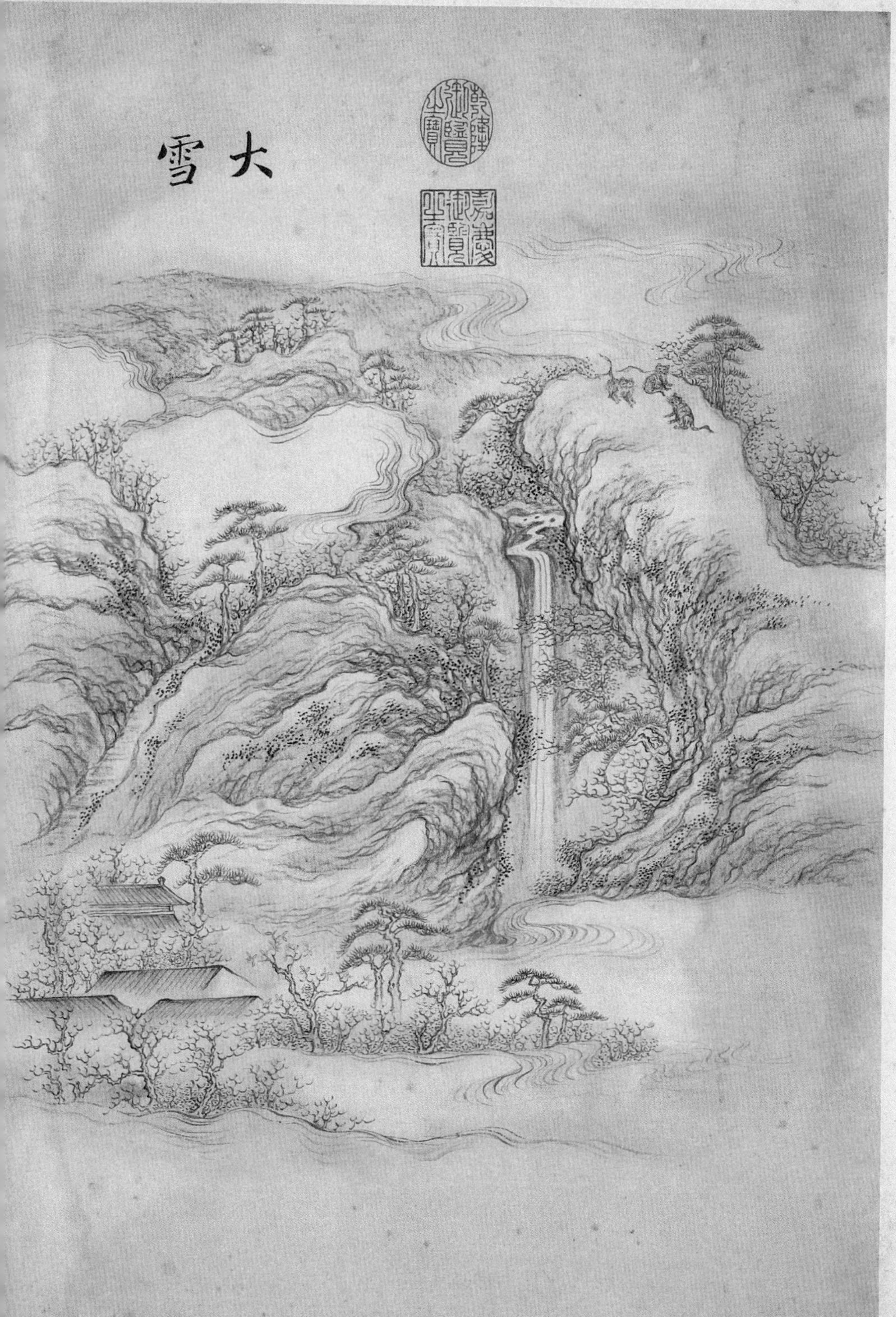


初候

## 鶡鴠不鳴

## Noisy Birds Fall Silent

二候

## 虎始交

## Tigers Begin Mating

三候

## 荔挺出

## Reeds Emerge Amid Snow

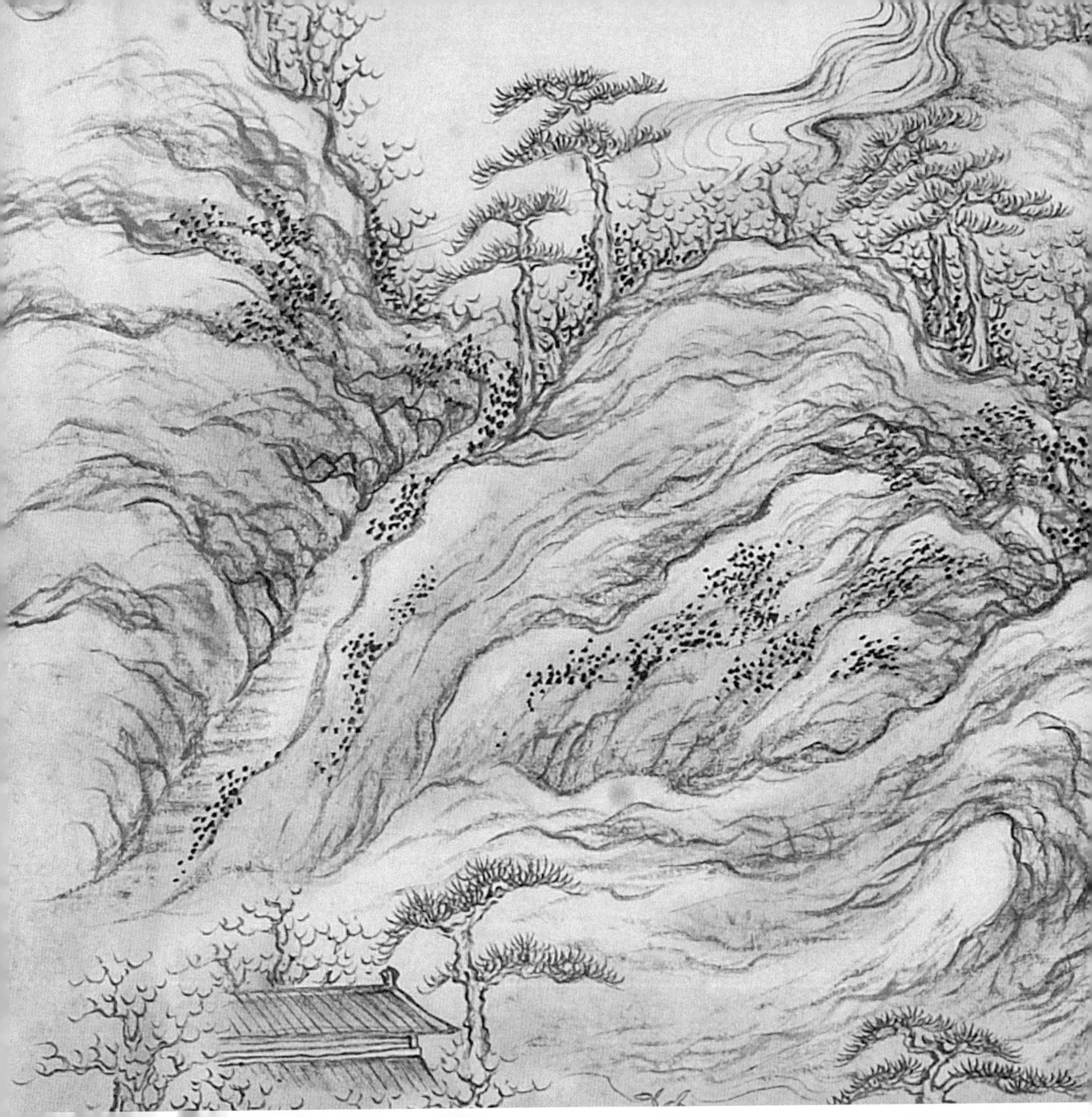

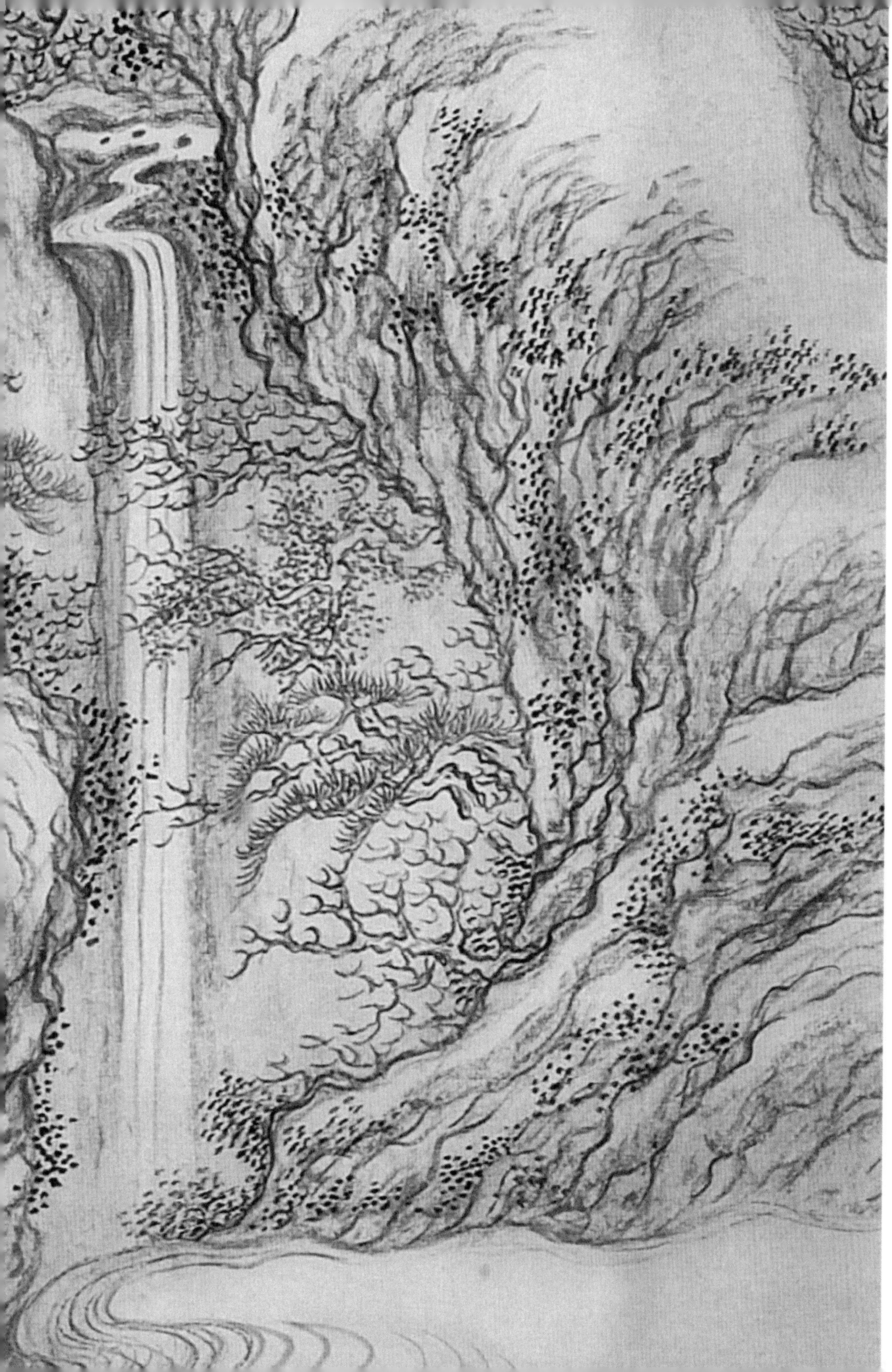

初候

# 鶡鴠不鳴

## Noisy Birds Fall Silent

The first whisper of Dàxuě is "Noisy Birds Fall Silent" (鶡鴠不鳴 Hé Dàn Bù Míng):
As the big snow blankets the earth, even the typically vocal and resilient birds known as "鶡鴠" or harsh birds cease their calls. The cold and frosty environment of Dàxuě mutes the natural soundscape, reflecting the profound impact of the season' s chill on wildlife. This sign symbolizes the deepening of winter and the pervasive silence that accompanies the coldest time of the year.

二候

# 虎始交
## Tigers Begin Mating

The Second Sign of Dàxuě "Tigers Begin Mating" (虎始交 Hǔ Shǐ Jiāo): In the deep cold of Dàxuě, tigers commence their mating rituals. This behavior is a powerful testament to the persistence of life even in the harshest conditions. The act of mating during this time reflects the cycle of life continuing unabated, with the raw vitality and strength of tigers symbolizing the spark of yang energy waiting to re-emerge after the cold has passed.

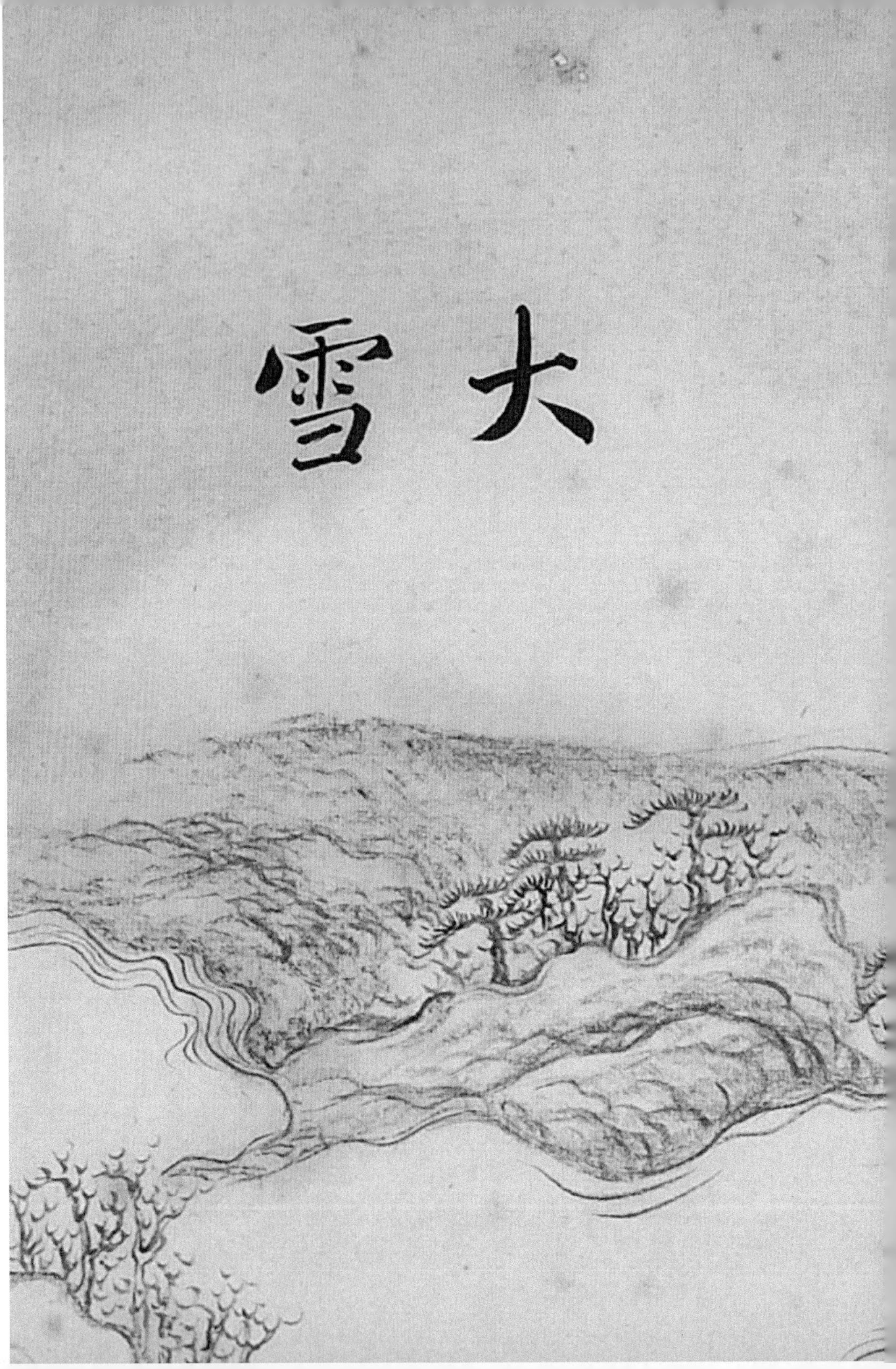

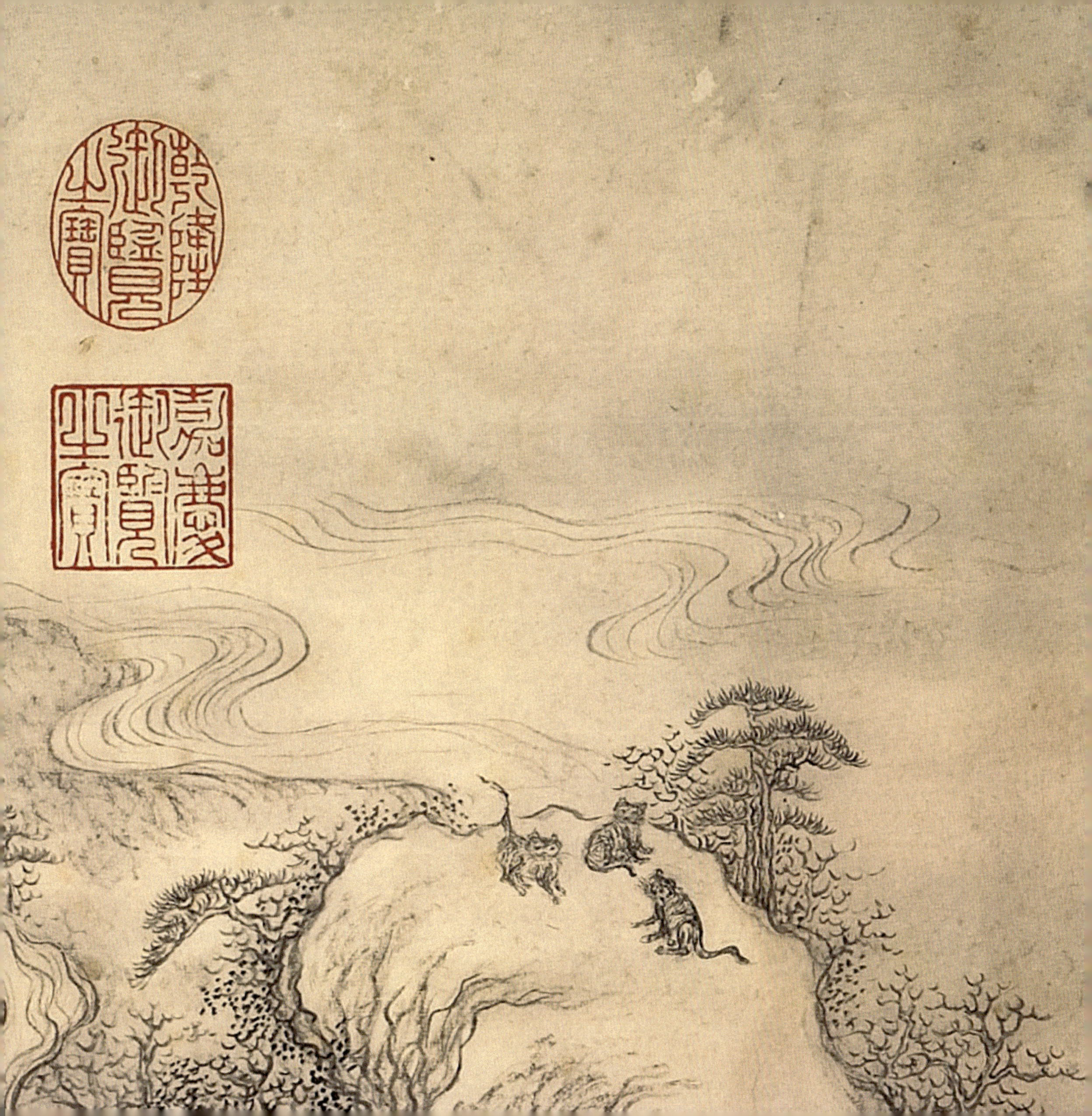
乾隆御覽之寶
嘉慶御覽之寶

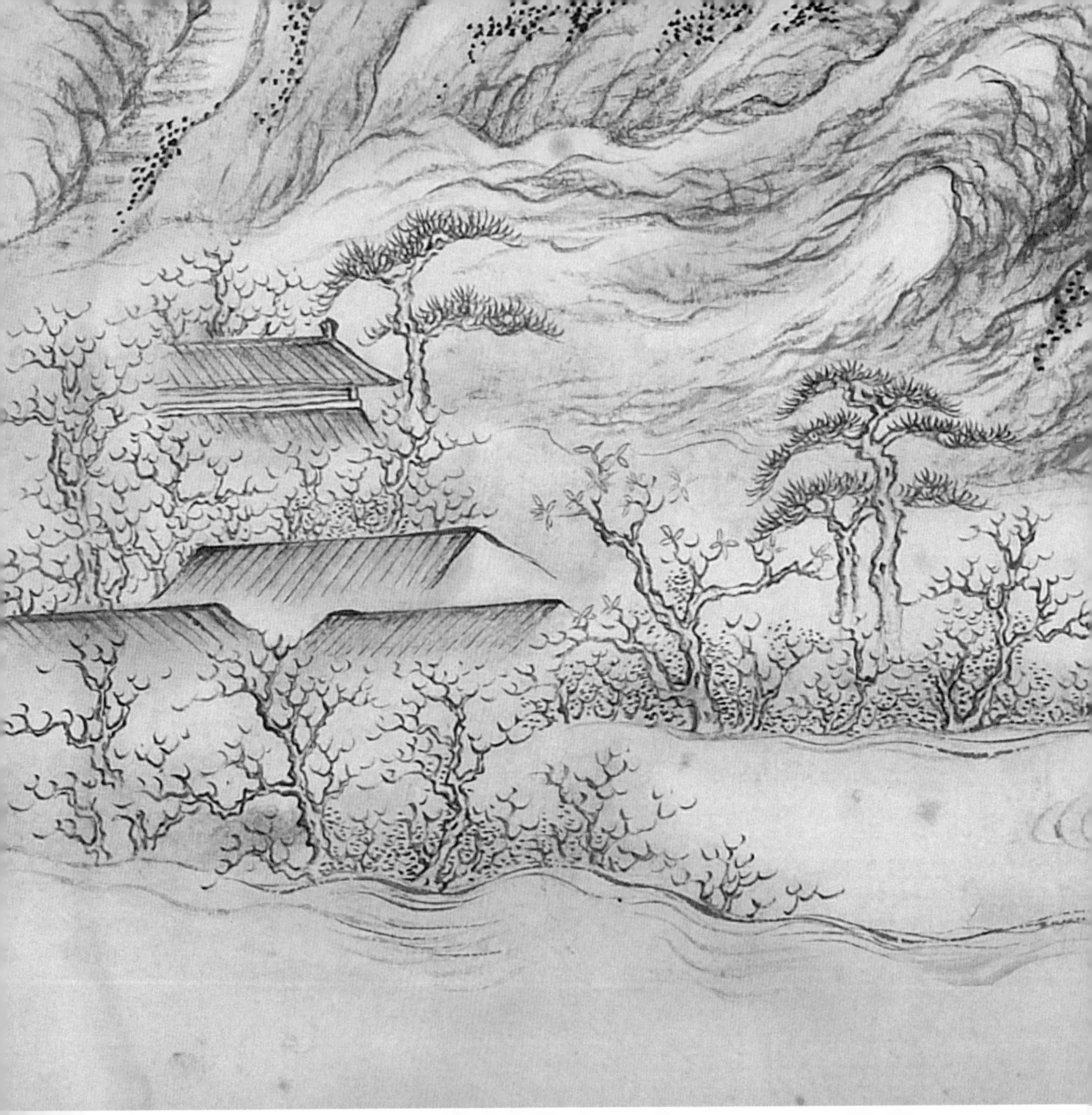

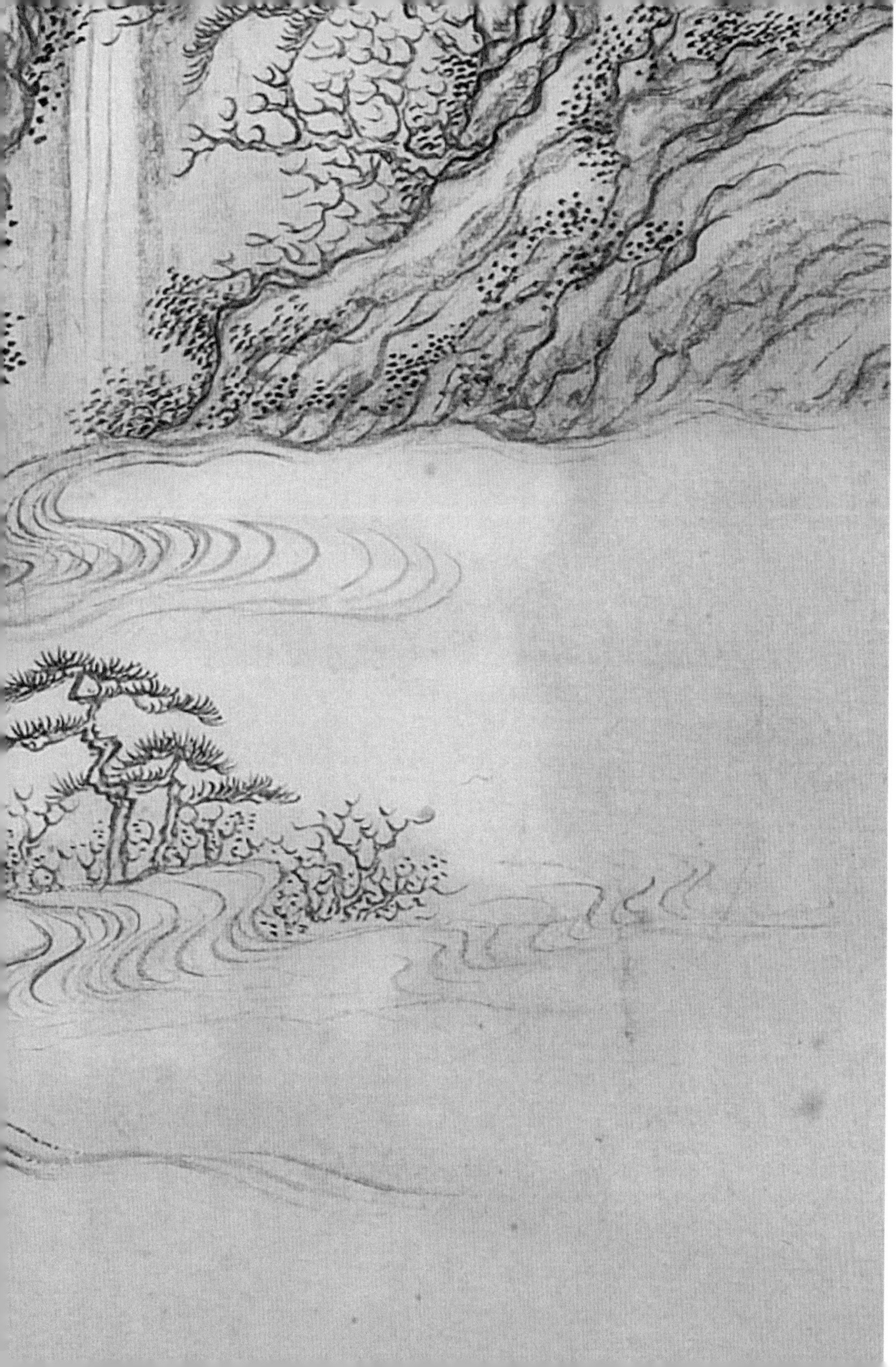

三候

# 荔挺出

## Reeds Emerge Amid Snow

The third hint of Dàxuě is "Reeds Emerge Amid Snow" (荔挺出 Lì Tǐng Chū).
Amid the snow-covered landscape, certain reeds known as "荔" manage to pierce through and stand tall against the blanket of white. This sign is a testament to the resilience and endurance of life, with these reeds symbolizing hope and the persistence of nature's will to survive and thrive despite the freezing conditions.

# 至

Zhì

Dōng

冬

# THE WINTER SOLSTICE

December 22 (± 1)

# 冬至

## The Winter Solstice

Dōngzhì : The Winter Solstice, is observed as the shortest day of the year, symbolizing the depth of winter and a pivotal point in the lunar calendar. It marks a period of reflection and celebration, acknowledging the longest night and the gradual return of light thereafter. Traditionally seen as one of the most significant moments of the year, Dōngzhì represents a turning point, leading to the gradual resurgence of warmth and light afterward.

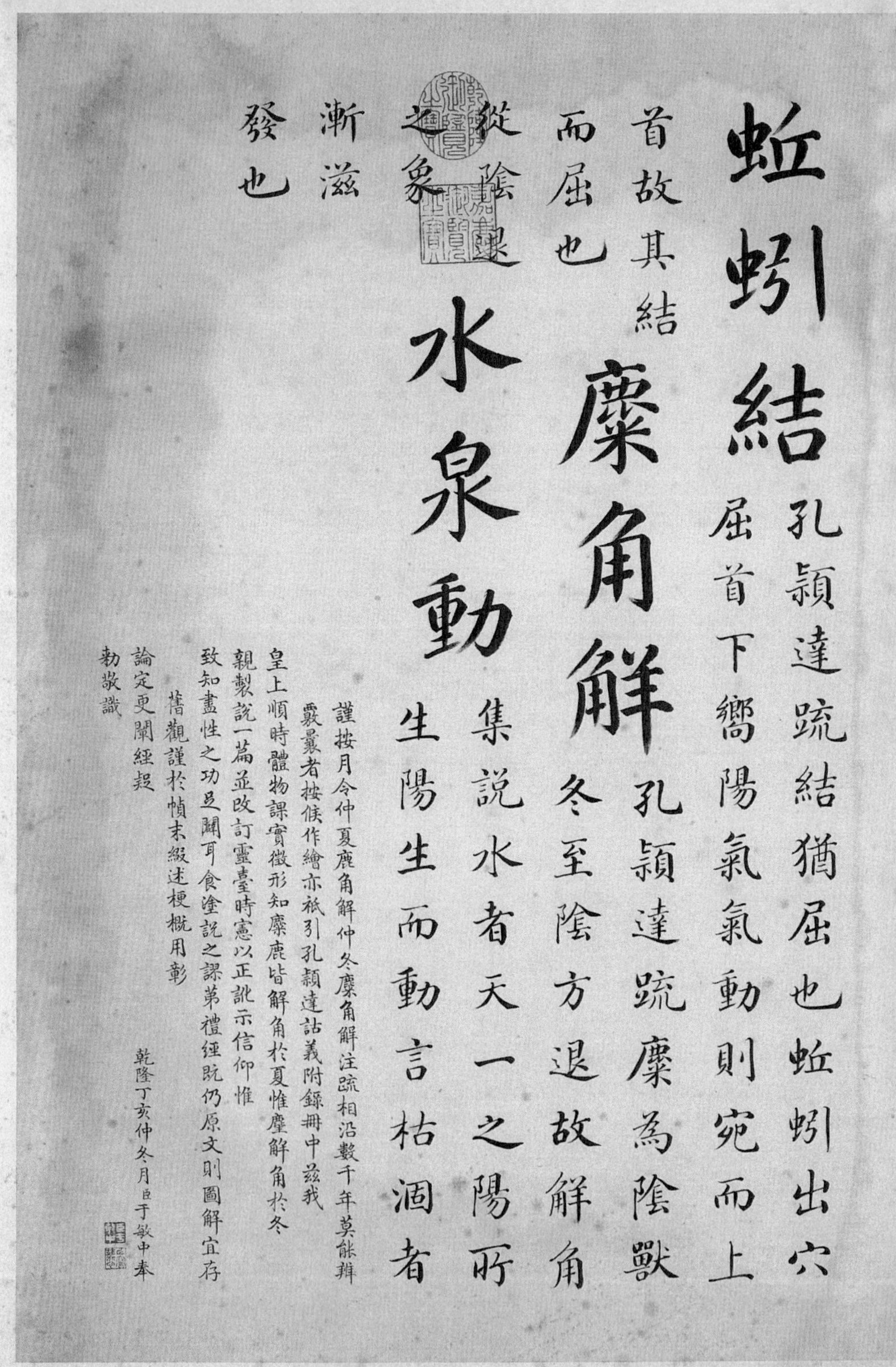

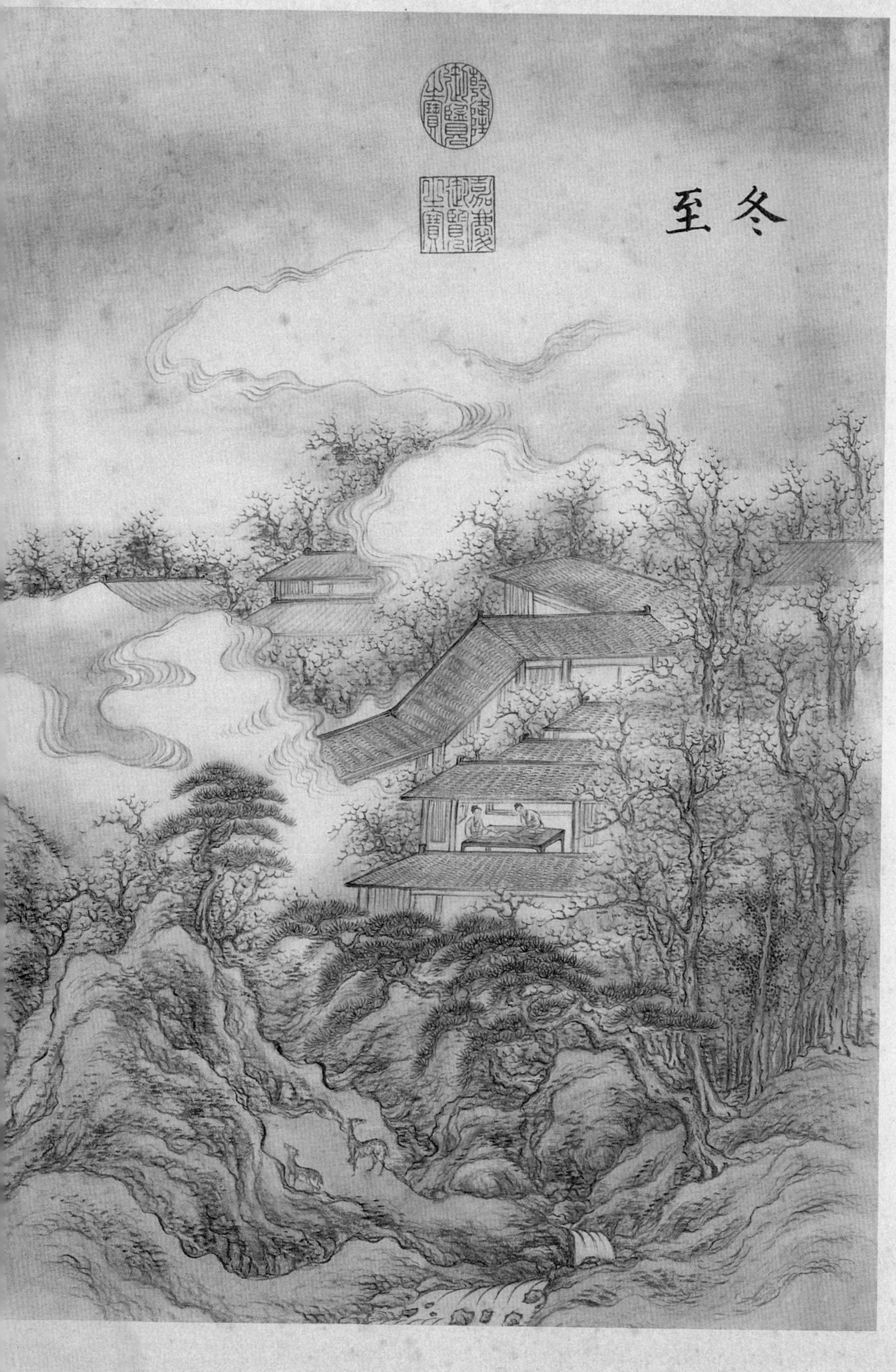


初候

**蚯蚓結**

Earthworms Ball Up

二候

**麋角解**

Elk Antlers Shed

三候

**水泉動**

Underground Springs Stir

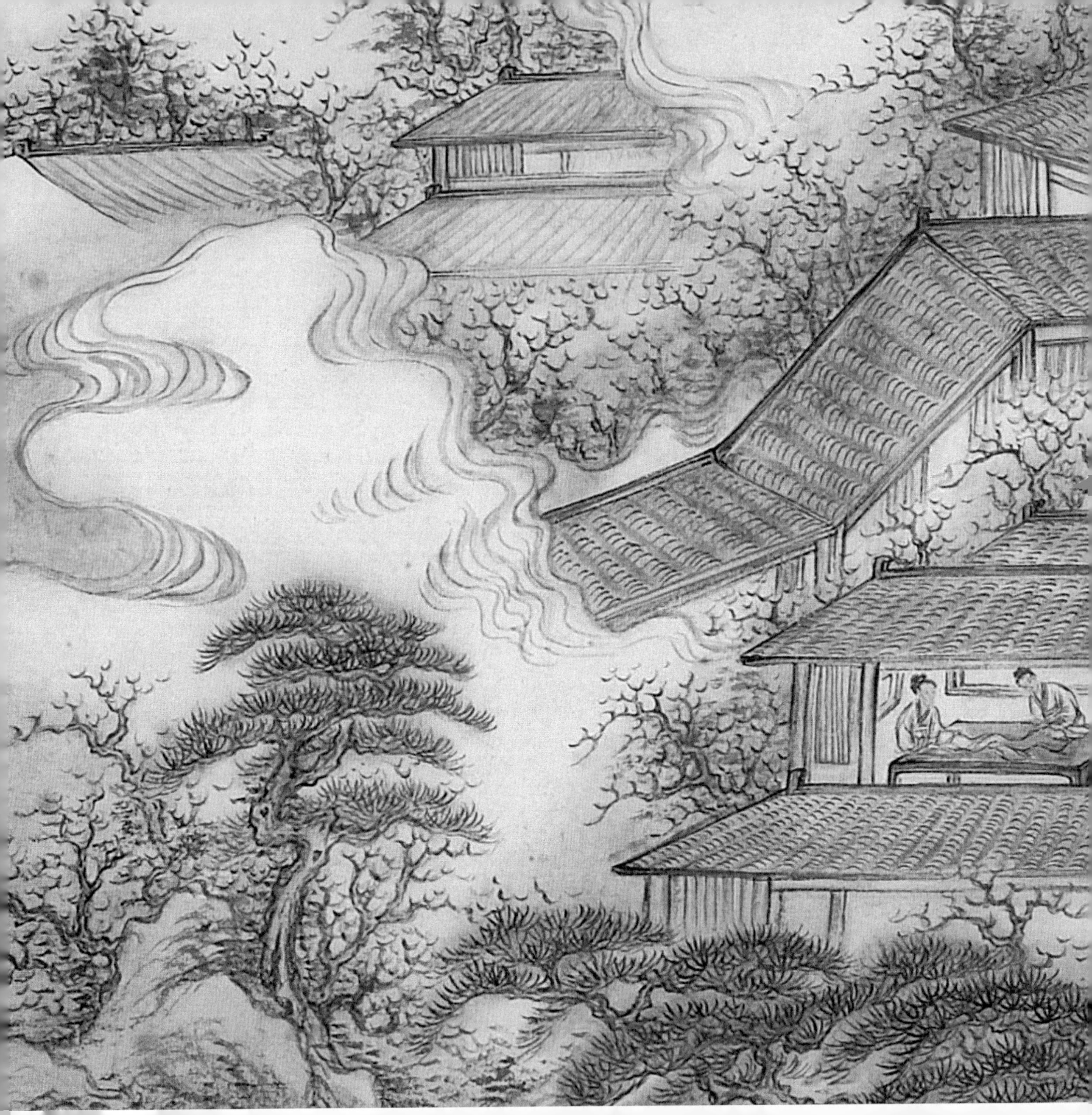

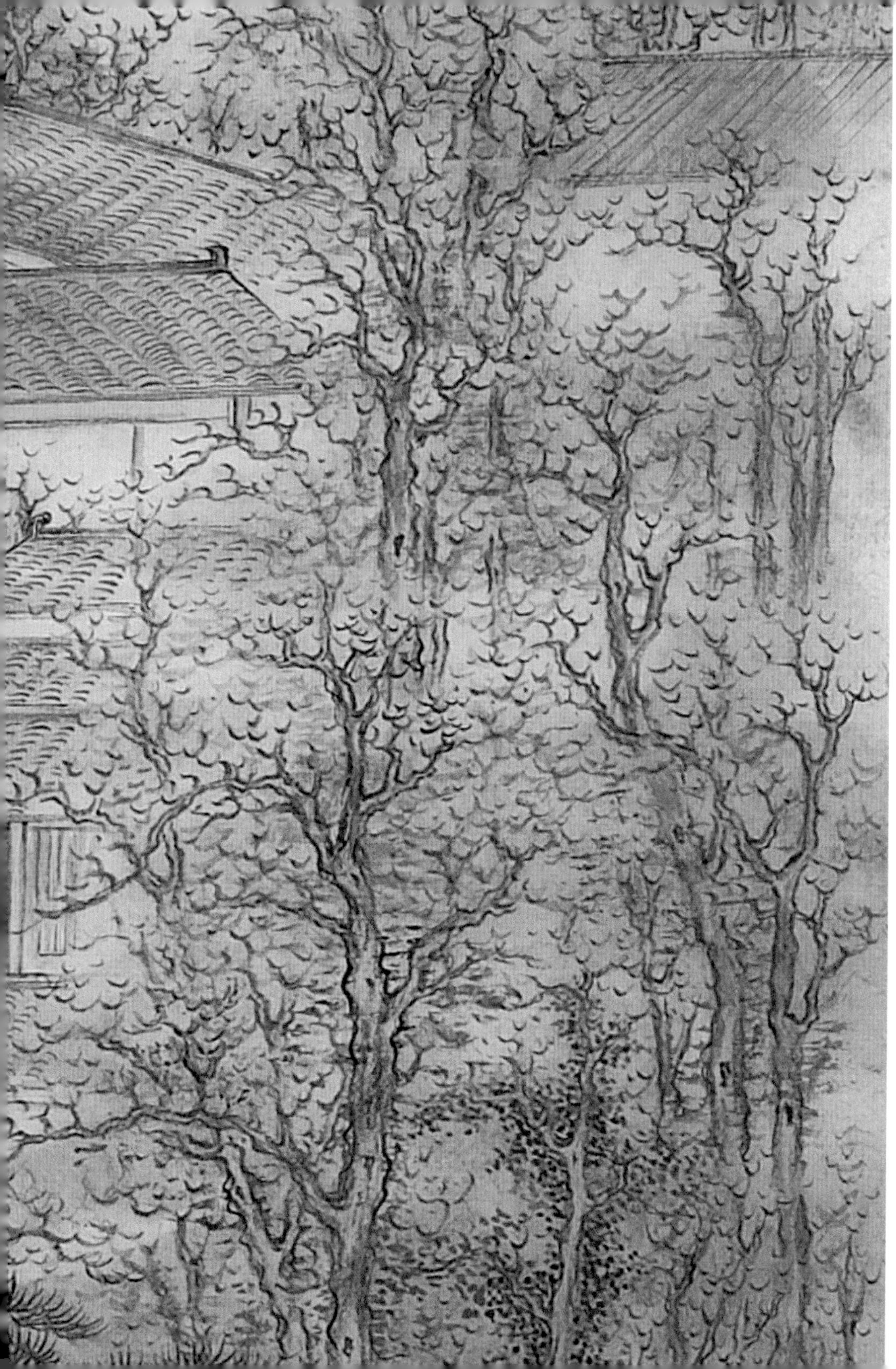

初候

# 蚯蚓結

## Earthworms Ball Up

The initial marker of Dōngzhì is "Earthworms Ball Up" (蚯蚓結 Qiū Yǐn Jié).

During Dōngzhì, earthworms engage in a survival strategy by balling up into knots underground. This behavior allows them to conserve heat and moisture during the coldest part of the year. It symbolizes the depth of winter, a time when even the creatures beneath the earth must adapt to survive. This period marks the peak of the cold season, with all of nature bracing for the extreme chill and awaiting the eventual thaw.

二候

# 麋角解

## Elk Antlers Shed

The second hint is "Elk Antlers Shed" (麋角解 Mí Jiǎo Jiě) refers to the shedding of antlers by the Mí, a creature similar to but distinct from deer, often associated with water and yin energy, as opposed to the yang energy of mountain deer. This event typically occurs as the Mí sense a retreat in the pervasive yin energy of winter, marking a significant moment in their annual cycle. It's a natural adaptation to the changing energies of the environment, reflecting the deep connection between living creatures and the cyclic flow of Yin and Yang.

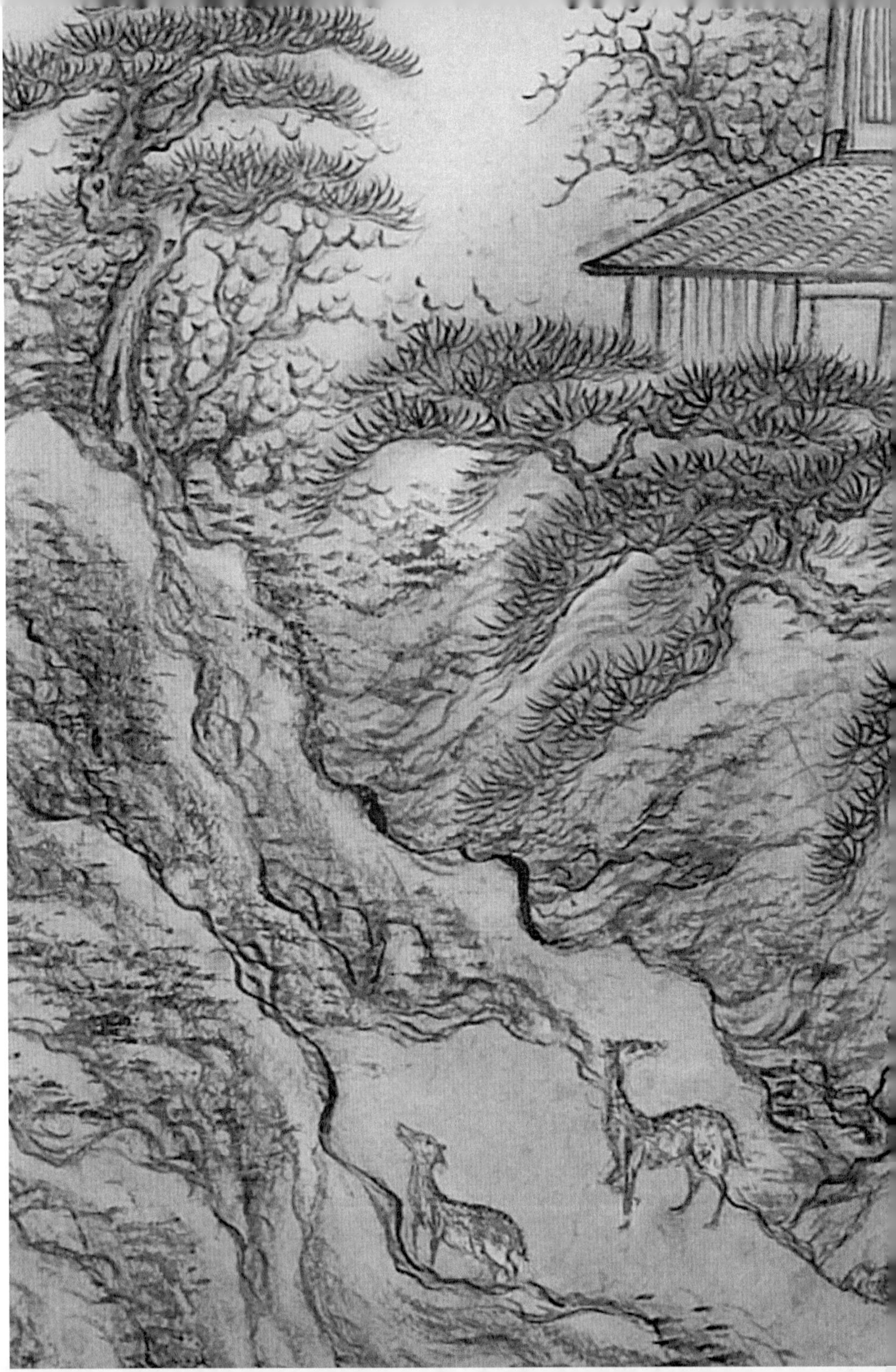

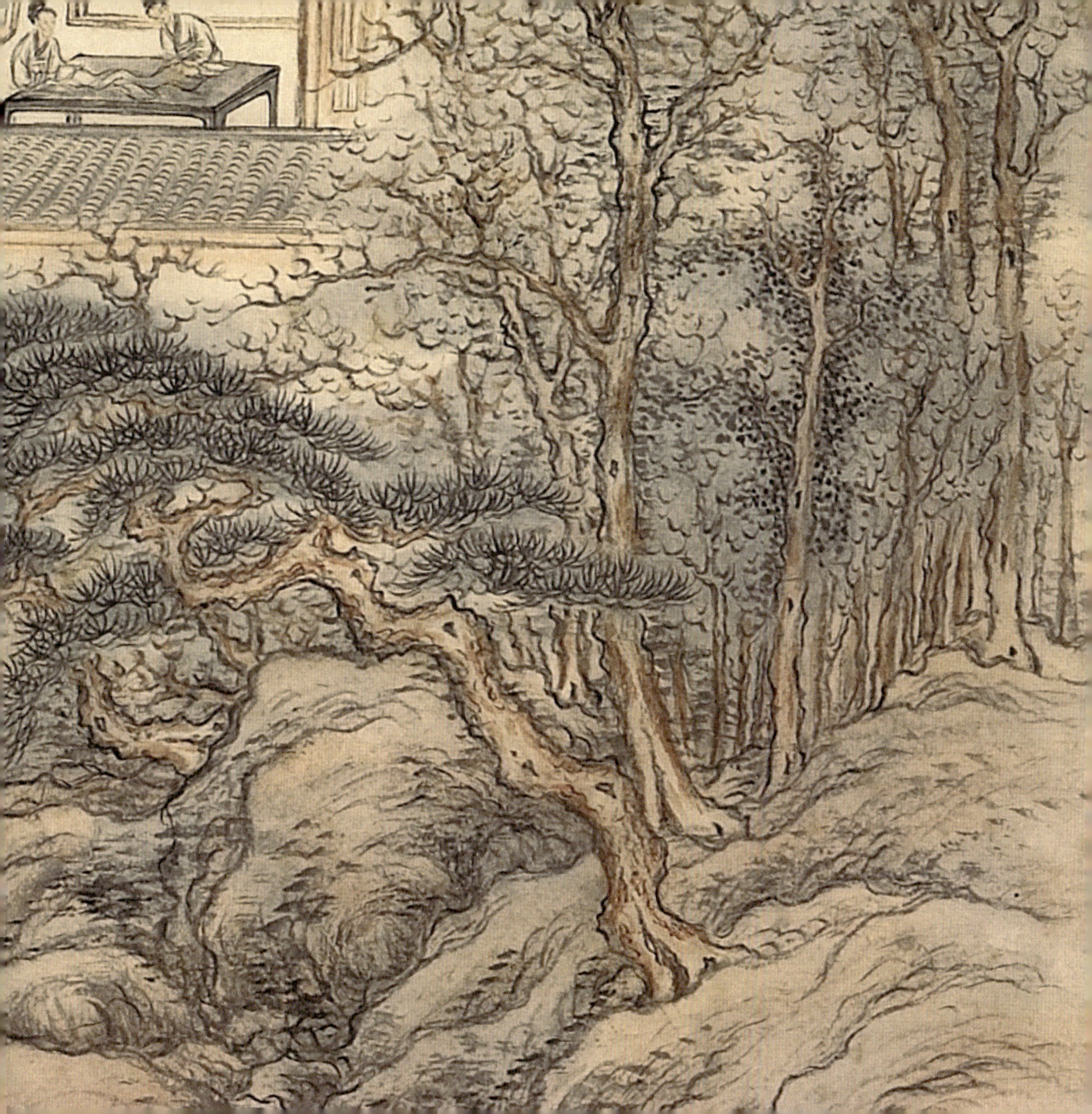

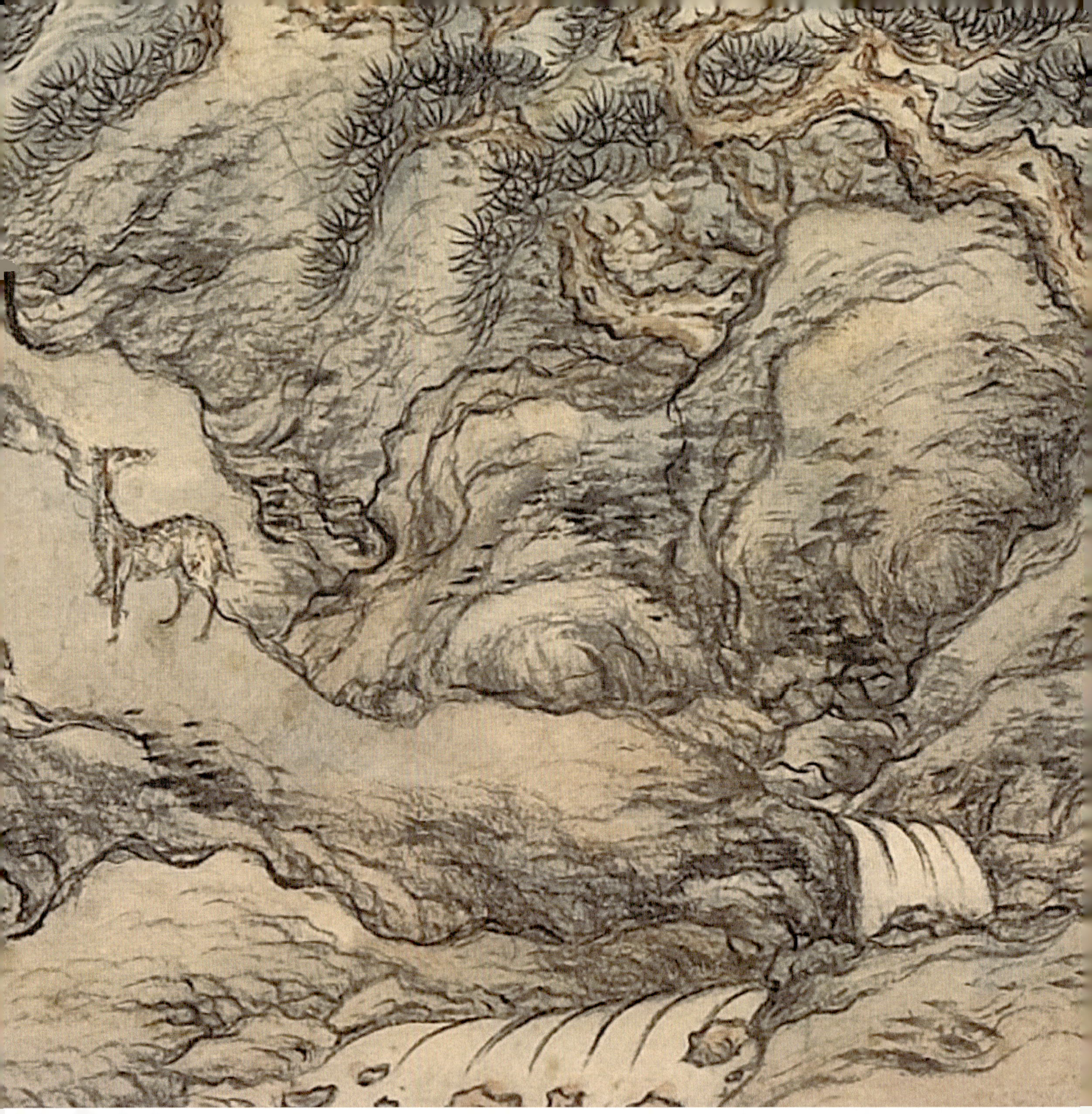

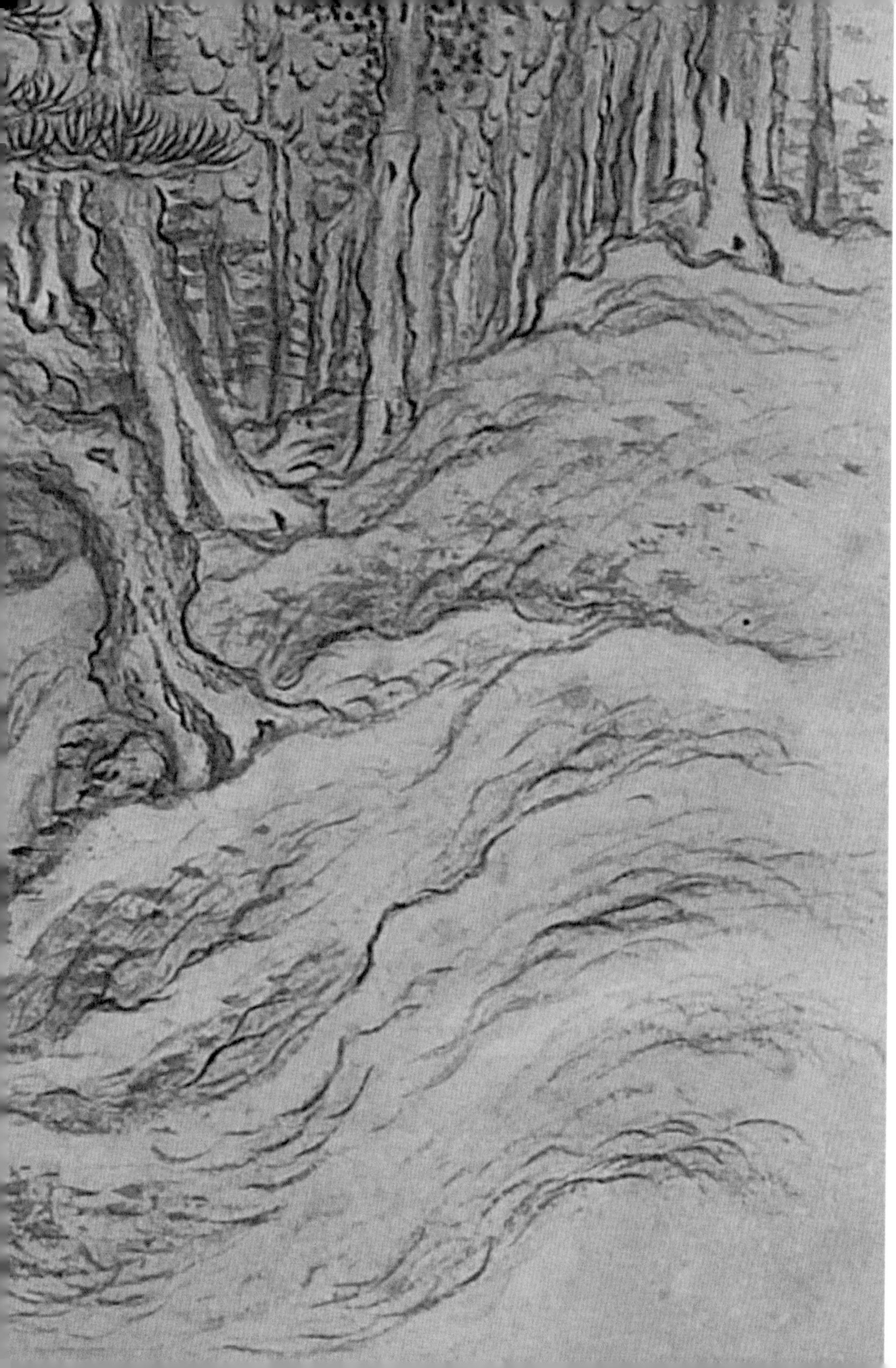

三候

# 水泉動

## Underground Springs Stir

The third sign of Dōngzhì is "Underground Springs Stir" (水泉動 Shuǐ Quán Dòng). As the deepest part of winter is marked by Dōngzhì, the underground springs, protected from the freezing surface, continue to flow. This movement is attributed to the Yang energy beneath the earth, ensuring that life persists even in the coldest moments. The stirring of the springs is a sign of the enduring vitality of nature and the promise of renewal and rejuvenation, as the hidden Yang energy maintains a subtle but essential flow of life.

寒

Hán

Xiǎo

小

# LESSER COLD

January 6 (± 1)

# 小寒

## Lesser Cold

Xiǎohán : “Lesser Cold,” corresponds to the summer’ s “Xiǎoshǔ (Lesser Heat)”, which means the initial intensification of cold, leading up to “Greater Cold.” It indicates a time when cold intensifies gradually, with the term itself capturing the increasing chill. This period marks the beginning of the coldest phase, where the accumulated chill starts to permeate, and the environment becomes most frigid. It’ s the time when ice is at its hardest and most durable, historically the moment for ancient officials to harvest and store ice. In agricultural practices, it’ s crucial to undertake appropriate measures to protect crops from the biting cold, ensuring their survival and preparation for the coming spring. This term emphasizes readiness and resilience in the face of deepening winter.

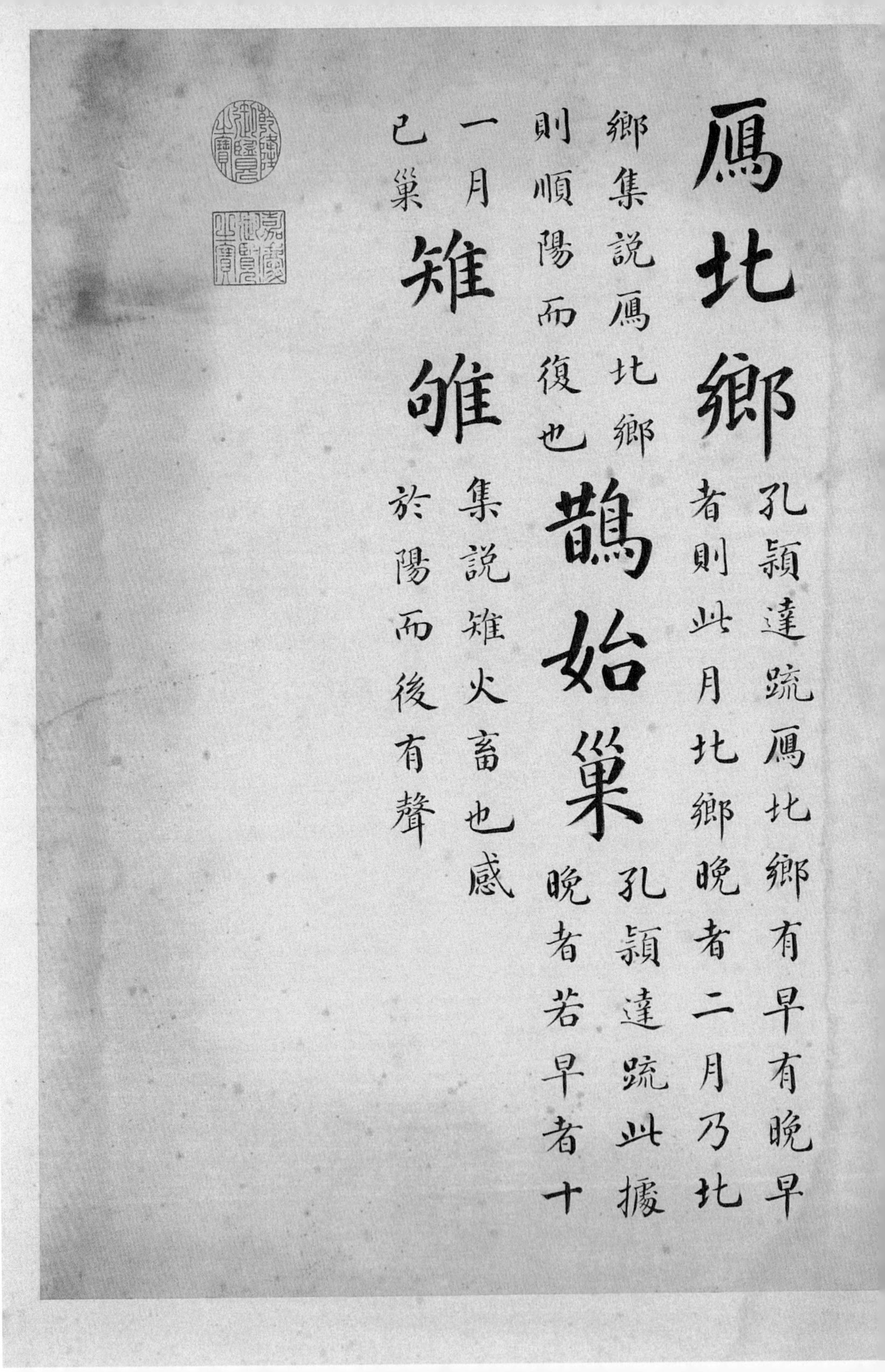

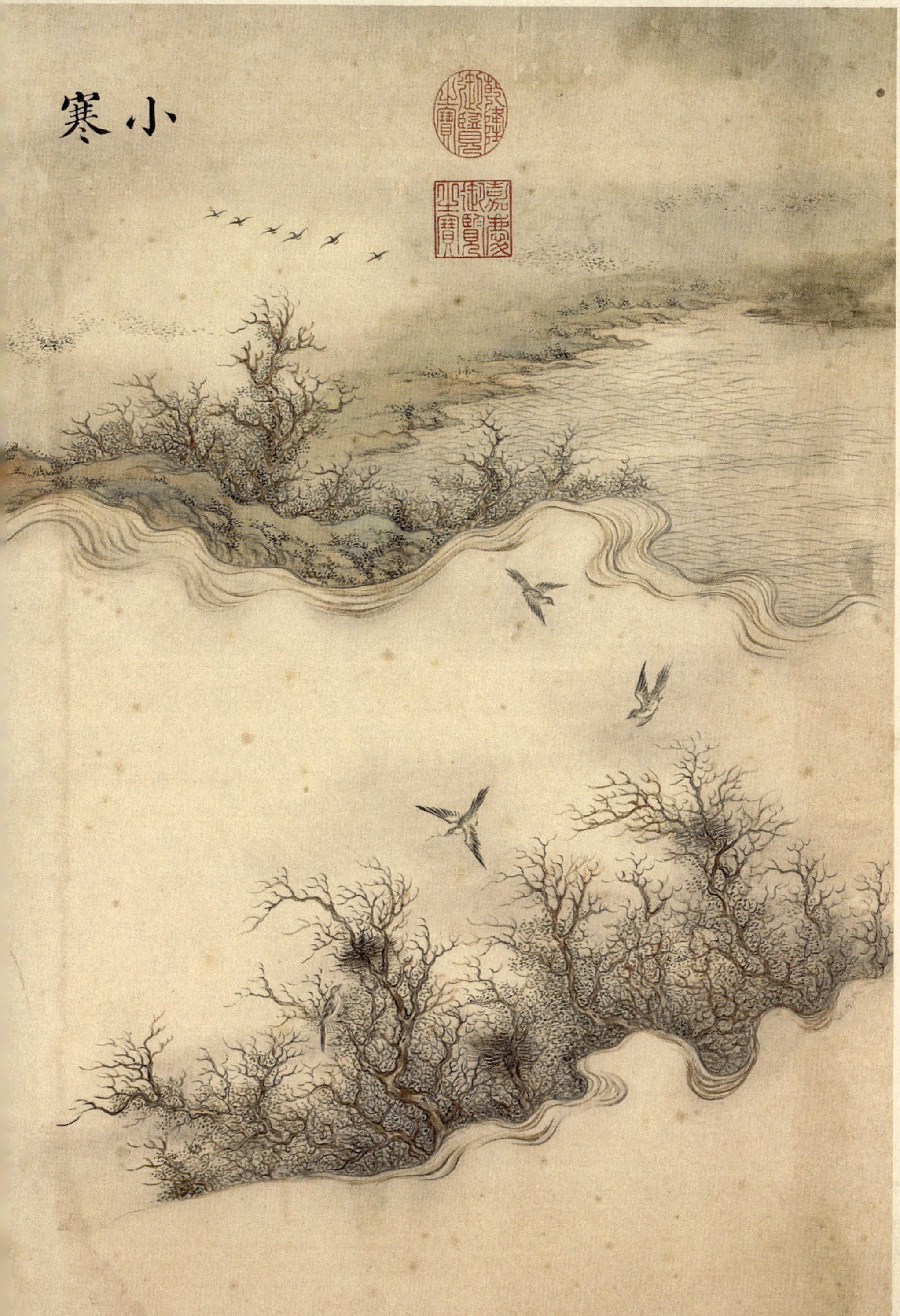


初候

## 雁北鄉

## Geese Begin Northward Journey

二候

## 鵲始巢

## Magpies Start Building Nests

三候

## 雉雊

## Pheasants Vocalize

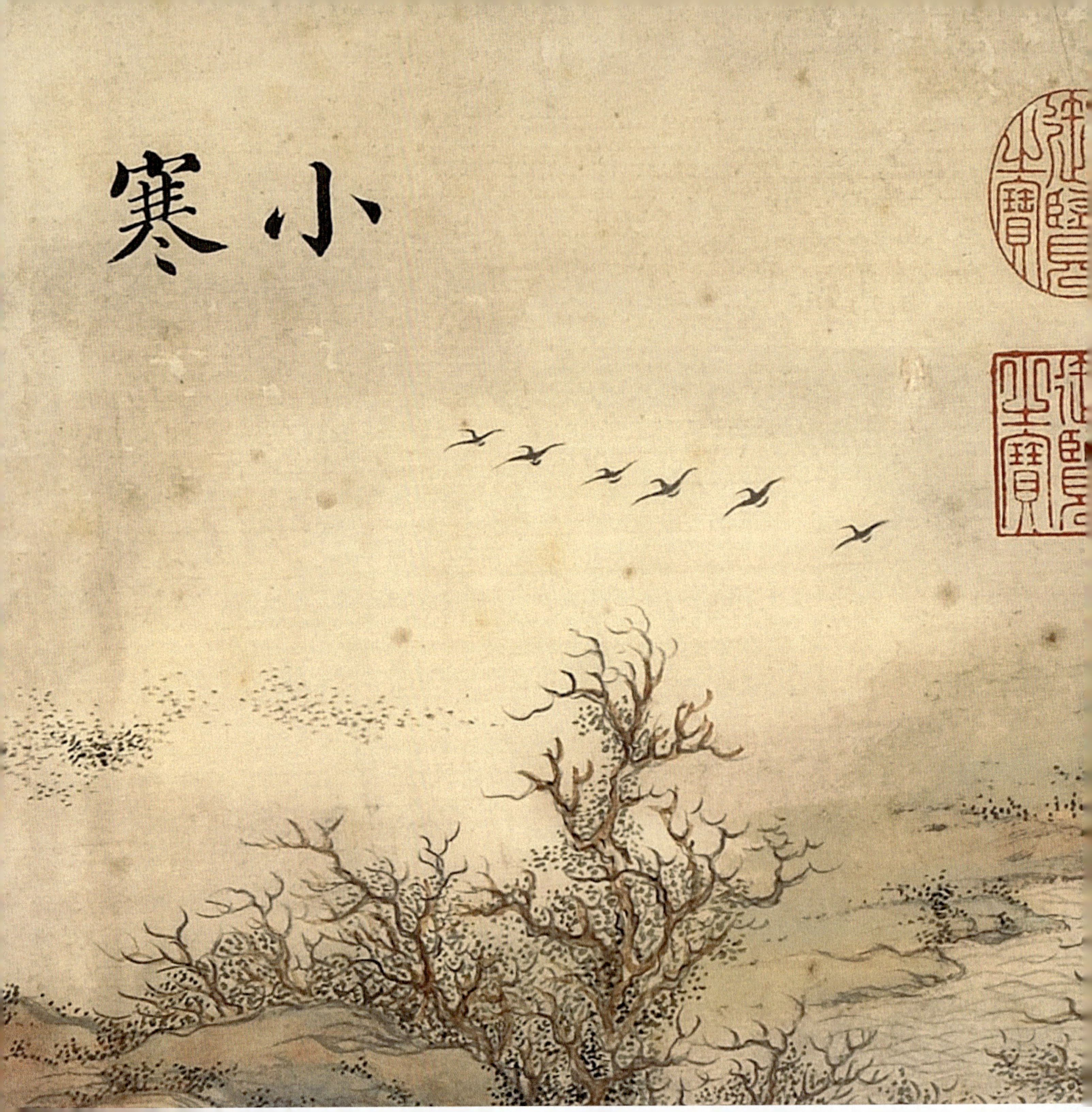
小寒

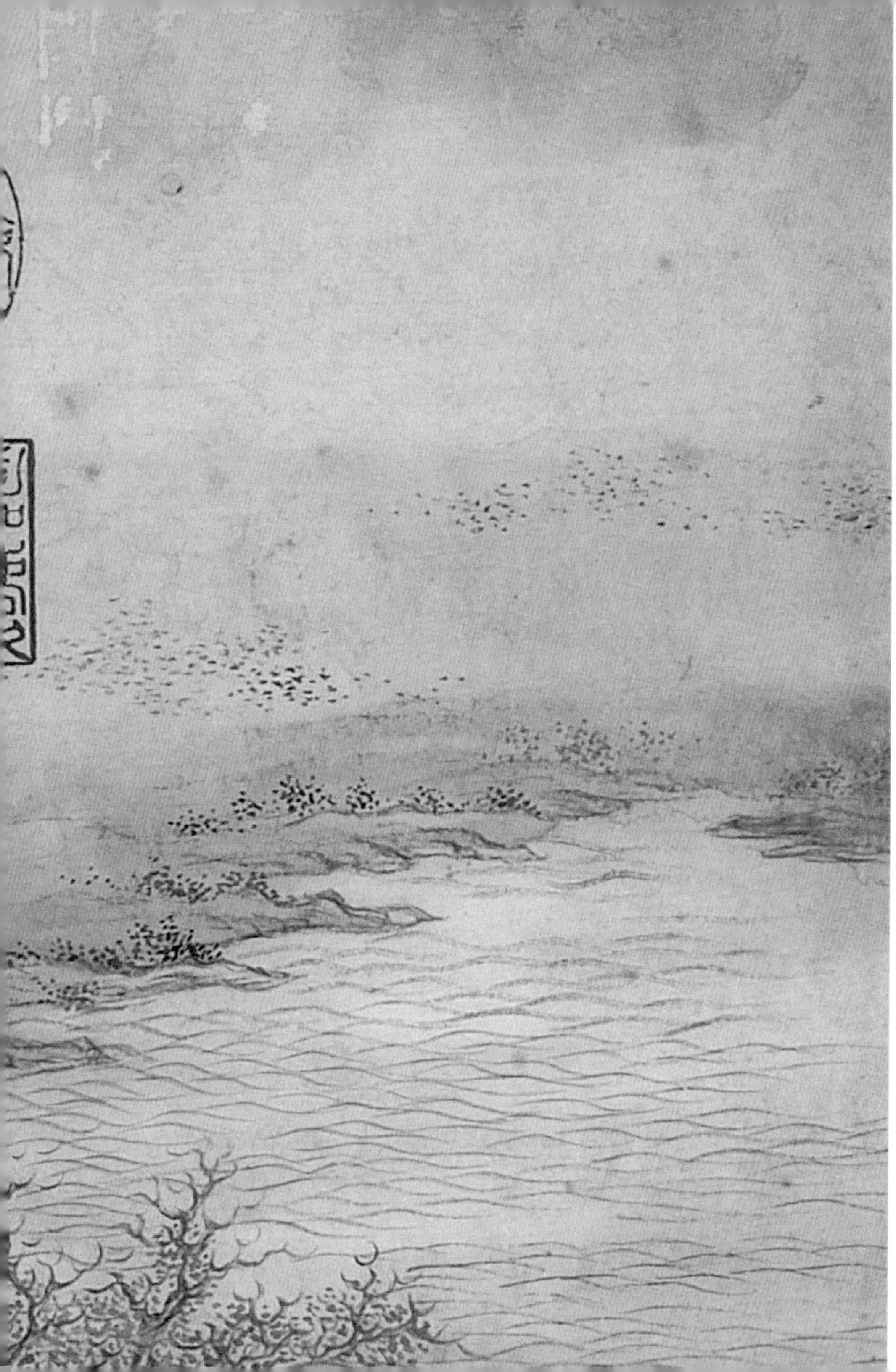

初候

# 雁北鄉

## Geese Begin Northward Journey

The first marker of Xiǎohán is "Geese Begin Northward Journey" (雁北鄉 Yàn Běi Xiāng): During Xiǎohán, some geese start their northward journey, anticipating the gradual resurgence of Yang energy. This behavior symbolizes the natural world's attunement to the subtle shifts in Yin and Yang, with the geese's migration indicating the early signs of the coming warmth and renewal of spring. The geese heading north is a testament to the cyclical nature of life and the instinctual knowledge embedded in these creatures, guided by the changing energies of the cosmos.

二候

# 鵲始巢

## Magpies Start Building Nests

The second cue of Xiǎohán is "Magpies Start Building Nests" (鵲始巢 Què Shǐ Cháo). As the cold persists but the days begin to slowly lengthen, magpies initiate the construction of their nests. This activity is a forward-looking gesture, preparing for the new generation in anticipation of spring' s return. The magpies, with their affinity for sunlight and active energy, embody the stirring of Yang within the still-dominant cold, heralding the eventual revival of warmth and growth.

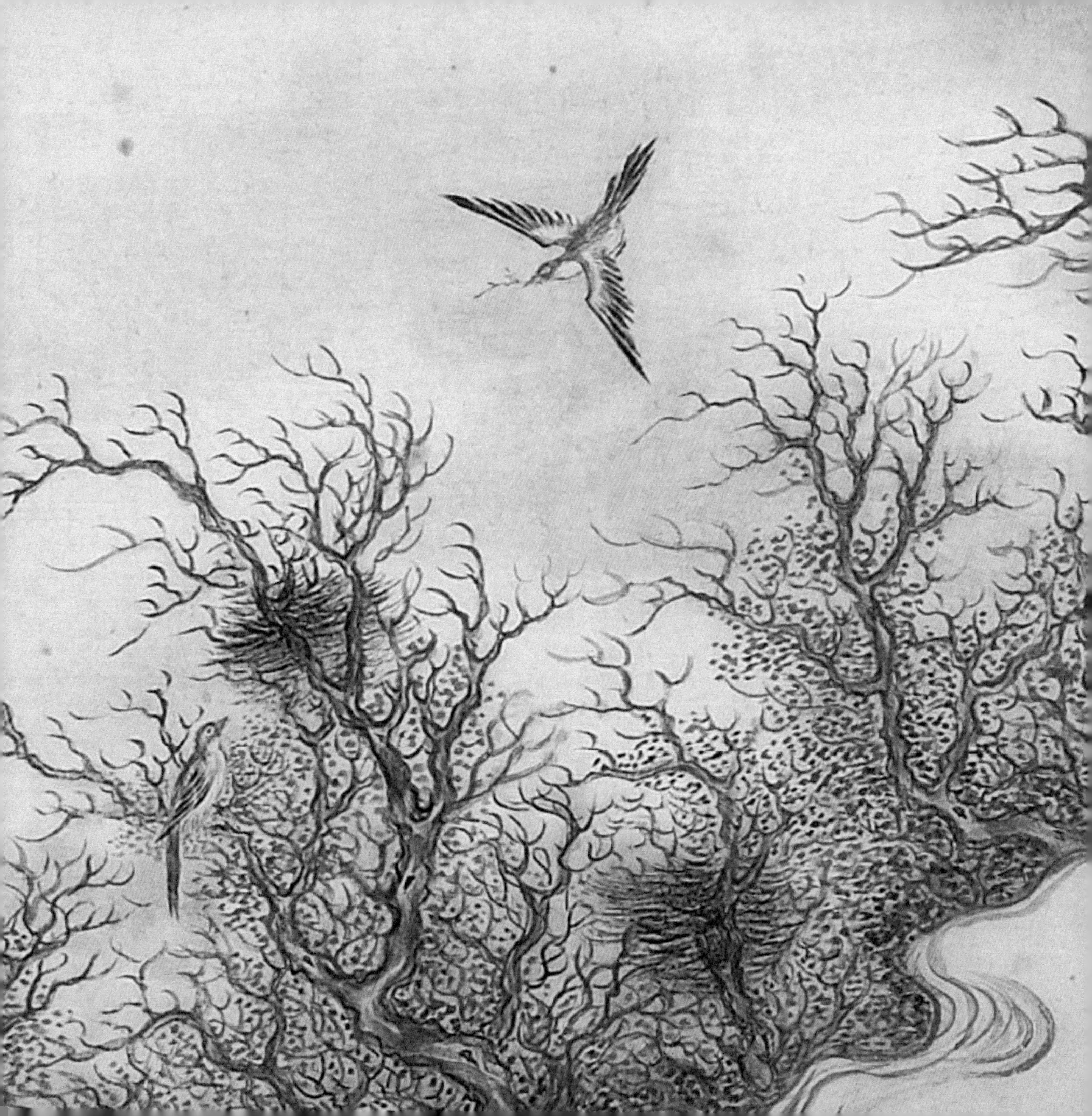

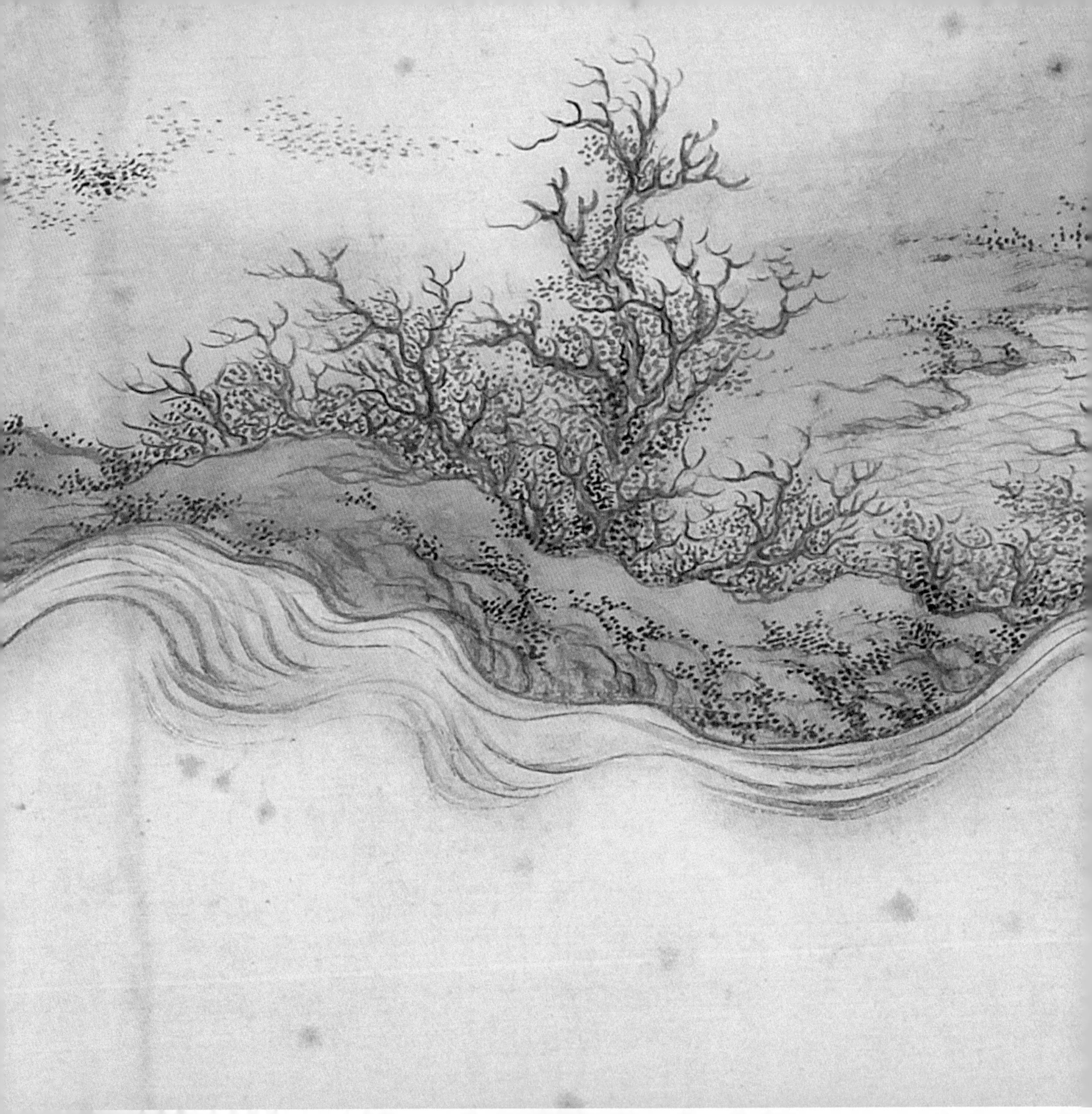

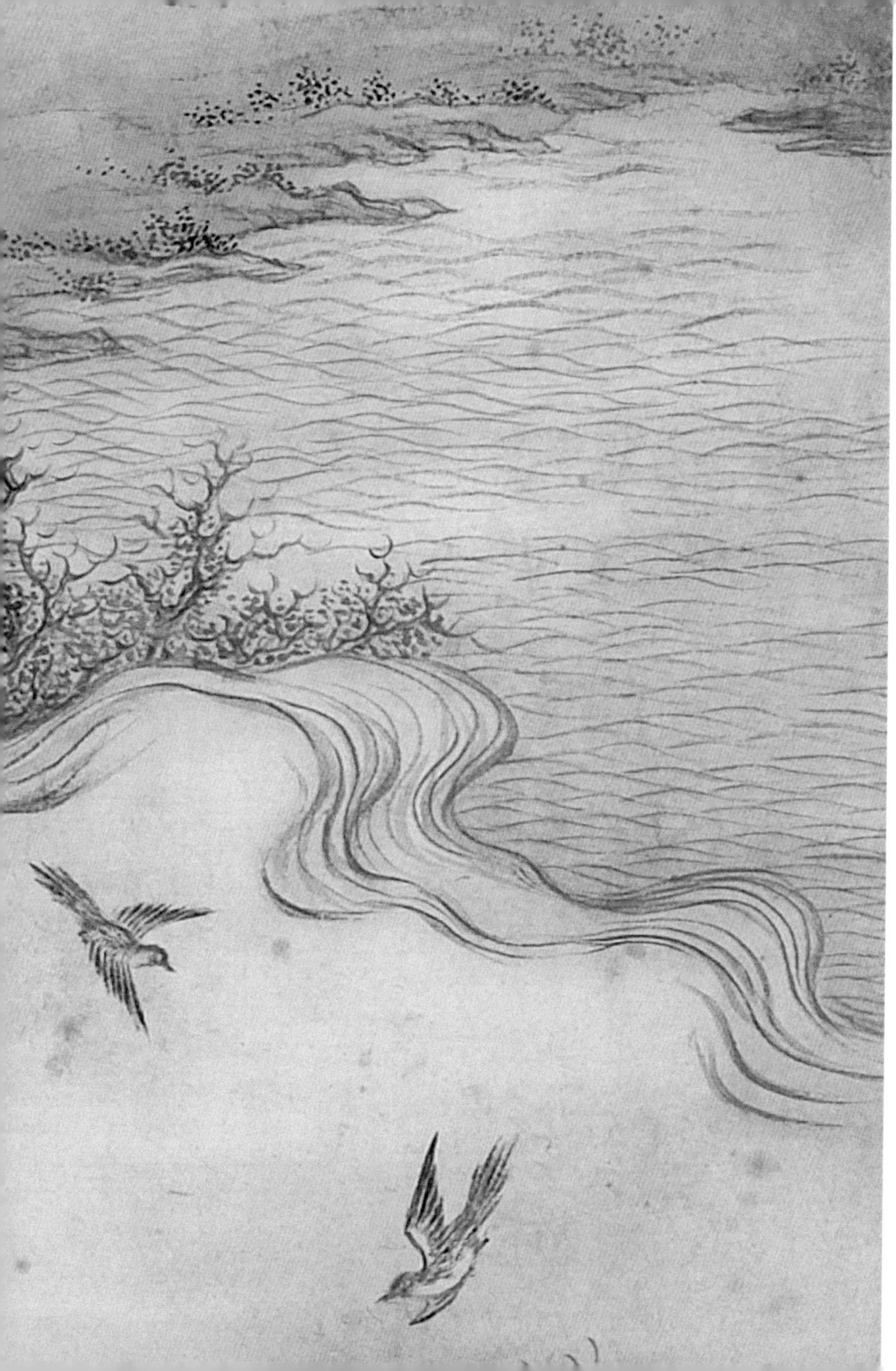

三候

# 雉雊

## Pheasants Vocalize

The third sign of Xiǎohán is “Pheasants Vocalize” (雉雊 Zhì Qú). As the cold season matures, pheasants begin their vocal calls, punctuating the winter silence. This sign represents not just the persistence of life through the cold but also the anticipation of spring’s return. The pheasants’ calls are seen as an expression of the gradual awakening of Yang energy, a natural response to the incremental increase of daylight and the promise of a new cycle of growth and vitality.

# 寒

Hán

Dà

大

# GREATER COLD

January 21 (± 1)

# 大寒

## Greater Cold

Dàhán : Representing the extreme of winter' s cold, marking the coldest days of the year. It' s a time characterized by intense cold and preparation for the Lunar New Year. As people engage in customs and traditions to welcome the new year, they also take measures to protect against the severe cold, such as covering crops with non-woven fabrics for insulation. Dàhán embodies the depth of winter and the anticipation of renewal, as the community comes together in warmth and festivity amidst the chill.

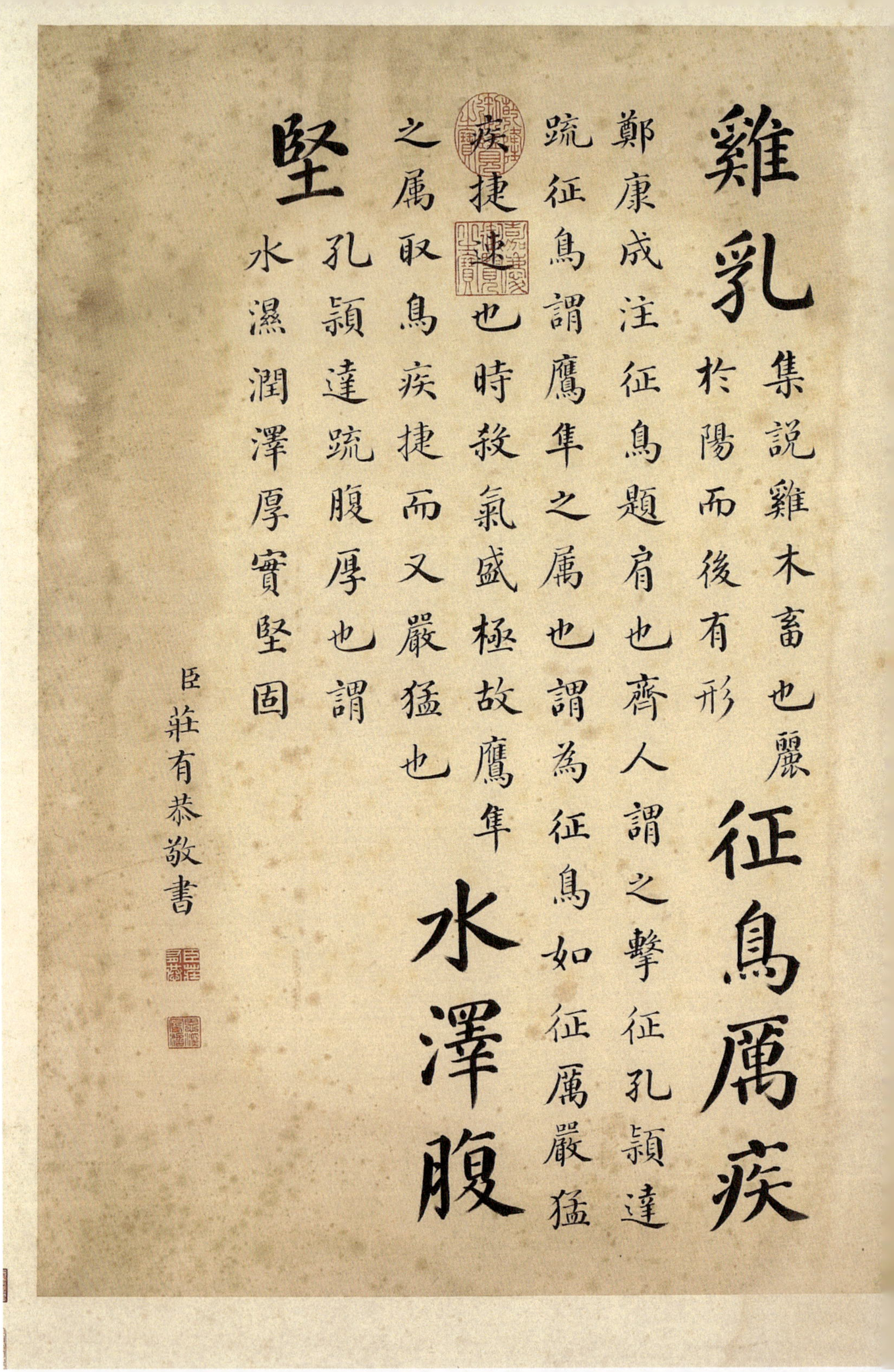

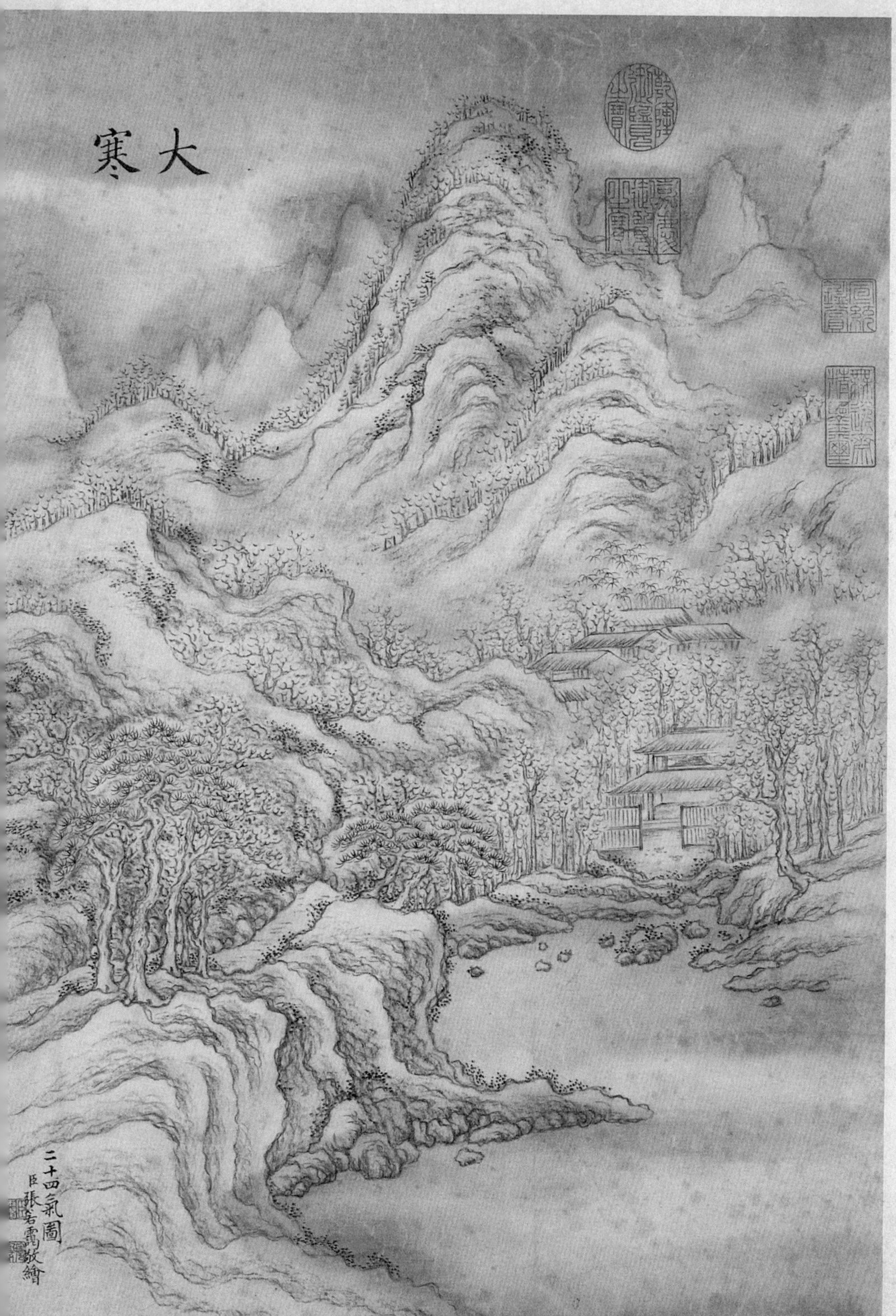


初候

# 雞乳

Hens Begin Brooding

二候

# 徵鳥厲疾

Predatory Birds Intensify Hunting

三候

# 水澤腹堅

Ice Solidifies Throughout Water Bodies

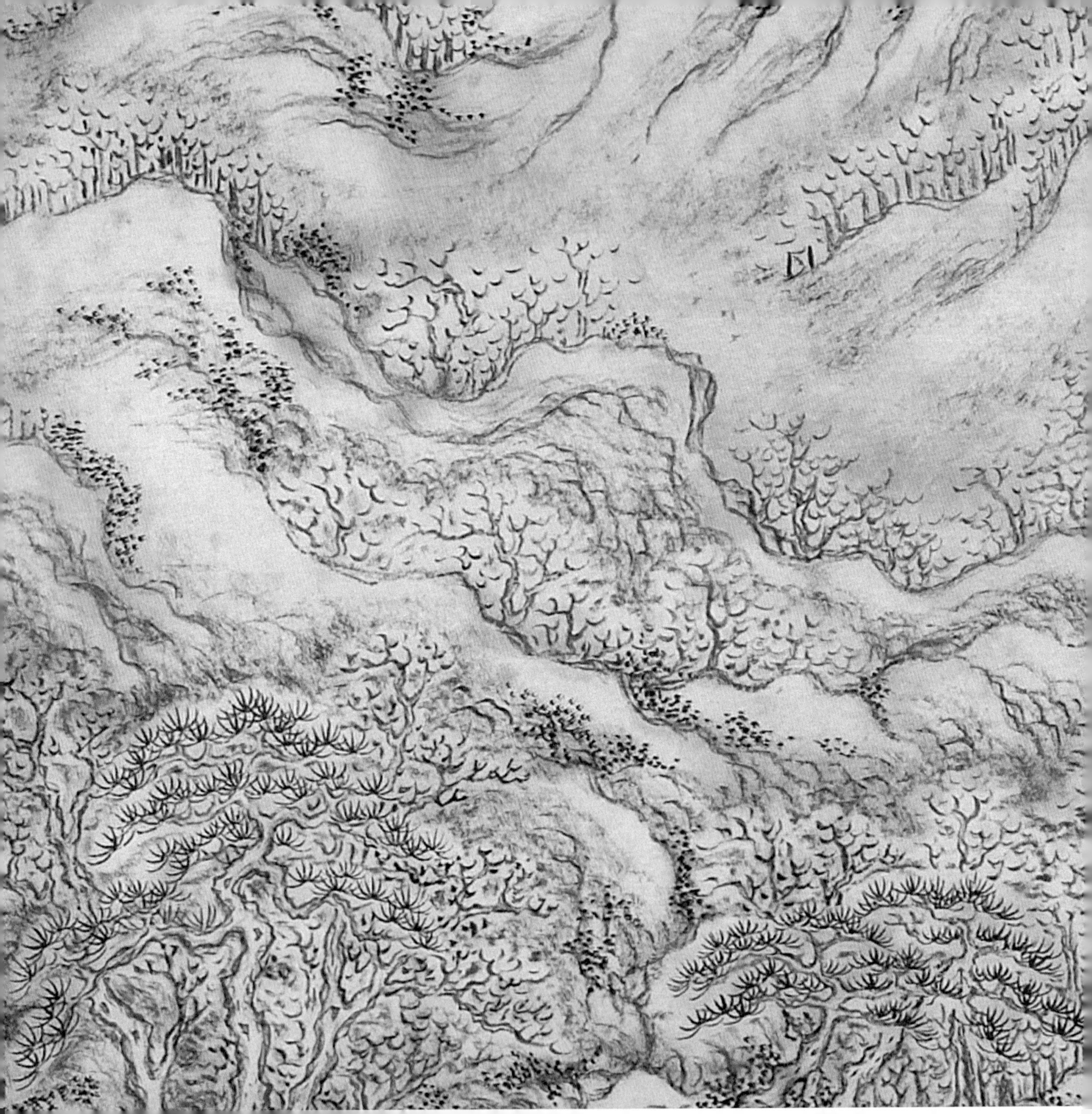

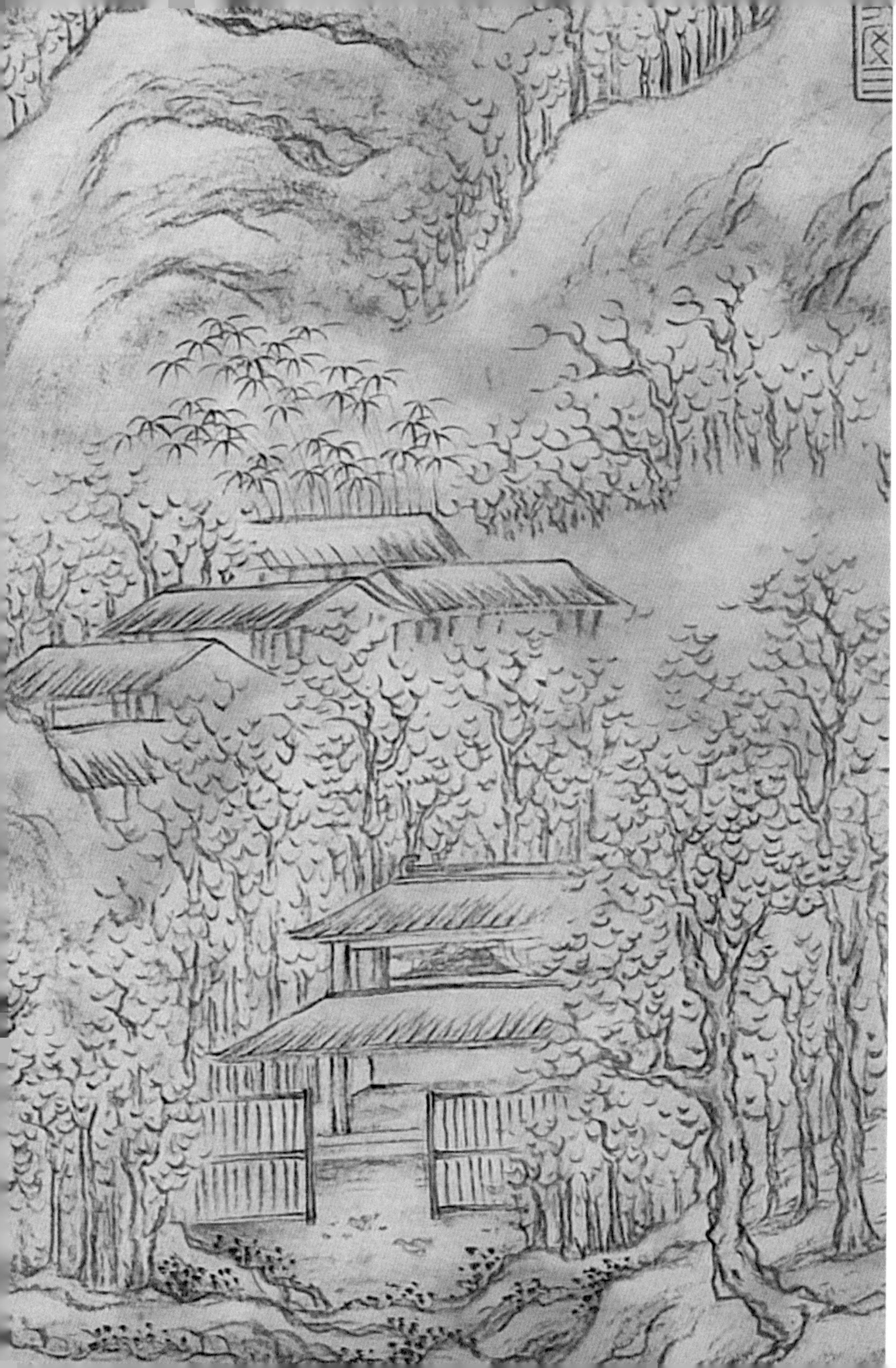

初候

# 雞乳

## Hens Begin Brooding

The first sign of Dàhán is "Hens Begin Brooding" (雞乳 Jī Rǔ). As the great cold of Dàhán sets in, hens commence the brooding process, sheltering and warming their eggs in preparation for the next generation. This sign symbolizes the continuation of life and the nurturing spirit amidst the coldest time of the year.

二候

# 徵鳥厲疾

## Predatory Birds Intensify Hunting

The second signal of Dàhán is “Predatory Birds Intensify Hunting” (徵鳥厲疾 Zhēng Niǎo Lì Jí). During this period, predatory birds like eagles and hawks escalate their hunting activities, driven by the need to survive the brutal cold and scarcity of food. The term “徵鳥厲疾” reflects the intensified efforts and fierce survival instincts of these raptors as they soar through the frigid skies, seeking sustenance. It’s a testament to the raw and relentless side of nature, where each creature adapts and strives in the face of winter’s challenges.

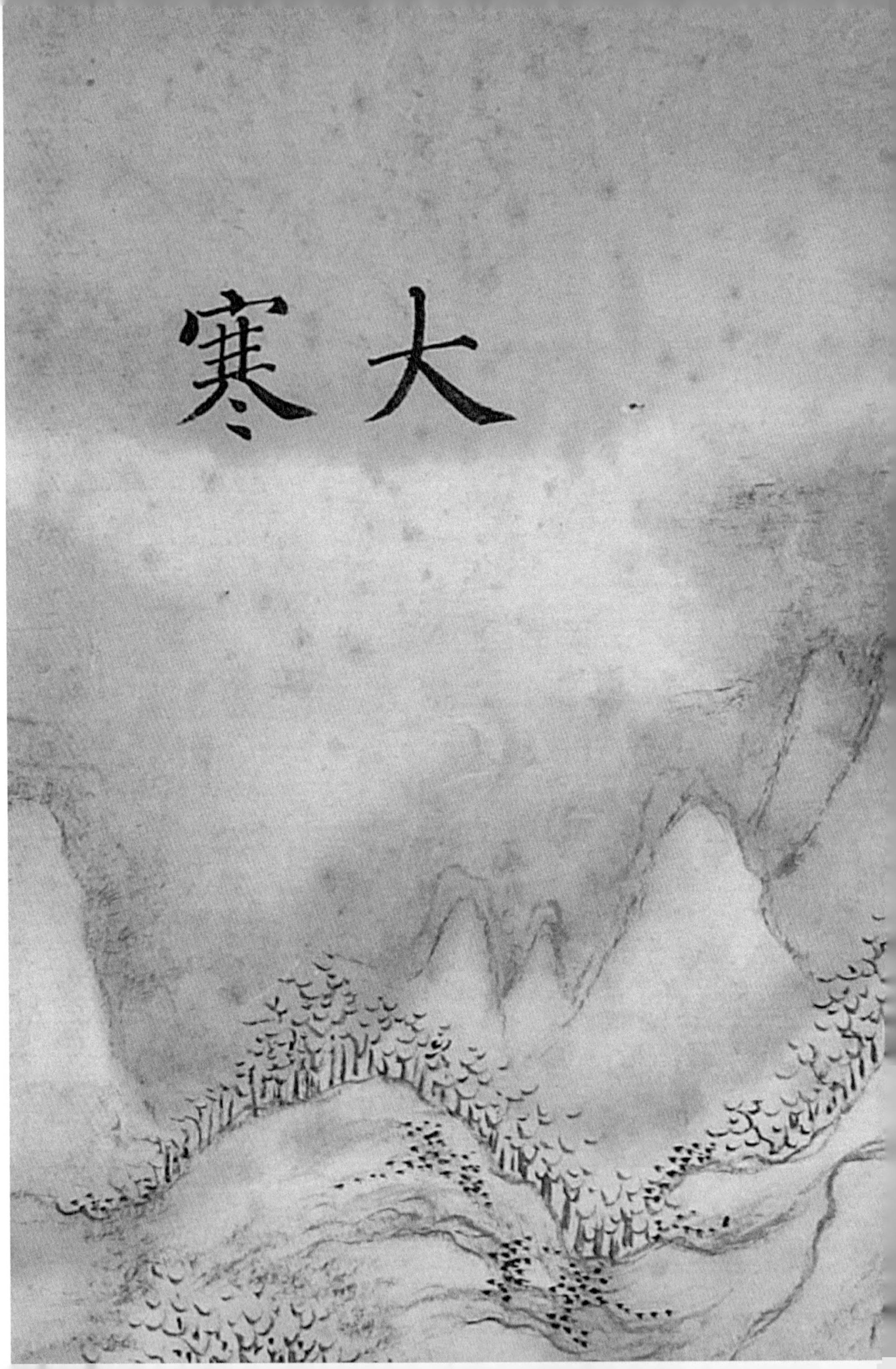

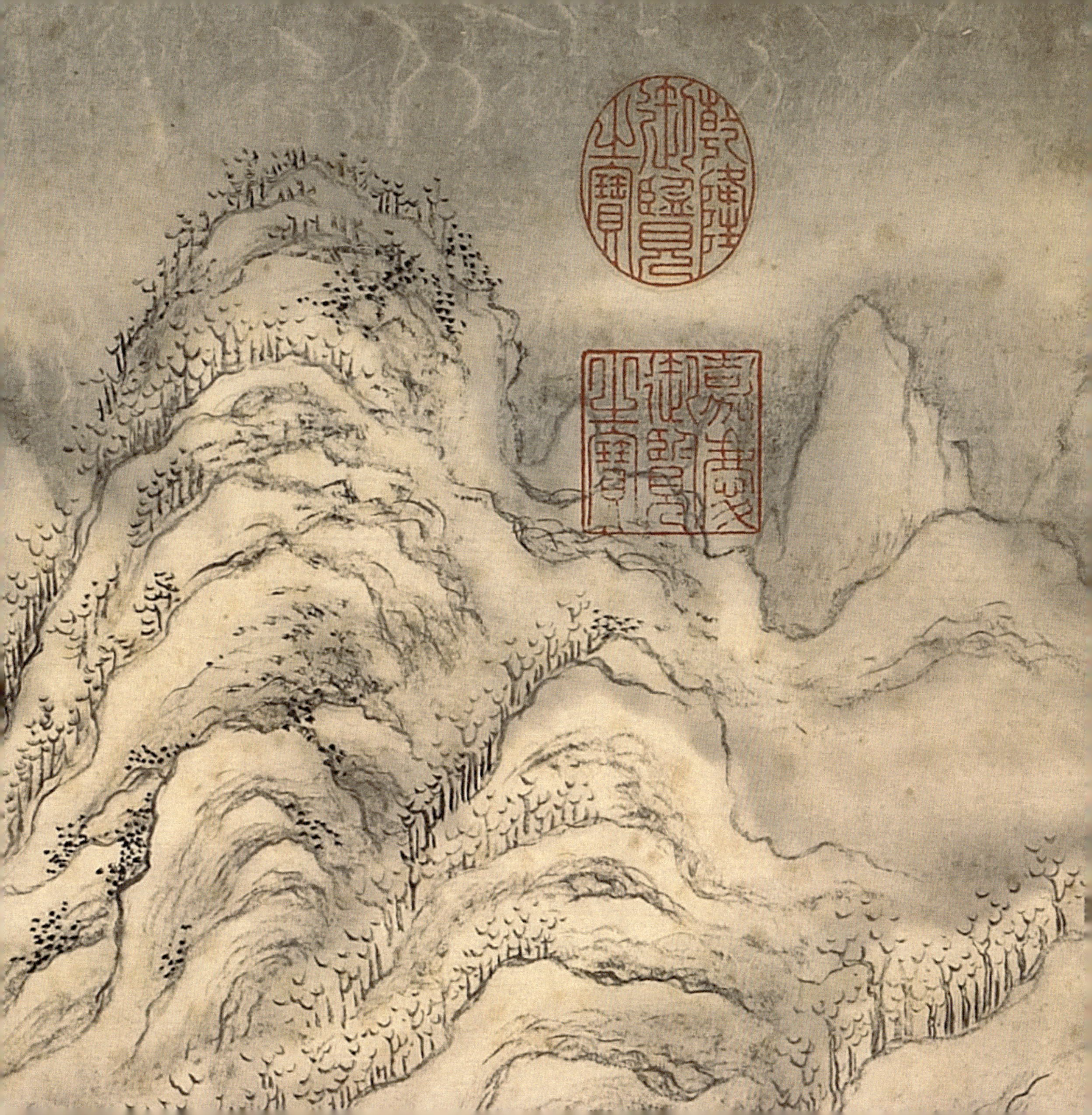
乾隆御覽之寶
嘉慶御覽之寶

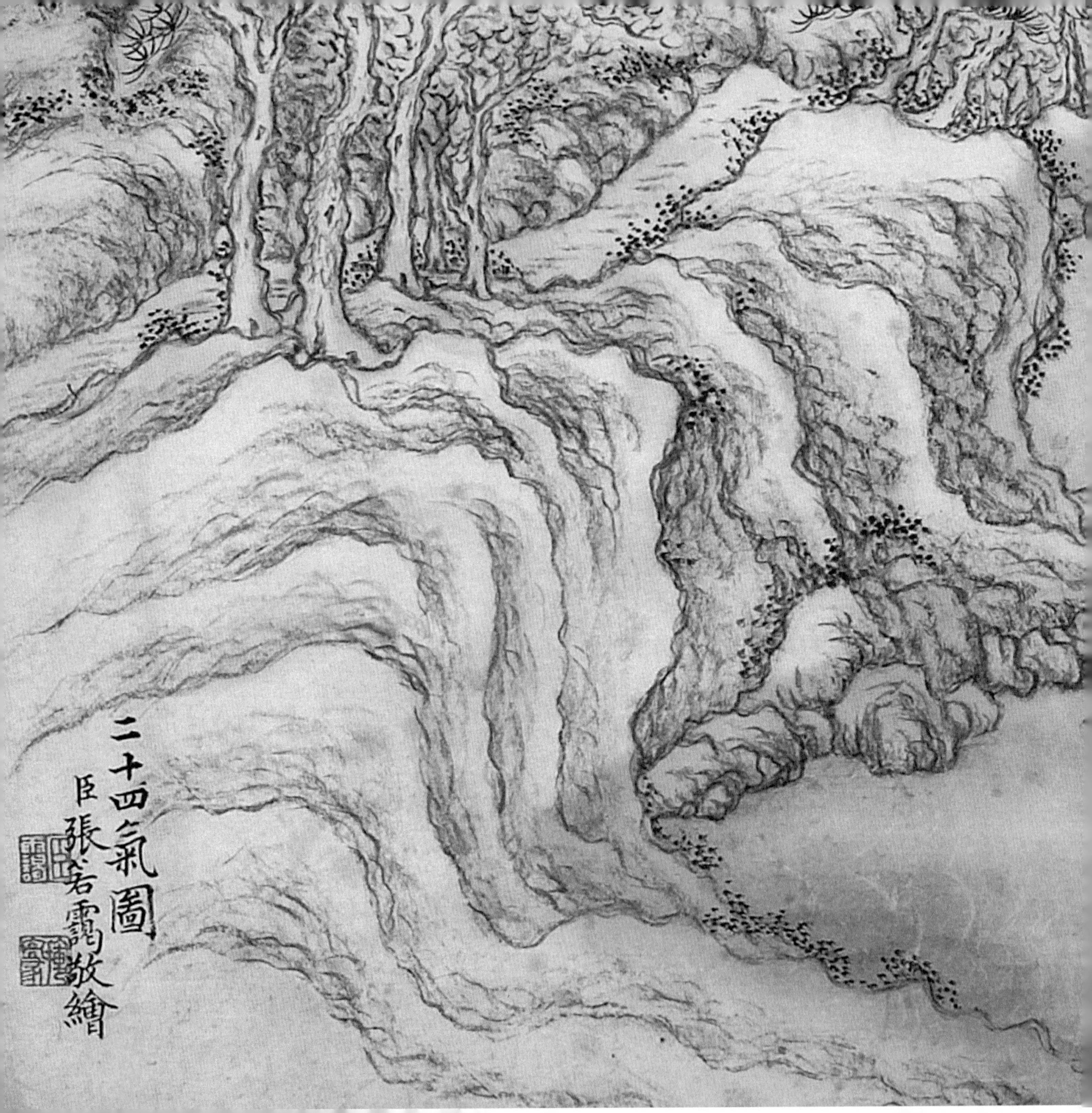
二十四氣圖
臣張若靄敬繪

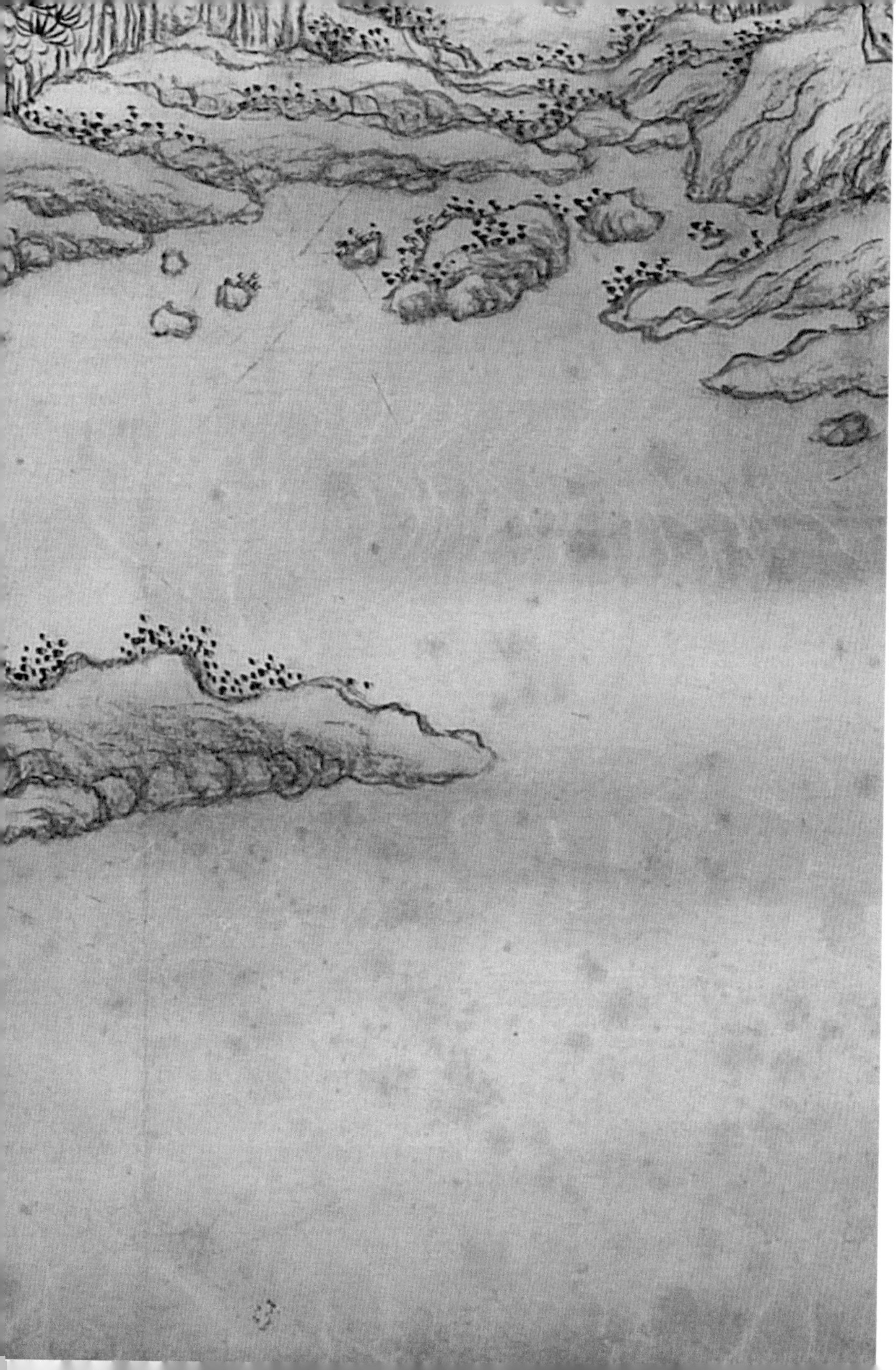

三候

# 水澤腹堅

## Ice Solidifies Throughout Water Bodies

The third hint of Dàhán is "Ice Solidifies Throughout Water Bodies" (水澤腹堅 Shuǐ Zé Fù Jiān). In the depth of Dàhán, the cold reaches its zenith, and bodies of water become encased in thick, hard ice. This sign marks the culmination of winter's grip, with rivers, lakes, and ponds solidifying into ice. It's a time when the landscape is transformed into a still, frozen tableau, reflecting the profound impact of the season's chill. The solidity of the ice over water bodies symbolizes the complete dominion of winter, a period of stillness and conservation in anticipation of the next year's thaw and renewal.

# 附錄

**Appendix**

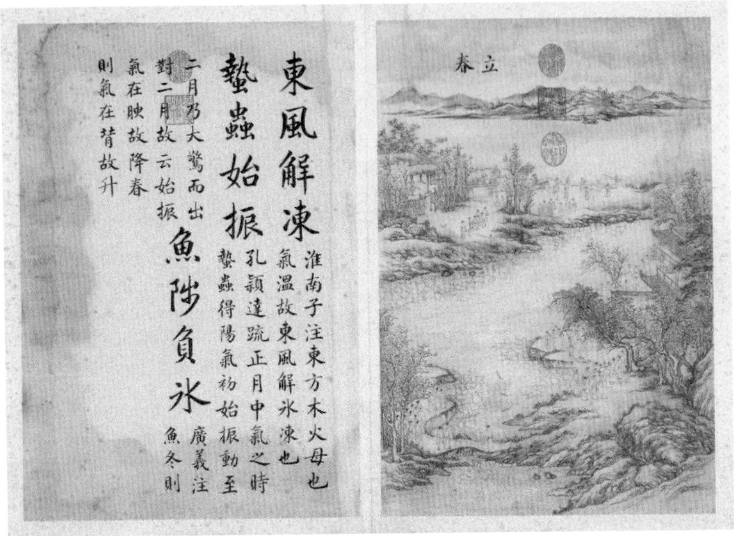


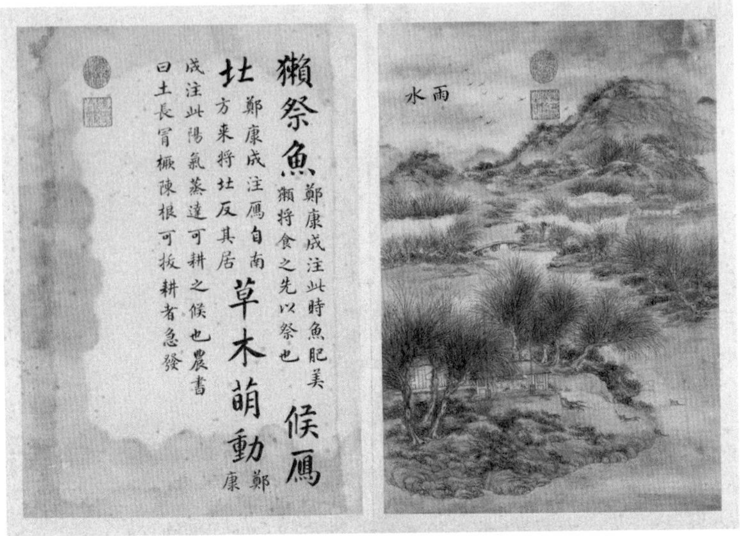


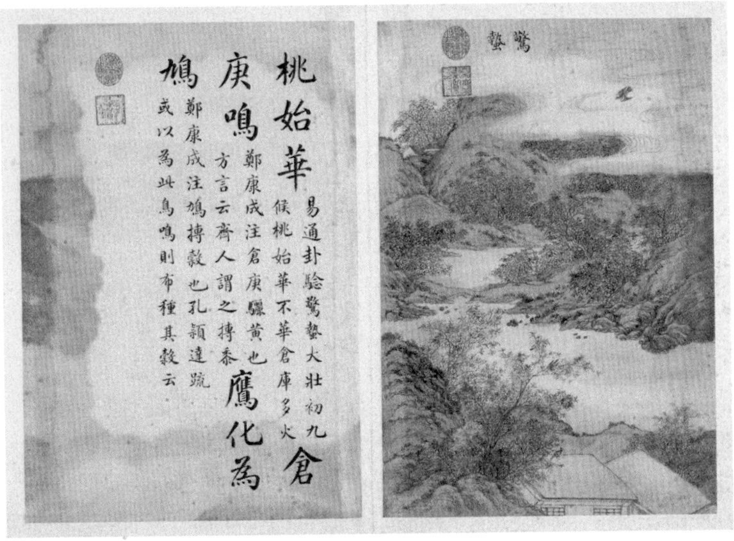


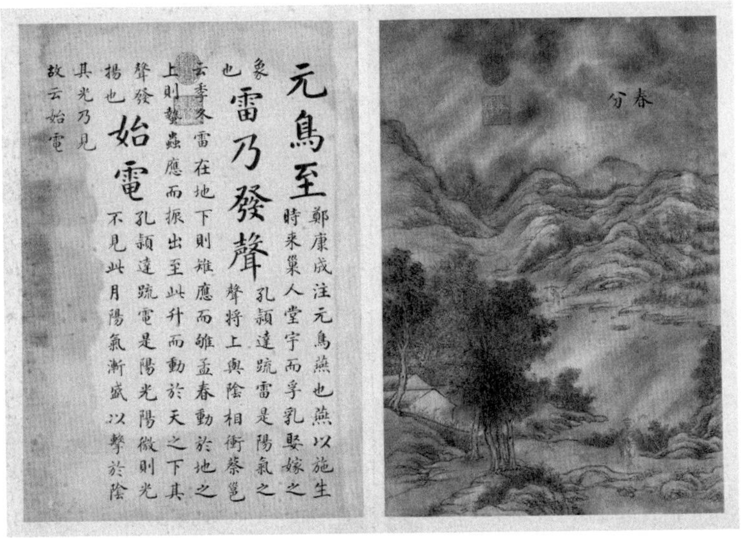


| **立春**<br>(Lìchūn)<br>February 4 | **雨水**<br>(Yǔshuǐ)<br>February 19 |
|---|---|
| **驚蟄**<br>(Jīngzhé)<br>March 6 | **春分**<br>(Chūnfēn)<br>March 21 |

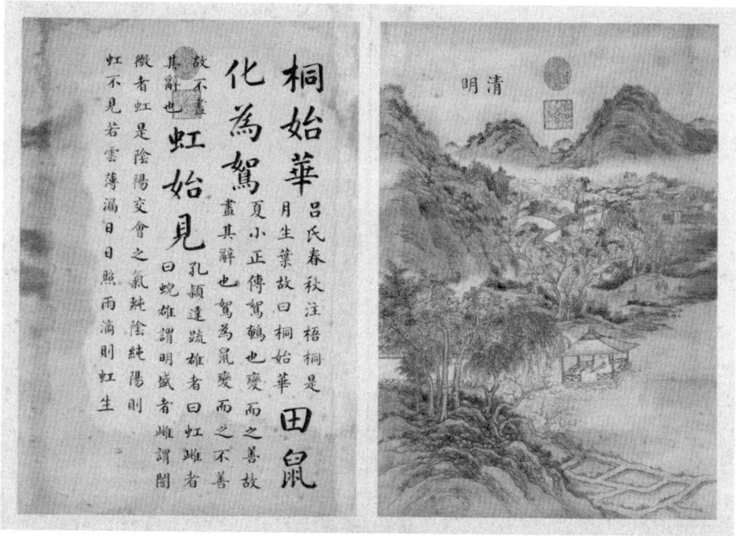


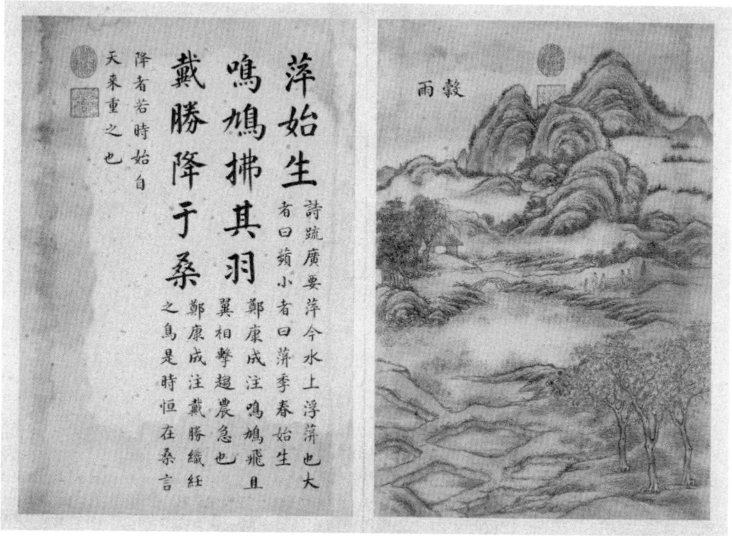


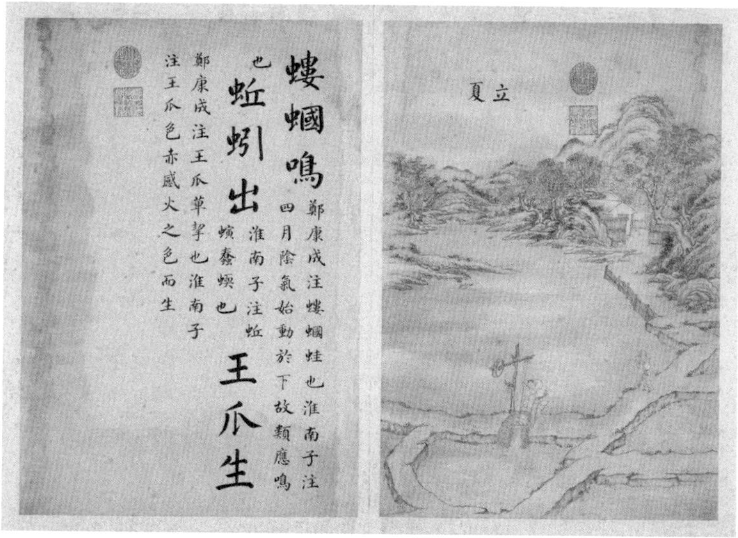


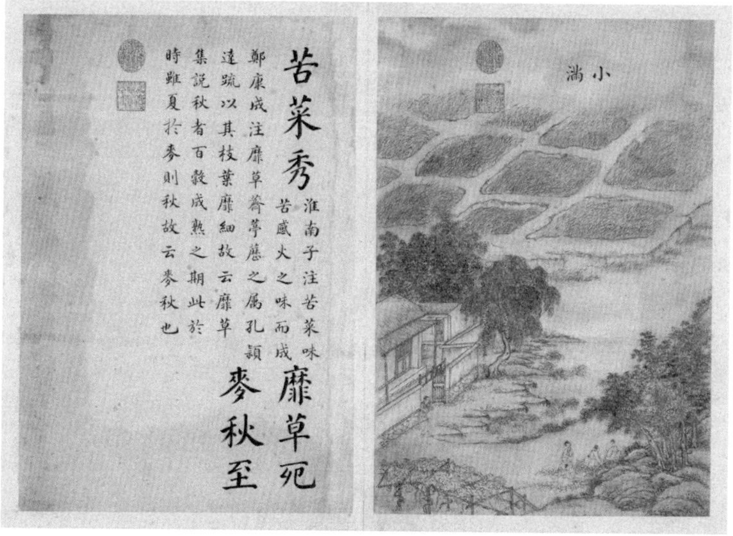


| | |
|---|---|
| **清明**<br>(Qīngmíng)<br>April 5 | **穀雨**<br>(Gǔyǔ)<br>April 20 |
| **立夏**<br>(Lìxià)<br>May 6 | **小滿**<br>(Xiǎomǎn)<br>May 21 |

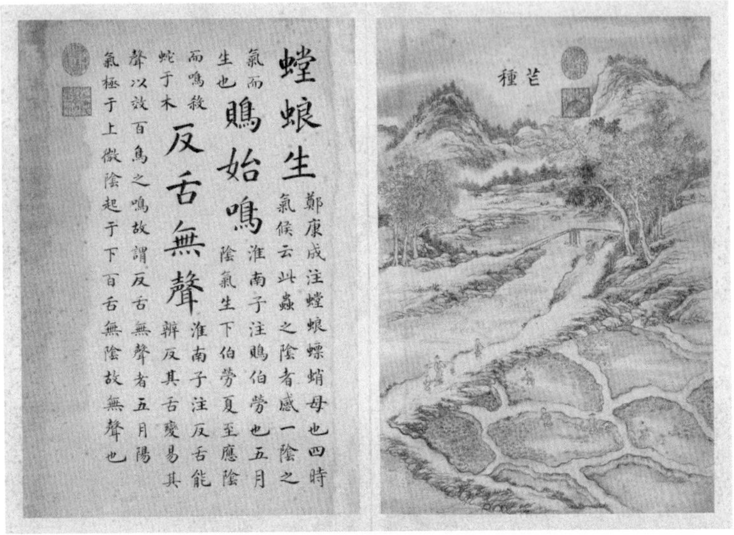


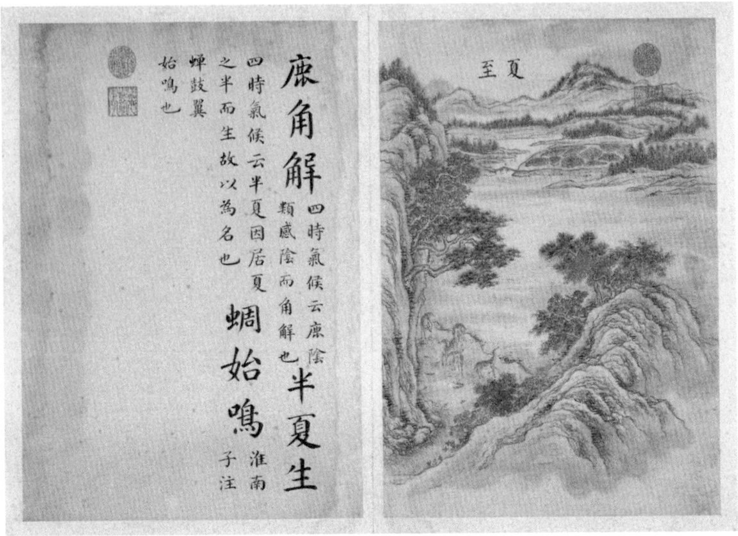


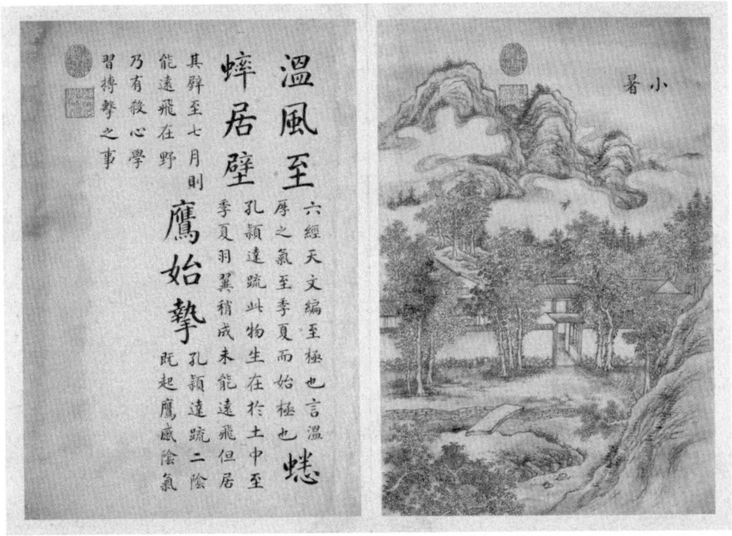


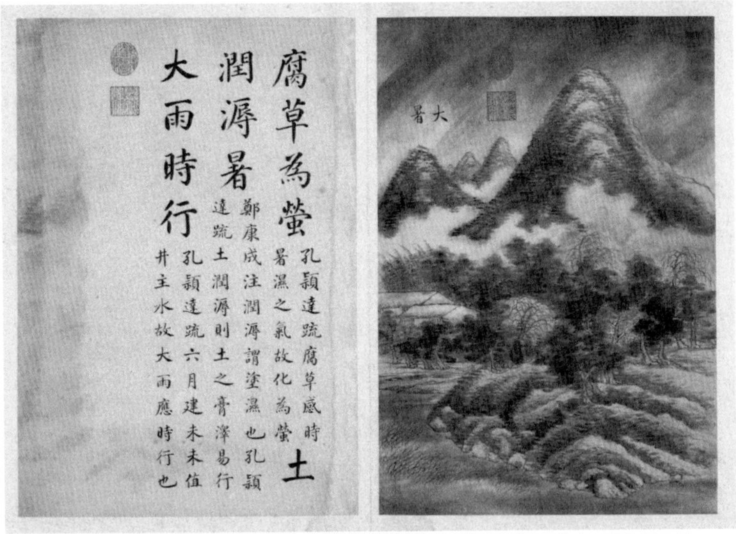


| | |
|---|---|
| **芒種**<br>(Mángzhòng)<br>June 6 | **夏至**<br>(Xiàzhì)<br>June 21 |
| **小暑**<br>(Xiǎoshǔ)<br>July 7 | **大暑**<br>(Dàshǔ)<br>July 23 |

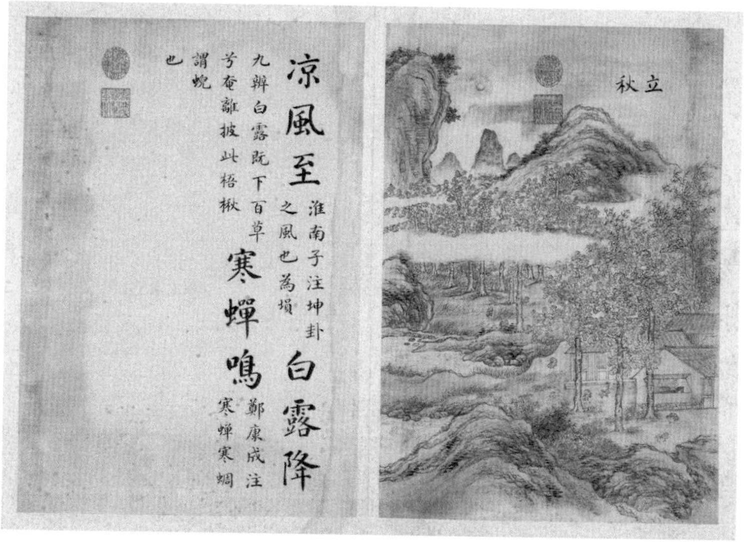


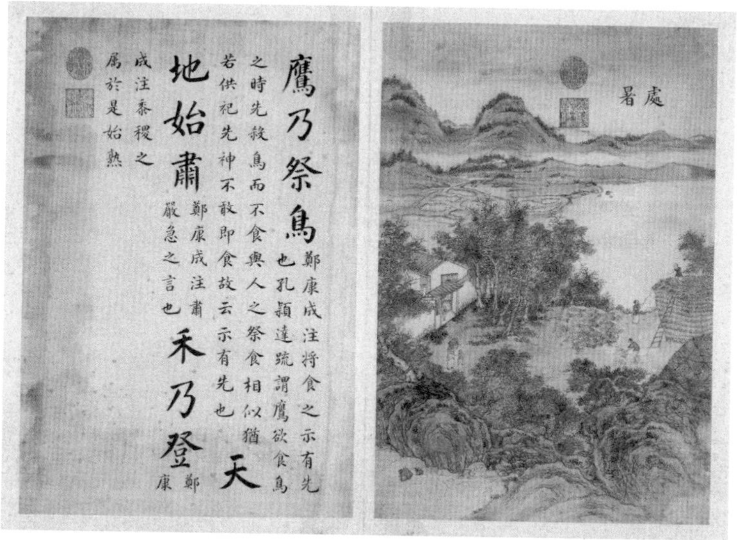


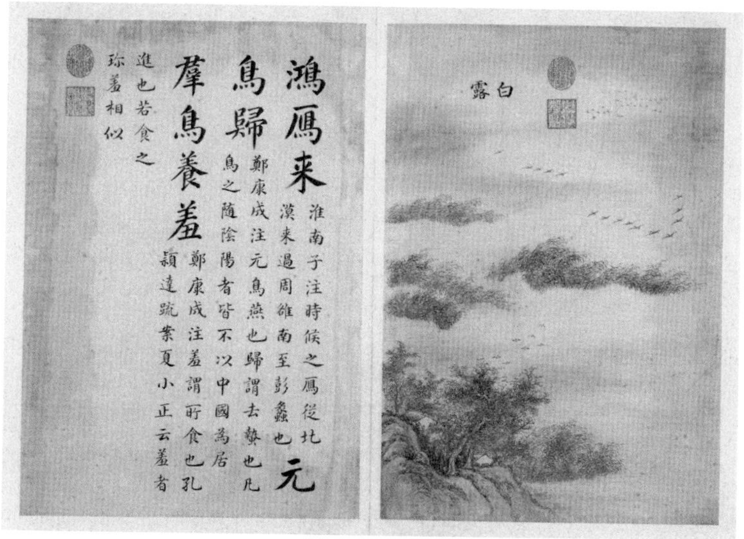


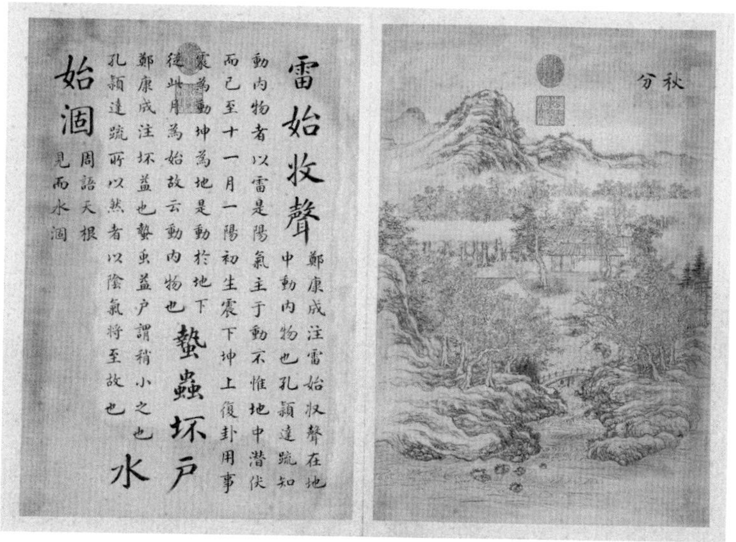


| | |
|---|---|
| **立秋**<br>(Lìqiū)<br>August 8 | **處暑**<br>(Chǔshǔ)<br>August 23 |
| **白露**<br>(Báilù)<br>September 8 | **秋分**<br>(Qiūfēn)<br>September 23 |

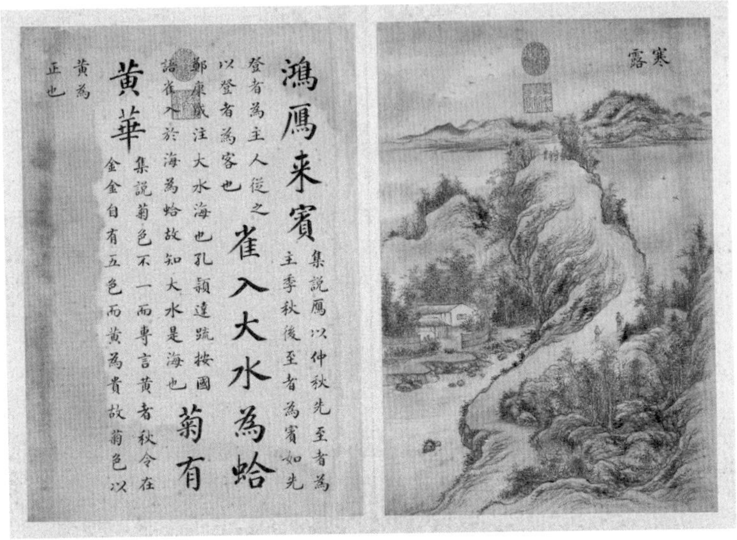


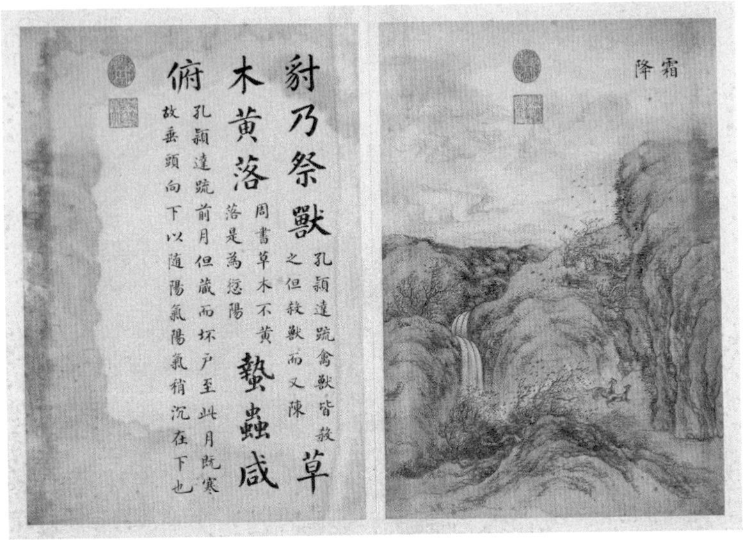


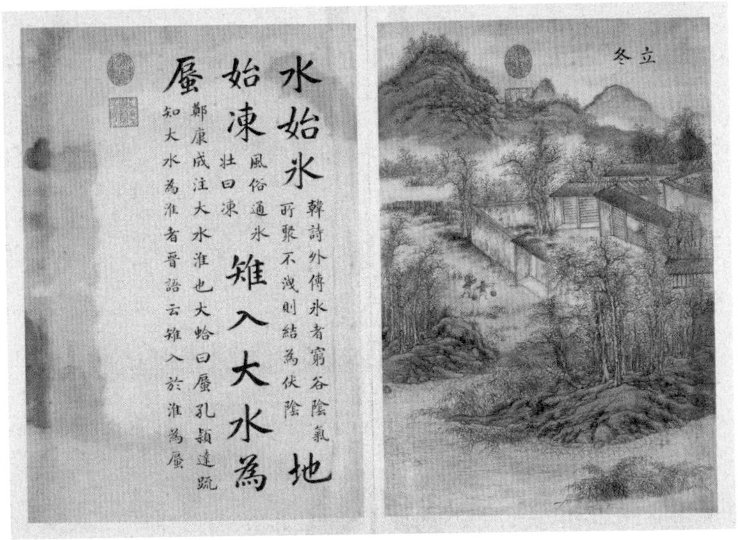


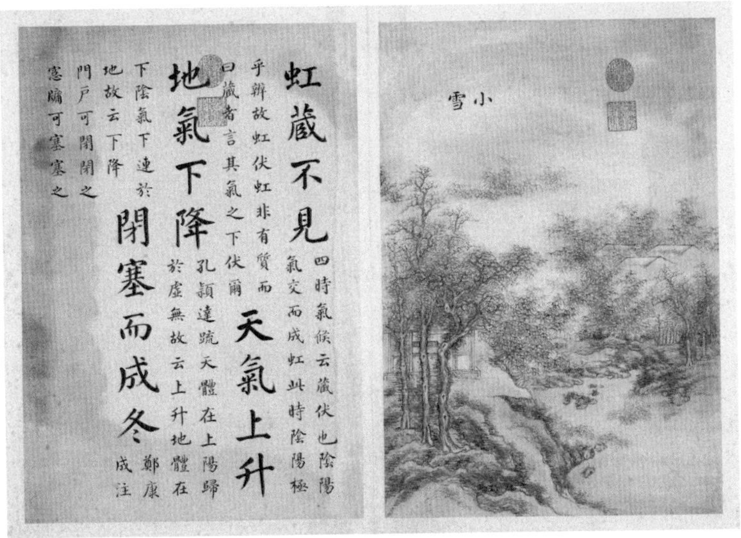


| **寒露** (Hánlù) October 8 | **霜降** (Shuāngjiàng) October 23 |
|---|---|
| **立冬** (Lìdōng) November 7 | **小雪** (Xiǎoxuě) November 22 |

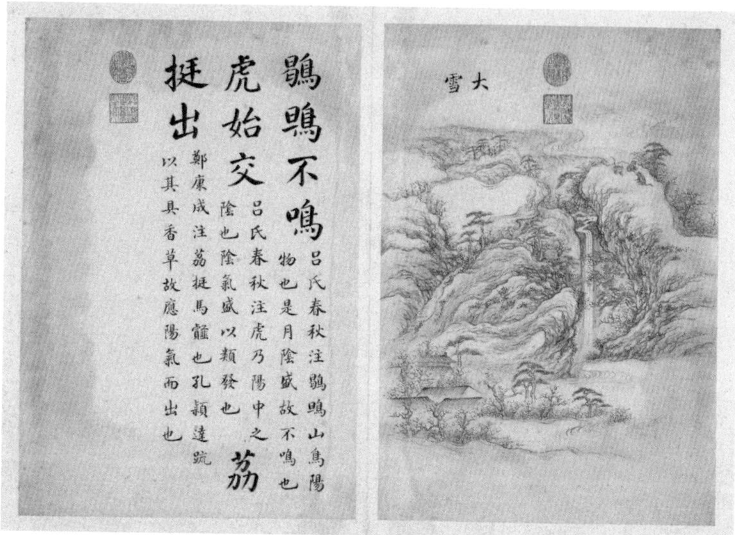


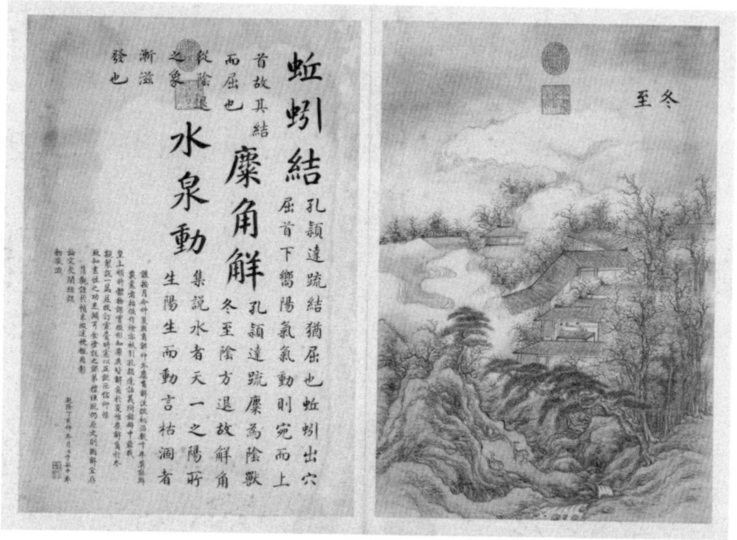


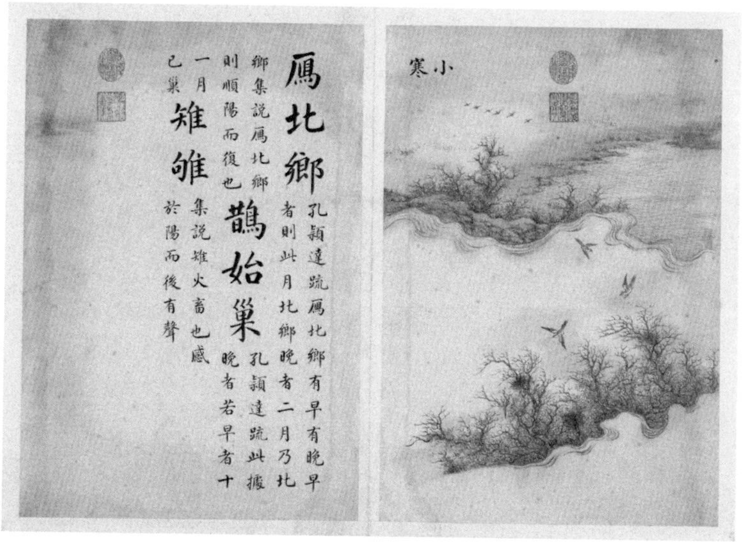


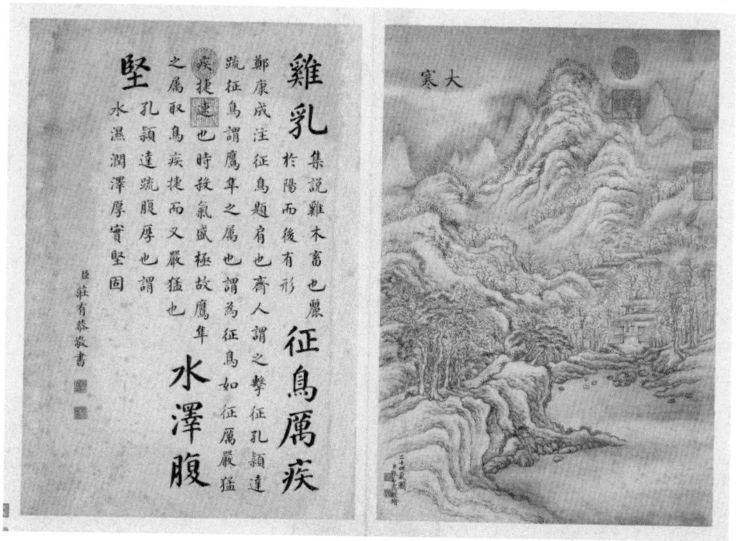


| | |
|---|---|
| **大雪**<br>(Dàxuě)<br>December 7 | **冬至**<br>(Dōngzhì)<br>December 22 |
| **小寒**<br>(Xiǎohán)<br>January 6 | **大寒**<br>(Dàhán)<br>January 21 |

# 註釋

## Notes

*The Gregorian dates associated with each of the 24 solar terms represent the central day most commonly recognized in traditional East Asian calendars. Historically, these terms were devised to closely follow the solar year, marking the subtle shifts in climate and agricultural cycles. Due to the precision of ancient astronomical knowledge, the actual occurrence of each solar term may vary slightly from year to year, typically by a margin of one day earlier or later. This variation is a reflection of the dynamic nature of the celestial movements that these terms are meant to capture. As readers journey through the seasons depicted in these pages, it's worth noting the enduring legacy of how ancient cultures harmonized their lives with the rhythm of the cosmos.

Gregorian Calendar: It is a solar calendar based primarily on the Earth's revolutions around the Sun, leading to a year of approximately 365 days. The calendar is divided into 12 months with a fixed number of days in each month.

Lunisolar Calendar: This type of calendar, such as the traditional Chinese, Hebrew, and Hindu calendars, is based on both lunar phases and the solar year. It typically consists of 12 lunar months, each beginning with a new moon. To align with the solar year, an extra month is added periodically.

**Paintings in this book were created by Zhang Ruohai during the 10th to 13th years of the Qianlong era of the Qing Dynasty.

Twenty-four openings, approximately 63x42.5 centimeters in size, colored on paper.

Currently housed in the National Palace Museum, Taipei (Collection numbers 故畫 003636 - 故畫 003647).

**Seventy-Two Nature Gifts in A Year:**
Time-Honored Artistry Melding with the Wisdom of the Twenty-Four Solar Terms

Yumeng (ILSA) Li

Published by PEAR & MULBERRY
https://www.pear-mulberry.com/



ISBN: 978-1-961660-10-6



PRINTED IN THE UNITED STATES OF AMERICA

First Edition



PEAR & MULBERRY
30 N Gould St
WY 82801, Sheridan
United States of America
pear.mulberry@gmail.com

Made in United States
Orlando, FL
04 November 2026

90199221R00162